AF358687

Nutritional Status and Interventions for Patients with Cancer

Nutritional Status and Interventions for Patients with Cancer

Editors

Nuno Borges
Fernando Mendes
Diana Martins

Basel • Beijing • Wuhan • Barcelona • Belgrade • Novi Sad • Cluj • Manchester

Editors
Nuno Borges
University of Porto
Porto
Portugal

Fernando Mendes
Polytechnic University
of Coimbra
Coimbra
Portugal

Diana Martins
Polytechnic University
of Coimbra
Coimbra
Portugal

Editorial Office
MDPI
St. Alban-Anlage 66
4052 Basel, Switzerland

This is a reprint of articles from the Special Issue published online in the open access journal *Nutrients* (ISSN 2072-6643) (available at: https://www.mdpi.com/journal/nutrients/special_issues/A80ILZV7GU).

For citation purposes, cite each article independently as indicated on the article page online and as indicated below:

Lastname, A.A.; Lastname, B.B. Article Title. *Journal Name* **Year**, *Volume Number*, Page Range.

ISBN 978-3-7258-1233-2 (Hbk)
ISBN 978-3-7258-1234-9 (PDF)
doi.org/10.3390/books978-3-7258-1234-9

Contents

About the Editors

Nuno Borges

Nuno Borges is a Nutritionist and currently an Associate Professor at the Faculty of Nutrition and Food Sciences, University of Porto, Portugal. He is presently researching the relationships between muscular function, nutritional status, and health outcomes in some diseases. Research interests also include pharmacology and drug–nutrient interactions.

Fernando Mendes

Fernando Mendes (FM) is a Coordinator Professor at the Biomedical Laboratory Sciences Scientific-Pedagogical Unit at Polytechnic University of Coimbra being an effective member of the Health & Technology Research Center, and a researcher collaborator at the Center for Innovative Biomedicine and Biotechnology at the University of Coimbra. He has published 30 full-text papers, 4 books, and 6 book chapters, more than 20 abstracts in periodic journals, more than 120 abstracts in conference proceedings, and 45 communications by invitation. Being co-author of more than 200 posters in international scientific meetings. He was a supervisor of more than 60 bachelor's degree students, 6 master's theses, and 5 research projects in innovation and entrepreneurship, e is currently the supervisor of 2 PhD thesis. FM has 3 patents. In collaboration with Brazil, as an invited researcher, has more than 20 financed research projects. Lately, he has given higher priority to the study of topics related to cancer treatment, namely using radiotherapy and immunotherapy. One of the subtopics that he has been studying aims to better understand the role of the immune system and the immune checkpoint inhibitors in cancer treatment, in prostate, bladder, and penile cancer among other cancers, which may contribute to increasing patients' life expectancy. He has been a reviewer of several journals, having reviewed more than 160 manuscripts. Has been awarded 10 prizes and distinctions.

Diana Martins

Diana Martins is an Associate Professor in the Biomedical Laboratory Sciences at Coimbra Health School. She is an effective member of the Health & Technology Research Center (H&TRC) research center and a collaborator in the Center for Innovative Biomedicine and Biotechnology (CIBB) at the University of Coimbra. A PhD in Pathology and Molecular Genetics at University of Porto and a Post doctoral grant in cancer research allow Diana to study the molecular mechanisms in breast and gastric cancer, focusing on biomarkers for cancer diagnosis and prognosis. Diana has published 36 full-text papers, 3 book chapters and 1 patent. Research interests include cancer, microenvironment, microbiome and cancer biomarkers.

Article

Validation of the Visual/Verbal Analogue Scale of Food Ingesta (Ingesta-VVAS) in Oncology Patients Undergoing Chemotherapy

Hanneke A. H. Wijnhoven [1,*], Loïs van der Velden [1,2], Carolina Broek [1,2], Marleen Broekhuizen [1,2], Patricia Bruynzeel [3], Antoinette van Breen [4], Nanda van Oostendorp [4] and Koen de Heer [2,5]

[1] Department of Health Sciences, Faculty of Science, Amsterdam Public Health Research Institute, Vrije Universiteit Amsterdam, 1081 HV Amsterdam, The Netherlands
[2] Department of Internal Medicine, Flevoziekenhuis, 1315 RA Almere, The Netherlands
[3] Department of Nutrition and Dietetics, Amsterdam UMC Locatie AMC, 1105 AZ Amsterdam, The Netherlands
[4] Dietetics Department, Flevoziekenhsuis, 1315 RA Almere, The Netherlands
[5] Department of Hematology, Amsterdam UMC Locatie AMC, 1105 AZ Amsterdam, The Netherlands
* Correspondence: hanneke.wijnhoven@vu.nl; Tel.: +31-(0)20-5989951

Citation: Wijnhoven, H.A.H.; van der Velden, L.; Broek, C.; Broekhuizen, M.; Bruynzeel, P.; van Breen, A.; van Oostendorp, N.; de Heer, K. Validation of the Visual/Verbal Analogue Scale of Food Ingesta (Ingesta-VVAS) in Oncology Patients Undergoing Chemotherapy. *Nutrients* **2022**, *14*, 3515. https://doi.org/10.3390/nu14173515

Academic Editor: Cheng-Chia Yu

Received: 18 July 2022
Accepted: 12 August 2022
Published: 26 August 2022

Publisher's Note: MDPI stays neutral with regard to jurisdictional claims in published maps and institutional affiliations.

Abstract: This study aimed to: (1) externally validate the Visual/Verbal Analogue Scale of food ingesta (ingesta-VVAS) that previously showed good discrimination between oncology patients who ingest more or less energy than required; (2) explore the discriminative properties of other questions. Dietitians performed 322 interviews in 206 adult oncology patients undergoing chemotherapy in two Dutch hospitals, including a 24-h dietary recall, assessment of the ingesta-VVAS and 12 additional questions related to reduced food intake. The ingesta-VVAS score was linearly associated with energy intake as % of Total Energy Expenditure (TEE) (standardized beta = 0.39, $p < 0.001$), with no differences between groups based on use of oral nutritional supplements, body mass index, in/outpatient setting or sex. The accuracy of the ingesta-VVAS score to predict low energy intake (<75% of TEE) was poor (Area Under the Receiver Operating Characteristic curve (AUC) = 0.668, 95% CI 0.603–0.733). The optimal multivariate model included the ingesta-VVAS score and a question on 'feeling sick' (AUC = 0.680, 95% CI 0.615–0.746). In conclusion, in our study the ingesta-VVAS discriminates poorly between oncology patients undergoing chemotherapy who ingest more or less energy than required. Adding a question on feeling sick only slightly improved model performance. Further external validation is warranted.

Keywords: nutrition; food intake; energy intake; cancer; screening; accuracy

1. Introduction

Oncology patients undergoing chemotherapy often experience periods of reduced energy intake due to a multitude of reasons, disease and treatment related [1–3]. If this lasts for more than a week, patients can lose significant body weight and may become malnourished. Weight loss and a poor nutritional status are associated with a worse response to cancer treatment, reduced quality of life and shorter survival [4–8]. Although a few studies on nutritional interventions showed a positive effect on treatment toxicity and survival, most were negative [9,10]. There is some evidence that nutritional interventions increase muscle mass and quality of life, and there is ample evidence that they attenuate the body weight loss due to chemotherapy [10]. Therefore, clinical guidelines recommend regular evaluation of nutritional intake and body weight in patients undergoing anticancer treatment [11,12]. These guidelines are increasingly implemented. In the Netherlands for example, 80% of adult hospital patients are screened for risk of malnutrition at admission [13] and malnourished or patients at risk are referred for nutritional counseling.

However, oncology patients undergoing chemotherapy treatment may also experience a reduction in energy intake after malnutrition screening. To facilitate the screening for a reduced energy intake during chemotherapy, Guerdoux-Ninot et al. [3] investigated the validity of the Visual/Verbal Analogue Scale of food ingesta (ingesta-VVAS) developed by Thibault et al. [14] in 1762 medical oncology patients scheduled for chemotherapy treatment. They found that the tool was easy-to-use and discriminated well between oncology patients who ingest less or more than 25 kcal per kilogram per day (kcal/kg/day) (Area Under the Receiver-Operating Characteristics curve (AUC) = 0.804) [3]. Therefore, external validation is warranted.

Guerdoux et al. [3] examined the ingesta-VVAS in hospitalized medical oncology patients scheduled for chemotherapy treatment, who did not use oral nutritional supplements (ONS). However, 30% of adult cancer patients use ONS [15]. As ONS use does not obviate further nutritional intervention, the determination of the discriminative accuracy in a population of patients that use ONS is also relevant. Furthermore, as not all patients are hospitalized during chemotherapy treatment, it is also important to test the accuracy in outpatients. As in practice screening may be applied at several time points during a patient's treatment episode, the accuracy of the ingesta-VVAS should also be tested using multiple measurements within a patient. Finally, no previous study tested if the discriminative accuracy may be improved by adding one or more intake or symptom questions related to malnutrition. The aims of the current study were therefore to: (1) validate the ingesta-VVAS in an external population of adult cancer patients undergoing chemotherapy, including patients who use ONS and outpatients; and (2) explore discriminative properties of additional questions that we hypothesized would also predict reduced energy intake.

2. Materials and Methods

2.1. Participants

Data for this study were collected in 2019–2021 among oncology in- and outpatients aged ≥18 years receiving systemic anti-cancer therapy (chemotherapy, targeted therapy, immune therapy and/or monoclonal antibodies, hereafter referred to as: "chemotherapy") in the Flevoziekenhuis (Almere) and the Amsterdam University Medical Centre (AUMC), location AMC (Amsterdam), the Netherlands. Patients treated only with immune checkpoint inhibitors or monoclonal antibodies were not included, as this is unlikely to affect nutritional intake [16]. Exclusion criteria were: use of parenteral nutrition or tube feeding; adequate communication not possible (due to e.g., delirium, a language barrier, a deaf/mute condition or mental retardation); or dietary restrictions based on a doctor's prescription. Between December 2019–March 2020, eligible participants were approached face-to-face for participation in the study in both hospitals, after which data collection was stopped because of the COVID pandemic. Between February–May 2021, the data collection was resumed by telephone. Patients could participate more than once, with at least two weeks between measurements. A flow chart depicting patient inclusion and measurements is depicted in Figure 1. The study was carried out in accordance with the Declaration of Helsinki. The Medical Ethics Review Committee of the Academical Medical Center (METC AMC) reviewed the study protocol and provided a waiver for full review. Written informed consent was obtained from all participants.

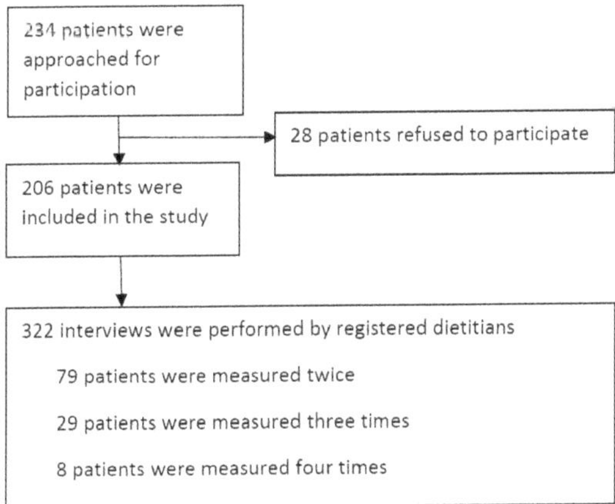

Figure 1. Flow chart of patient inclusion and measurements performed among oncology in- and outpatients aged ≥18 years receiving systemic anti-cancer therapy in two Dutch Hospitals between 2019–2020.

2.2. Data Collection

2.2.1. Procedures

Interviews were conducted by five registered dietitians within 3 weeks after administration of chemotherapy. Each interview included a 24-h dietary recall and assessment of 13 screening questions, including the ingesta-VVAS, and a few demographic questions. The order of assessing the 24 h recall and the screening questions was randomized.

2.2.2. Patient Characteristics

Patient characteristics were obtained from the hospital records or were assessed during the interview when not available. This included: sex; age; body weight (kg); body height (cm); setting (in- or outpatient); hospital; number of days since chemotherapy; type of cancer; type of treatment(s); and whether a dietitian was involved. Body mass index (BMI) was calculated as body weight (kg) divided by body height squared (m^2). After the resumption of the study in 2021, we added questions from the Short Nutritional Assessment Questionnaire (SNAQ) [17] and Malnutrition Universal Screening Tool (MUST) [18] to the interview.

2.2.3. Dietary Intake

Dietary intake was assessed by an in-depth 24-h dietary recall of 15–20 min. The experienced dietitians collected detailed information about food preparation methods and the ingredients used in compound dishes. The amount of each food product was estimated by using common size containers (e.g., glasses, cups, different types of spoons). Energy (kcal) and protein (gram) intake were based on data of the Dutch Food Composition Table (2019) [19]. For patients using ONS, energy and protein intake from ONS were recorded separately. ONS comprised both medical nutrition and commercially available protein powders and shakes.

Protein intake was expressed as grams per day (g/day) and in grams per kilogram body weight per day (g/kg BW/day). Energy intake was expressed as kcal per day (kcal/day), kcal per kilogram body weight per day (kcal/kg BW/day) and as percentage of Total Energy Expenditure (TEE) (%TEE). A lower than required intake was defined as an energy intake <75% of TEE [20]. TEE was calculated by the World Health Organization (WHO) formula [21] plus a 30% addition for physical activity/disease activity. When the BMI was >30 kg/m^2, the WHO formula was replaced by the Harris & Benedict 1918 formula [22,23]. We also defined low energy intake as a caloric intake <25 kcal/kg/day as was done by Guerdoux et al. [3].

2.2.4. Ingesta-VVAS

The ingesta-VVAS was assessed (in Dutch) with the following question: 'If you consider that, at times when you are in good health, you eat 10 out of 10, how much do you currently eat on a scale from 0 to 10?' 0 would mean eating "nothing at all" and 10 eating "as usual" [3]. We specified "currently" as pertaining to the past 24 h. During the face-to-face data collection most patients responded to the verbal question but a few patients preferred the visual scale as described by Guerdoux et al. [3].

2.2.5. Additional Questions

We included 12 additional screening questions, derived from the Patient-Generated Subjective Global Assessment Short Form (PG-SGA SF), a validated screening tool for malnutrition in chemotherapy outpatients [24]. Some questions were slightly adapted. We included a question on food intake: 'In the past 24 h, did you eat less compared to your usual intake before the cancer diagnosis? (yes/no). Additionally, we included 11 questions on symptoms related to reduced food intake (answer options: yes; no) in the past 24 h: 'Did you have problems eating?'; 'Were you nauseous?'; 'Did you vomit?'; 'Did you experience taste alterations?'; 'Did you have a painful mouth?'; 'Did you experience pain in your abdomen?'; 'Did you experience pain while swallowing?"; 'Did you experience being bothered by food smells?'; 'Did you feel full quickly?'; 'Did you feel fatigued?'; and 'Did you feel sick?'.

2.3. Statistical Analyses

Patient characteristics were presented as mean with standard deviation ($\pm$SD), median with interquartile range (IQR) or frequency (%), for only the first measurement (single cases), for all measurements combined, and stratified by ONS use. The association between the ingesta-VVAS score and energy intake as percentage of TEE was examined by a generalized estimating equations (GEE) model to adjust for repeated measures within individuals as well as by a linear model treating all measurements as independent. As this resulted in equivalent model parameters, we only presented the results of simpler linear regression model, also depicted by a boxplot. Results were also presented stratified by ONS use, BMI, setting (in-/outpatients) and sex (male/female). Interaction was tested by including interaction terms in the linear models.

To examine the external validity of the ingesta-VVAS in detecting an energy intake <75% of TEE, a logistic regression model and AUC curve were used. To illustrate the performance of the ingesta-VVAS in practice, sensitivity, specificity, positive predictive value (PPV), negative predictive values (NPV), and diagnostic accuracy (*n* correctly classified/total *n*) were calculated using the optimal cut-off ($\leq$7) derived by Guerdoux et al. [3] and, if different, the optimal cut-off (based on maximized sensitivity and specificity) in our sample, stratified by ONS use. Sensitivity analyses were performed with <25 kcal/kg/day as reference standard.

To explore the discriminative properties of the 12 additional questions and the Ingesta-VVAS score in predicting a low energy intake, the association between the response to each single question and an energy intake of <75% TEE was examined with a univariate logistic regression model. Subsequently, a forward and backward selection method was used to derive the most optimal multivariate prediction model [25]. A *p*-value < 0.05 was required for inclusion in the final multivariable model. The performance of the final model was assessed by the AUC.

A *p*-value (two-sided) less than 0.05 was considered statistically significant. Statistical analyses were performed by using Statistical Package for Social Sciences (SPSS) version 27 (IBM Corp., Armonk, NY, USA).

3. Results

Of the 234 patients approached, 28 patients (12%) did not want to participate, mostly due to feeling sick or having no interest. In total 322 measurements were performed in

206 patients of whom 79 were measured twice, 29 thrice and 8 four times (Figure 1). Table 1 shows the characteristics of the study sample, of only the first measurement (single cases) as well as all measurements combined, stratified by use of ONS. Mean age was 62 years (SD 12). Most patients were outpatients (81%) and received their chemotherapy 0–6 days ago. At their first measurement, 35% of patients had an energy intake <75% of TEE and in 40% the Ingesta-VVAS score was ≤7. ONS was used by 21% of patients of which 90% was seen by a dietitian. ONS users had a higher mean energy and protein intake than non-users, were slightly older, more often had cancer of gastrointestinal origin, and a lower BMI. Although ONS users more often had an ingesta-VVAS score ≤7 (55% versus 37%), they less often had an energy intake <75% of TEE (24% versus 34%).

Table 1. Characteristics of the included oncology patients undergoing chemotherapy, stratified by oral nutritional supplement (ONS) use.

Characteristics	First Measurement $n = 206$	All (Multiple) Measurements $n = 322$	ONS Use $n = 67$	No ONS Use $n = 255$
Age (years) [a]	62 ± 12	62 ± 12	65 ± 11	61 ± 12
Female sex [b]	114 (55%)	186 (58%)	33 (49%)	153 (60%)
Type of cancer [b]				
Gastro-intestinal	49 (24%)	78 (24%)	27 (40%)	51 (20%)
Breast	4 (21%)	78 (24%)	11 (16%)	67 (26%)
Hematological	63 (31%)	95 (30%)	16 (24%)	79 (31%)
Other	50 (24%)	71 (22%)	13 (20%)	58 (23%)
Setting: inpatient (vs. outpatient) [b]	39 (19%)	44 (14%)	12 (18%)	32 (13%)
Days since last chemotherapy [c]	3 (0–6)	3 (1–6)	3 (1–6)	4 (1–6)
Dietitian involved [b]	83 (40%)	136 (42%)	60 (90%)	76 (30%)
Body Mass Index (kg/m^2) [a]	26.7 ± 4.9	26.9 ± 4.9	25.8 ± 4.8	27.2 ± 4.9
Protein intake (g/kg/day)	0.95 ± 0.45	0.95 ± 0.43	1.18 ± 0.51	0.89 ± 0.39
Energy intake (kcal/day) [a,d]	1770 ± 684	1804 ± 673	2035 ± 711	1743 ± 651
Energy intake ONS (kcal/day) [a]	-		490 ± 235	-
Energy intake % of TEE [a,e]	87 ± 34	89 ± 33	102 ± 38	85 ± 31
Energy intake <75% of TEE [b,e]	72 (35%)	105 (33%)	16 (24%)	89 (34%)
Energy intake <25 kcal/kg/day [b]	127 (62%)	197 (61%)	32 (48%)	165 (65%)
ingesta-VVAS score [f]				
0	3 (2%)	3 (1%)	1 (2%)	2 (1%)
1	0	0	0	0
2	0	0	0	0
3	3 (2%)	3 (1%)	1 (2%)	2 (1%)
4	13 (6%)	15 (5%)	6 (9%)	9 (4%)
5	16 (8%)	27 (8%)	7 (10%)	20 (8%)
6	17 (8%)	35 (11%)	12 (18%)	23 (9%)
7	30 (15%)	49 (15%)	10 (15%)	39 (15%)
8	41 (20%)	62 (19%)	14 (21%)	48 (19%)
9	16 (8%)	25 (8%)	5 (8%)	20 (8%)
10	67 (33%)	103 (32%)	11 (16%)	192 (36%)
ingesta-VVAS score [c,f]	8 (6–10)	8 (6–10)	7 (6–8)	8 (7–10)
ingesta-VVAS score ≤ 7 [b,f]	82 (40%)	132 (41%)	37 (55%)	95 (37%)
MUST score [b,g]				
0 (low malnutrition risk)	87 (69)	172 (72)	15 (33)	157 (81)
1 (medium malnutrition risk)	22 (17)	37 (15)	16 (35)	21 (11)
≥2 (high malnutrition risk)	18 (13)	31 (13)	15 (33)	16 (8)
SNAQ score [b,h]				
0, 1 (no malnutrition)	82 (65)	164 (68)	15 (33)	149 (77)
2 (moderate malnutrition)	7 (6)	14 (6)	4 (9)	10 (5)
≥3 (severe malnutrition)	38 (30)	62 (26)	27 (59)	35 (18)

Values are expressed as: [a] mean ± SD, [b] n (%), [c] median (25–75% IQR), [d] calculated by a 24-h recall and includes intake from (regular) food and ONS; [e] Total Energy Expenditure (TEE) is calculated by the World Health Organization 1985 formula [21] (BMI ≤ 30 kg/m^2) or Harris & Benedict 1918 formula [23] (BMI > 30 kg/m^2), both with an addition of 30% for physical activity/disease activity; [f] Ingesta-Verbal/Visual Analogue Scale (Ingesta-VVAS) ranges from 0 (I eat nothing at all) to 10 (as usual); [g] MUST: malnutrition universal screening tool, added to data collection in 2021; [h] SNAQ: short nutritional assessment questionnaire, added to data collection in 2021. SD, standard deviation; IQR, interquartile range; ONS, oral nutritional supplements; BMI, body mass index.

We found a positive linear association (standardized beta (β) = 0.39, $p < 0.001$) between the ingesta-VVAS score and energy intake as percentage of TEE in the total study sample including all measurements (n = 322) (Figure 2). There was no significant interaction by ONS use (P interaction = 0.116; β = 0.52 for ONS users; β = 0.42 for non-ONS users), BMI (P interaction = 0.130; β = 0.43 for BMI $\leq$ 25 kg/m^2; β = 0.39 for BMI > 25 kg/m^2), setting (P interaction = 0.844; β = 0.37 for outpatients; β = 0.46 for inpatients), or sex (P interaction = 0.106; β = 0.31 for males; β = 0.43 for females).

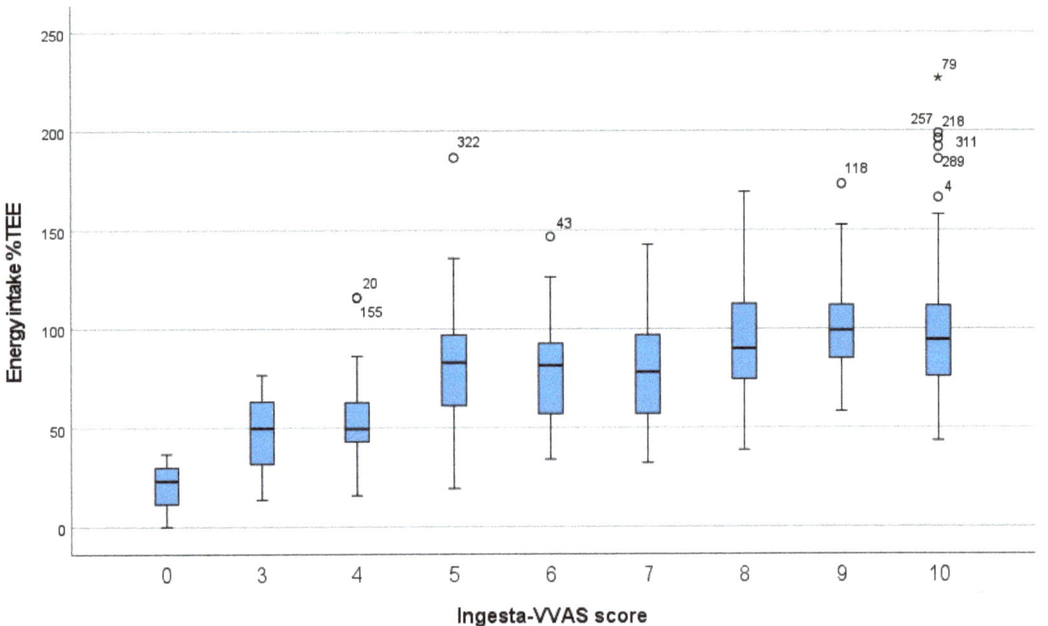

Figure 2. Association (n = 322) between the Ingesta-Verbal/Visual Analogue Scale (ingesta-VVAS) and energy intake, calculated by a 24-h recall and expressed as percentage of Total Energy Expenditure (TEE), calculated by the World Health Organization 1985 formula [21] (BMI $\leq$ 30 kg/m^2) or Harris & Benedict 1918 formula [23] (BMI > 30 kg/m^2), both with an addition of 30% for physical activity/disease activity. BMI, body mass index. ° outlier (3rd quartile + 1.5* interquartile range). * outlier (3rd quartile + 3* interquartile range).

In the logistic regression model, a 1-point lower ingesta-VVAS score was associated with an 1.38 higher odds (95% confidence interval (CI) 1.23–1.57) of an energy intake <75% of TEE. The AUC was 0.668 (95% CI 0.603–0.733, Figure 3). In our sample a cut-off of $\leq$7 maximized sensitivity plus specificity, irrespective of ONS use. Table 2 shows that at a cut-off point $\leq$7, 66.1% of patients was classified correctly and that sensitivity and specificity were respectively 61.0% (95% CI 51.4–69.9%); and 68.7% (95% CI 62.3–74.6%). Sensitivity analyses with the reference standard of <25 kcal/kg/day resulted in slightly weaker associations (Odds Ratio 1.27, 95% CI 1.12–1.44).

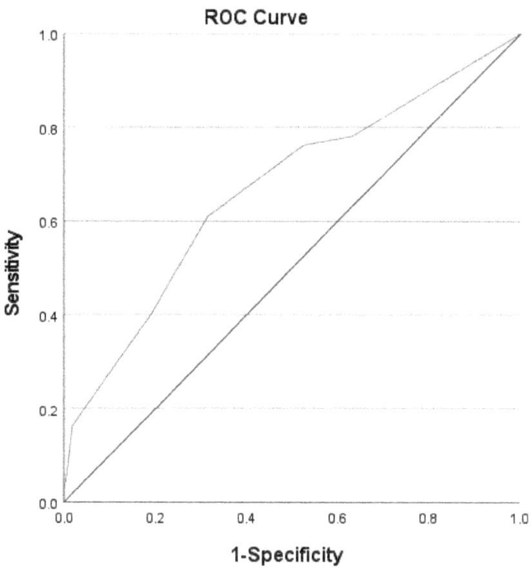

Figure 3. Receiver Operating Characteristic (ROC) curve (*n* = 322) of the Ingesta-verbal/Visual analogue Scale (ingesta-VVAS) to predict energy intake, calculated by a 24-h recall, <75% of Total Energy Expenditure (TEE) (Area under the ROC curve = 0.668, 95% CI 0.603–0.733). TEE is calculated by the World Health Organization 1985 formula [21] (BMI $\leq$ 30 kg/m^2) or Harris & Benedict 1918 formula [23] (BMI > 30 kg/m^2), both with an addition of 30% for physical activity/disease activity. CI, confidence interval.

Table 2. Performance of the Ingesta-verbal/Visual analogue Scale (ingesta-VVAS) to predict energy intake <75% of Total Energy Expenditure (TEE) (*n* = 322).

	Energy Intake <75% of TEE	Energy Intake $\geq$75% of TEE	Total (*n*)
Ingesta-VVAS score $\leq$ 7	64	68	132
Ingesta-VVAS score > 7	41	149	190
Total (*n*)	105	217	322

Sensitivity 61.0% (95% CI 51.4–69.9%); specificity 68.7% (95% CI 62.3–74.6%); positive predictive value 48.5% (95% CI 40.0–57.0%); negative predictive value 78.4% (95% CI 72.2–83.9%); Diagnostic accuracy = (149 + 64)/322 = 66.1%. Energy intake is calculated by a 24-h recall. TEE is calculated by the World Health Organization 1985 formula [21] (BMI $\leq$ 30 kg/m^2) or Harris & Benedict 1918 formula [23] (BMI > 30 kg/m^2), both with an addition of 30% for physical activity/disease activity. CI, confidence interval.

Based on univariate logistic regression models (Table 3), seven questions were found to be associated with an energy intake < 75% of TEE besides the ingesta-VVAS score. The final multivariable model consisted of two predictors: the 'ingesta-VVAS score' and 'feeling sick' (Table 4). The AUC of this model was 0.680 (95% CI 0.615–0.746).

Table 3. Associations between each candidate predictor and energy intake <75% of Total Energy Expenditure (TEE) (n = 322).

Question (Candidate Predictor)	Energy Intake <75% of TEE (n = 105)	Energy Intake ≥75% of TEE (n = 217)	Odds Ratio (95% CI)	p-Value
Eating less	66 (63%)	78 (36%)	3.02 (1.86–4.89)	<0.001
Problems eating	33 (31%)	31 (14%)	2.75 (1.57–4.82)	<0.001
Nausea	26 (25%)	40 (18%)	1.46 (0.83–2.55)	0.189
Vomiting	6 (6%)	2 (1%)	6.51 (1.29–32.85)	0.023
Taste alterations	52 (50%)	84 (38%)	1.58 (0.99–2.53)	0.055
Painful mouth	19 (18%)	22 (10%)	1.96 (1.01–3.81)	0.047
Painful abdomen	22 (21%)	46 (21%)	0.99 (0.56–1.75)	0.960
Pain swallowing	6 (6%)	8 (4%)	1.58 (0.54–4.69)	0.407
Bothered by food smells	11 (11%)	24 (11%)	0.94 (0.44–2.00)	0.875
Feeling full quickly	69 (66%)	91 (42%)	2.65 (1.63–4.31)	<0.001
Feeling fatigued	79 (75%)	158 (73%)	1.14 (0.67–1.94)	0.643
Feeling sick	39 (37%)	38 (18%)	2.78 (1.64–4.72)	<0.001

Values are expressed as n (%). Energy intake is calculated by a 24-h recall. TEE is calculated by the World Health Organization 1985 formula [21] (BMI ≤ 30 kg/m^2) or Harris & Benedict 1918 formula [23] (BMI > 30 kg/m^2), both with an addition of 30% for physical activity/disease activity.

Table 4. Final multivariate model for prediction of energy intake <75% of Total Energy Expenditure (TEE) (n = 322).

Predictors	Regression Coefficient	Odds Ratio (95% CI)	p-Value
Constant	−1.59		
Ingesta-VVAS score [a] (reversed)	0.29	1.33 (1.17–1.51)	<0.001
Feeling sick	0.65	1.92 (1.09–3.91)	0.024

Energy intake is calculated by a 24-h recall. TEE is calculated by the World Health Organization 1985 formula [21] (BMI ≤ 30 kg/m^2) or Harris & Benedict 1918 formula [23] (BMI > 30 kg/m^2), both with an addition of 30% for physical activity/disease activity. [a] Ingesta-Verbal/Visual Analogue Scale (ingesta-VVAS) reversed so that a lower score is associated with an 1.33 higher odds on energy intake <75% of TEE.

4. Discussion

The main aim of this study was to externally validate the Visual/Verbal Analogue Scale of food ingesta (ingesta-VVAS) as a tool to discriminate between oncology patients undergoing chemotherapy who ingest less than 75% of their required energy or more. In a study sample of 206 adult oncology in- and outpatients including 322 measurements, we found that the ingesta-VVAS score was linearly associated (β = 0.39, p < 0.001) with energy intake as percentage of required, with no differences between groups based on use of oral nutritional supplements, body mass index or in/outpatient setting. However, the accuracy of the ingesta-VVAS score to predict low energy intake (<75% of TEE) was poor. If in a hypothetical cohort of 100 patients, 35 have a low energy intake, on average 14 patients (95% CI 11–17) will be missed using the ingesta-VVAS, while 20 patients (95% CI 16–25) would be incorrectly classified as having a low energy intake.

As a secondary aim, we explored the discriminative properties of 12 additional questions and found that the ingesta-VVAS was retained in the final model together with a question on "feeling sick". Adding this question only slightly improved model performance.

The ingesta-VVAS was first examined by Thibault et al. [14] in 114 undernourished or at risk patients in two French University hospitals. Both the 10-point verbal (ρ = 0.66) and visual (ρ = 0.74) scale correlated well with energy intake based on a 3-day dietary records. Stronger correlations were found for inpatients (ρ = 0.73) compared to outpatients (ρ = 0.32), for those with a BMI < 19 kg/m^2 (ρ = 0.78) compared to those with a BMI ≥ 25 kg/m^2 (ρ = 0.39), and for those malnourished based on the Nutritional Risk Index (NRI) [26] (ρ = 0.82) compared to those who were not malnourished (ρ = 0.11). Subsequently, a larger study was performed by Guerdoux et al. [3] among 1762 medical oncology patients who were hospitalized for more than 48 h and scheduled for chemotherapy treatment, and did not use artificial nutrition or ONS. They found that in 95% of patients it was feasible to use the verbal form of the ingesta-VVAS. In this French population, the ingesta-VVAS score correlated well with mean daily energy intake based on a 24-h recall about food intake the day before the hospitalization (ρ = 0.67), with no major differences between subgroups based on previous weight loss, BMI and NRI. The discriminative accuracy was good.

Our study differed in a number of respects from the study by Guerdoux et al. [3]. First, our study was performed in the Netherlands. Although we are not aware of national differences in response to questions on food intake, this could explain differences in performance. It is difficult to compare between both studies the percentage of patients at risk of malnutrition due to variation in screening instruments applied. The percentages with a major nutritional risk score according to the NRI [26] in the Guerdoux study and a high malnutrition risk score according to the MUST [18] in our study were comparable (13%). The median ingesta-VVAS score was higher in our study (8 compared to 6 observed by Guerdoux et al. [3]), even though energy intake was comparable (62% < 25 kcal/kg/day compared to 67% [3]). The on average higher ingesta-VVAS score at the same energy intake level in our study is reflected by the lower sensitivity and positive predictive value at a cut-off of 7 in our study. Second, we also included outpatients and patients using ONS, and third, in 2021 we switched from verbal interviews (*n* = 82) to telephone interviews (*n* = 240). However, correlations were not modified by clinical setting or ONS use and we also did not observe interaction by verbal/telephone interview (P interaction = 0.54; β = 0.49 for 2020 verbal interviews; β = 0.34 for 2021 telephone interviews), nor was there a difference in mean ingesta-VVAS score or energy intake between verbal or telephone interviews (data not shown). However, it cannot be excluded that this played a role.

When exploring the discriminative properties of 12 additional questions on (symptoms related to) reduced food intake, we found that the Ingesta-VVAS was retained in the final model together with a question on "feeling sick". The individual questions on 'eating less', 'problems eating', 'taste alterations', 'vomiting', 'painful mouth', 'feeling full quickly' and 'feeling sick', were associated with a lower energy intake. Many patients reported 'feeling fatigued' but this was not associated with a lower energy intake. The finding that the ingesta-VVAS was retained in the final model with only one additional question shows that it captures the predictive validity of virtually all other questions. Although adding the question on 'feeling sick' improved model performance, this was not a large improvement.

Besides the large study of Guerdoux et al. [3] and a recent study by the group of Thibault et al. that evaluated the accuracy of the visual analogue scale for food intake as a screening test for malnutrition in primary care [27], we are not aware of external validation studies in a cohort of oncology patients. Thus, our study, the first in a non-French speaking country, thereby provides an important contribution to the external validation of the ingesta-VVAS. Strengths of our study were the inclusion of patients using ONS and outpatients, allowing us to examine the validity of the ingesta-VVAS in these (sub)groups as well. Moreover, we included multiple measurements per patients as in clinical practice screening may also take place on different days. Of 206 patients, 79 were measured at least twice. The strength of associations was consistent for the first and second measurement (β = 0.42 for measurement one, β = 0.36 for measurement two) suggesting that the ingesta-VVAS can be used to screen patients multiple times. Additionally, adjustment for repeated observations did not alter the results. The confidence interval of the AUC (95% CI 0.603–0.733) shows that our study was not underpowered and the sample size large enough to justify the conclusions.

Several limitations need to be mentioned. First, we used a single 24 h recall to assess energy intake. This validates the ingesta-VVAS as a dietary snapshot to be able to detect an acute drop in energy intake. However, multiple 24 h recalls are preferred when measuring long-term exposure to dietary intake factors and for example incidence of chronic diseases [28]. For use in such a long-term setting our validation method is not optimal. We chose to repeat the original validation method by Guerdoux et al. [3] Second, using a single 24 h recall increases the probability of recall bias and may result in lower accuracy of dietary intake assessment and thereby weaken observed correlations. Therefore, all interviews were performed by experienced dietitians and we repeated the noted food intake at the end of the recall to minimize recall bias.

5. Conclusions

Our study showed poor accuracy of the ingesta-VVAS in detecting low energy intake in oncology patients undergoing chemotherapy as measured by a single 24-h dietary recall, whether patients were in- or outpatients, used ONS or not, and whether the ingesta-VVAS was assessed by telephone or to face-to-face. Adding a question on feeling sick only slightly improved model performance. More external validation studies are necessary before the ingesta-VVAS can be implemented in clinical practice.

Author Contributions: H.A.H.W. and K.d.H. designed the study. H.A.H.W. performed the statistical analyses. L.v.d.V., C.B., M.B., P.B., A.v.B. and N.v.O. contributed to the design of the study and performed the data collection. H.A.H.W. and K.d.H. drafted the original paper. All authors revised the manuscript critically for important intellectual content. All authors have read and agreed to the published version of the manuscript.

Funding: This research received no external funding.

Institutional Review Board Statement: Ethical review and approval were waived for this study by medical ethics committee of the Amsterdam UMC locations AMC (W19_406) because the participants in our study only underwent procedures that are considered standard of care.

Informed Consent Statement: Informed consent was obtained from all subjects involved in the study.

Data Availability Statement: The data presented in this study are available on request from the corresponding author. The data are not publicly available due to privacy restrictions.

Acknowledgments: We thank the study participants for their participation in the study.

Conflicts of Interest: The authors declare no conflict of interest.

References

1. Bozzetti, F. Screening the nutritional status in oncology: A preliminary report on 1000 outpatients. *Support. Care Cancer* **2009**, *17*, 279–284. [CrossRef]
2. Coa, K.I.; Epstein, J.B.; Ettinger, D.; Jatoi, A.; McManus, K.; Platek, M.E.; Price, W.; Stewart, M.; Teknos, T.N.; Moskowitz, B. The Impact of Cancer Treatment on the Diets and Food Preferences of Patients Receiving Outpatient Treatment. *Nutr. Cancer* **2015**, *67*, 339–353. [CrossRef]
3. Guerdoux-Ninot, E.; Flori, N.; Janiszewski, C.; Vaillé, A.; de Forges, H.; Raynard, B.; Baracos, V.E.; Thezenas, S.; Senesse, P. Assessing dietary intake in accordance with guidelines: Useful correlations with an ingesta-Verbal/Visual Analogue Scale in medical oncology patients. *Clin. Nutr.* **2019**, *38*, 1927–1935. [CrossRef]
4. Di Fiore, A.; Lecleire, S.; Gangloff, A.; Rigal, O.; Benyoucef, A.; Blondin, V.; Sefrioui, D.; Quiesse, M.; Iwanicki-Caron, I.; Michel, P.; et al. Impact of nutritional parameter variations during definitive chemoradiotherapy in locally advanced oesophageal cancer. *Dig. Liver Dis.* **2014**, *46*, 270–275. [CrossRef]
5. Topkan, E.; Parlak, C.; Selek, U. Impact of Weight Change During the Course of Concurrent Chemoradiation Therapy on Outcomes in Stage IIIB Non-Small Cell Lung Cancer Patients: Retrospective Analysis of 425 Patients. *Int. J. Radiat. Oncol.* **2013**, *87*, 697–704. [CrossRef]
6. Bullock, A.F.; Greenley, S.; McKenzie, G.A.G.; Paton, L.W.; Johnson, M.J. Relationship between markers of malnutrition and clinical outcomes in older adults with cancer: Systematic review, narrative synthesis and meta-analysis. *Eur. J. Clin. Nutr.* **2020**, *74*, 1519–1535. [CrossRef]
7. Andreyev, H.J.N.; Norman, A.R.; Oates, J.; Cunningham, D. Why do patients with weight loss have a worse outcome when undergoing chemotherapy for gastrointestinal malignancies? *Eur. J. Cancer* **1998**, *34*, 503–509. [CrossRef]
8. Persson, C.; Glimelius, B. The relevance of weight loss for survival and quality of life in patients with advanced gastrointestinal cancer treated with palliative chemotherapy. *Anticancer Res.* **2002**, *22*, 3661–3668.
9. de van der Schueren, M.A.E.; Laviano, A.; Blanchard, H.; Jourdan, M.; Arends, J.; Baracos, V.E. Systematic review and meta-analysis of the evidence for oral nutritional intervention on nutritional and clinical outcomes during chemo(radio)therapy: Current evidence and guidance for design of future trials. *Ann. Oncol.* **2018**, *29*, 1141–1153. [CrossRef]

10. van der Werf, A.; Langius, J.A.E.; Beeker, A.; Tije, A.J.T.; Vulink, A.J.; Haringhuizen, A.; Berkhof, J.; van der Vliet, H.J.; Verheul, H.M.W.; de van der Schueren, M.A.E. The effect of nutritional counseling on muscle mass and treatment outcome in patients with metastatic colorectal cancer undergoing chemotherapy: A randomized controlled trial. *Clin. Nutr.* **2020**, *39*, 3005–3013. [CrossRef]

11. August, D.A.; Huhmann, M.B.; American Society for Parenteral and Enteral Nutrition. Clinical Guidelines: Nutrition Support Therapy during Adult Anticancer Treatment and in Hematopoietic Cell Transplantation. *J. Parenter. Enter. Nutr.* **2009**, *33*, 472–500. [CrossRef]

12. Arends, J.; Bachmann, P.; Baracos, V.; Barthelemy, N.; Bertz, H.; Bozzetti, F.; Fearon, K.; Hütterer, E.; Isenring, E.; Kaasa, S.; et al. ESPEN guidelines on nutrition in cancer patients. *Clin. Nutr.* **2017**, *36*, 11–48. [CrossRef]

13. Kruizenga, H.; Van Keeken, S.; Weijs, P.; Bastiaanse, L.; Beijer, S.; Waal, G.H.-D.; Jager-Wittenaar, H.; Jonkers-Schuitema, C.; Klos, M.; Remijnse-Meester, W.; et al. Undernutrition screening survey in 564,063 patients: Patients with a positive undernutrition screening score stay in hospital 1.4 d longer. *Am. J. Clin. Nutr.* **2016**, *103*, 1026–1032. [CrossRef]

14. Thibault, R.; Goujon, N.; Le Gallic, E.; Clairand, R.; Sébille, V.; Vibert, J.; Schneider, S.M.; Darmaun, D. Use of 10-point analogue scales to estimate dietary intake: A prospective study in patients nutritionally at-risk. *Clin. Nutr.* **2009**, *28*, 134–140. [CrossRef]

15. Hébuterne, X.; Lemarié, E.; Michallet, M.; De Montreuil, C.B.; Schneider, S.; Goldwasser, F. Prevalence of Malnutrition and Current Use of Nutrition Support in Patients with Cancer. *J. Parenter. Enter. Nutr.* **2014**, *38*, 196–204. [CrossRef]

16. Schirrmacher, V. From chemotherapy to biological therapy: A review of novel concepts to reduce the side effects of systemic cancer treatment (Review). *Int. J. Oncol.* **2019**, *54*, 407–419. [CrossRef]

17. Kruizenga, H.M.; Seidell, J.C.; de Vet, H.C.W.; Wierdsma, N.J. Development and validation of a hospital screening tool for malnutrition: The short nutritional assessment questionnaire (SNAQ). *Clin. Nutr.* **2005**, *24*, 75–82. [CrossRef]

18. Stratton, R.J.; Hackston, A.; Longmore, D.; Dixon, R.; Price, S.; Stroud, M.; King, C.; Elia, M. Malnutrition in hospital outpatients and inpatients: Prevalence, concurrent validity and ease of use of the 'malnutrition universal screening tool' ('MUST') for adults. *Br. J. Nutr.* **2004**, *92*, 799–808. [CrossRef]

19. RIVM. NEVO Online Version 2019/6.0. Available online: https://www.rivm.nl/documenten/voedingsstoffen-in-nevo-online-2019 (accessed on 1 April 2021).

20. Blum, D.; European Palliative Care Research Collaborative; Omlin, A.; Fearon, K.; Baracos, V.; Radbruch, L.; Kaasa, S.; Strasser, F. Evolving classification systems for cancer cachexia: Ready for clinical practice? *Support. Care Cancer* **2010**, *18*, 273–279. [CrossRef]

21. *Energy and Protein Requirements. Report of a Joint FAO/WHO/UNU Expert Consultation*; World Health Organization Technical Report Series; no. 724; Istituto Scotti Bassani: Rome, Italy, 1985; pp. 1–206.

22. Kruizenga, H.M.; Hofsteenge, G.H.; Weijs, P.J. Predicting resting energy expenditure in underweight, normal weight, overweight, and obese adult hospital patients. *Nutr. Metab.* **2016**, *13*, 85. [CrossRef]

23. Harris, J.A.; Benedict, F.G. A Biometric Study of Human Basal Metabolism. *Proc. Natl. Acad. Sci. USA* **1918**, *4*, 370–373. [CrossRef]

24. Abbott, J.; Teleni, L.; McKavanagh, D.; Watson, J.; McCarthy, A.L.; Isenring, E. Patient-Generated Subjective Global Assessment Short Form (PG-SGA SF) is a valid screening tool in chemotherapy outpatients. *Support. Care Cancer* **2016**, *24*, 3883–3887. [CrossRef]

25. Peduzzi, P.; Concato, J.; Kemper, E.; Holford, T.R.; Feinstein, A.R. A simulation study of the number of events per variable in logistic regression analysis. *J. Clin. Epidemiol.* **1996**, *49*, 1373–1379. [CrossRef]

26. Buzby, G.P.; Williford, W.O.; Peterson, O.L.; Crosby, L.O.; Page, C.P.; Reinhardt, G.F.; Mullen, J.L. A randomized clinical trial of total parenteral nutrition in malnourished surgical patients: The rationale and impact of previous clinical trials and pilot study on protocol design. *Am. J. Clin. Nutr.* **1988**, *47*, 357–365. [CrossRef]

27. Bouëtté, G.; Esvan, M.; Apel, K.; Thibault, R. A visual analogue scale for food intake as a screening test for malnutrition in the primary care setting: Prospective non-interventional study. *Clin. Nutr.* **2021**, *40*, 174–180. [CrossRef]

28. Shim, J.-S.; Oh, K.; Kim, H.C. Dietary assessment methods in epidemiologic studies. *Epidemiol. Health* **2014**, *36*, e2014009. [CrossRef]

Article

Association of Low Handgrip Strength with Chemotherapy Toxicity in Digestive Cancer Patients: A Comprehensive Observational Cohort Study (FIGHTDIGOTOX)

Pierre Martin [1,2,*], Damien Botsen [1,2], Mathias Brugel [2], Eric Bertin [3], Claire Carlier [1,2], Rachid Mahmoudi [4], Florian Slimano [5], Marine Perrier [2] and Olivier Bouché [2]

[1] Department of Medical Oncology, Godinot Cancer Institute, 51100 Reims, France
[2] Department of Gastroenterology and Digestive Oncology, Université de Reims Champagne-Ardenne, CHU Reims, 51100 Reims, France
[3] Department of Nutrition, Endocrinology and Diabetology, CHU Reims, 51100 Reims, France
[4] Department of Internal Medicine and Geriatrics, Université de Reims Champagne-Ardenne, VieFra, CHU Reims, 51100 Reims, France
[5] Department of Pharmacy, Université de Reims Champagne-Ardenne, CHU Reims, 51100 Reims, France
* Correspondence: pierre.martin@chu-reims.fr; Tel.: +33-6-70-11-46-10

Abstract: In the FIGHTDIGO study, digestive cancer patients with dynapenia experienced more chemotherapy-induced neurotoxicities. FIGHTDIGOTOX aimed to evaluate the relationship between pre-therapeutic handgrip strength (HGS) and chemotherapy-induced dose-limiting toxicity (DLT) or all-grade toxicity in digestive cancer patients. HGS measurement was performed with a Jamar dynamometer. Dynapenia was defined according to EWGSOP2 criteria (<27 kg (men); <16 kg (women)). DLT was defined as any toxicity leading to dose reduction, treatment delay, or permanent discontinuation. We also performed an exploratory analysis in patients below the included population's median HGS. A total of 244 patients were included. According to EWGSOP2 criteria, 23 patients had pre-therapeutic dynapenia (9.4%). With our exploratory median-based threshold (34 kg for men; 22 kg for women), 107 patients were dynapenic (43.8%). For each threshold, dynapenia was not an independent predictive factor of overall DLT and neurotoxicity. Dynapenic patients according to EWGSOP2 definition experienced more hand-foot syndrome ($p = 0.007$). Low HGS according to our exploratory threshold was associated with more all-grade asthenia ($p = 0.014$), anemia ($p = 0.006$), and asthenia with DLT ($p = 0.029$). Pre-therapeutic dynapenia was not a predictive factor for overall DLT and neurotoxicity in digestive cancer patients but could be a predictive factor of chemotherapy-induced anemia and asthenia. There is a need to better define the threshold of dynapenia in cancer patients.

Keywords: digestive system neoplasms; dose-limiting toxicity; dynapenia; muscle strength; sarcopenia; frailty; clinical nutrition; malnutrition

Citation: Martin, P.; Botsen, D.; Brugel, M.; Bertin, E.; Carlier, C.; Mahmoudi, R.; Slimano, F.; Perrier, M.; Bouché, O. Association of Low Handgrip Strength with Chemotherapy Toxicity in Digestive Cancer Patients: A Comprehensive Observational Cohort Study (FIGHTDIGOTOX). *Nutrients* **2022**, *14*, 4448. https://doi.org/10.3390/nu14214448

Academic Editors: Nuno Borges, Diana Martins and Fernando Mendes

Received: 21 September 2022
Accepted: 20 October 2022
Published: 22 October 2022

Publisher's Note: MDPI stays neutral with regard to jurisdictional claims in published maps and institutional affiliations.

1. Introduction

Digestive cancers are among the most common spectrum of cancer in the world [1]. Anticancer agents have potential acute and chronic toxicities which may require treatment dose adaptations. Identifying predictive factors could help physicians to prevent the occurrence of chemotherapy-induced dose-limiting toxicity (DLT).

Low lean body mass and sarcopenia have been shown to predict anticancer drug toxicity in patients with breast or colorectal cancer [2–4]. Sarcopenia was primarily defined as the age-related progressive and generalized loss of skeletal muscle mass [5]. The European Working Group on Sarcopenia in Older People (EWGSOP) extended the definition of sarcopenia as the association of low muscle mass, plus low muscle strength or low physical performance, occurring in various diseases [6]. Many studies have evaluated sarcopenia using measurements of muscle mass quantity or quality with whole-body imaging methods, especially computed tomography [7]. Nevertheless, these methods are costly, time-consuming, irradiating, and not adapted for routine clinical practice. In 2019, EWGSOP2 recommended using the handgrip strength (HGS) measurement to screen sarcopenia [5]. Loss of muscle strength, also named dynapenia [8], has been defined by EWGSOP2 consensus as HGS < 27 kg in men and <16 kg in women, based on the geriatric part of a cohort study [5,9].

Nevertheless, a great heterogeneity of cut-off points is presented in the literature, and not adapted for cancer patients [9–11]. HGS has already proven its interest in the elderly, since the loss of HGS has been associated with more postoperative complications, increased length of hospitalization, higher rehospitalization rate, and poorer physical status [12,13]. In cancer patients, dynapenia has been associated with cancer-related fatigue [14], poor quality of life [15], postoperative complications [16], and mortality [17]. HGS could also be a reliable and effective tool to screen for malnutrition in digestive cancer patients [18].

The FIGHTDIGO study has demonstrated the feasibility and acceptability of HGS measurement using a JAMAR dynamometer in an outpatient cancer unit [19]. An ancillary analysis from a small sample of FIGHTDIGO patients suggested that patients with pre-therapeutic dynapenia (defined by using cut-off points of <30 kg in men and <20 kg in women) experienced more chemotherapy-induced dose-limiting neurotoxicity (DLN), but no difference in terms of other DLT [20]. Considering the new HGS cut-off points recommended by EWGSOP2 to define sarcopenia, additional studies are required to confirm dynapenia as a potential predictor of DLN or DLT.

The present FIGHTDIGOTOX study aimed to assess the relationship between pre-therapeutic HGS and chemotherapy-induced DLT and/or all-grade toxicity in digestive cancer patients treated in an outpatient cancer unit.

2. Materials and Methods

2.1. Design and Population

FIGHTDIGOTOX is a comprehensive observational retrospective monocentric cohort study including patients older than 18 years old, diagnosed with primary digestive cancer and receiving an intravenous anticancer drug in the Oncology Day-Hospital of the Reims university hospital in France. From November 2015 to December 2018, patients aged more than 65 years-old had an HGS measurement before initiation of chemotherapy in the prospective AgElOn study (NCT02807129). From November 2018 to March 2020, each newly admitted patient (waiting for a first anticancer drug infusion) was invited to perform HGS measurement as part of the routine practice.

Patients were excluded if they had a history of previous anti-cancer treatment, did not understand, or practice the HGS test, had any history of neuromuscular disorder, had received exclusive oral chemotherapy or immunotherapy, and/or had early stopped anticancer treatment (≤1 cycle) unrelated to adverse effects.

2.2. Outcomes

The primary objective was to study the association between pre-therapeutic dynapenia with chemotherapy induced all-grade toxicities and DLT. The secondary objective was to analyze the same association using an exploratory median-based HGS threshold to define dynapenia.

2.3. Ethical Approval

The study was conducted in accordance with the Helsinki Declaration. Informed written consent was obtained from each patient enrolled in the AgElOn trial. This trial was approved by the ethics committee (Committee for the Protection of Person EST I DIJON, 25 March 2016) and was registered on Clinicaltrials.gov (NCT02807129). Patients' records were anonymized prior to analysis. The database was constituted in accordance with the reference methodology MR004 of the French National Commission on Informatics and

Liberty (CNIL). As per French regulations concerning the retrospective study, no informed consent or additional ethical committee review was required.

2.4. Data Collection

Patients' characteristics of interest (including sex, age, tumor location, disease stage, comorbidities, anticancer drug regimen, concomitant radiotherapy, ECOG Performance Status (PS), Body Mass Index (BMI), G8 score in older patients (tool to identify elderly cancer patients who benefit from a comprehensive geriatric assessment)), and biological characteristics (serum albumin level, C-reactive protein (CRP), lymphocyte count and the modified Glasgow Prognostic Score (mGPS)) were retrospectively collected from medical records. The mGPS was calculated from serum CRP and albumin levels and is known to be an independent prognostic factor in oncology [21].

2.5. Handgrip Strength Measurement and Dynapenia Definition

HGS was measured with a Jamar hydraulic dynamometer which has already proven its reliability [22]. HGS measurement was performed in all patients at baseline before the administration of the antineoplastic treatment, either during the medical consultation or during the first hospital stay.

The HGS measurement protocol was previously described by Ordan et al. [18]. There were 5 possible handle positions, and position 2 is used in our daily practice. The test was performed with the dominant and non-dominant hand. Patients performed maximal isometric contraction within 3 s in both hands. A verbal motivation was given by the physician to access their best score. After the first measurement, a one-minute break was taken before the second measurement for each hand. The highest value from the four measurements was finally collected.

The HGS test value was defined according to different thresholds. First, initially planned analysis was performed using the newly validated EWGSOP2 criteria for dynapenia (HGS < 27 kg for men and <16 kg for women) [5]. Second, we defined dynapenia using an additional exploratory threshold, as an HGS below the sex-based median of our population. These definitions are designated as original (EWGSOP2) and exploratory, respectively.

2.6. Chemotherapy-Induced Dose-Limiting Toxicities (DLT), Dose-Limiting Neurotoxicity (DLN) and All-Grade Chemotherapy-Induced Toxicities

Data on chemotherapy-induced toxicities occurring during the first six months after the initiation of first-line chemotherapy were collected from each patient's electronic health record. All-grade toxicities during anticancer treatment were also collected (including dose-limiting). Chemotherapy-induced toxicity was graded according to the National Cancer Institute-Common Terminology Criteria for Adverse Events (NCI-CTCAE Version 5.0). Chemotherapy-induced DLT was defined as any toxicity leading to dose reduction (temporary or permanent), treatment delays, or permanent treatment discontinuation. Progressive disease as the cause of treatment discontinuation was not considered DLT. Pre-therapeutic dose adaptation was defined as an initial dose reduction by individual clinical appreciation considering patient profile (age, ECOG PS, organ failure, or malnutrition).

Toxicities were analyzed according to each chemotherapy side-effect profile. Neuropathy was only considered in patients receiving oxaliplatin, cisplatin, and docetaxel; hand-foot syndrome (HFS) and oral mucositis only in patients receiving 5-Fluorouracil (5FU)- or capecitabine-based chemotherapy regimens. Finally, nausea and vomiting were not considered in patients receiving 5FU or gemcitabine alone.

2.7. Statistical Analyses

Quantitative variables were expressed as mean and standard deviation or median and interquartile range(s) (IQR) and compared using the non-parametric Kruskal–Wallis test. Qualitative data were described by frequencies and percentages and compared with the Chi-square test or Fisher's exact test when appropriate. All *p*-values were two-sided, and a *p*-value ≤ 0.05 was considered significant. The tests were performed to compare all-grade and dose-limiting toxicities with the original and the exploratory dynapenia thresholds. An additional multivariate analysis was performed including significant patient characteristics in a stepwise regression multivariate analysis. All data were collected using EpiInfo 7.2.5.0 and analyzed using R Studio (R Core Team, 2022).

3. Results

3.1. Characteristics of Patients

A total of 322 medical records were screened and 244 patients were included (Figure 1). The characteristics of the included population are described in Table 1.

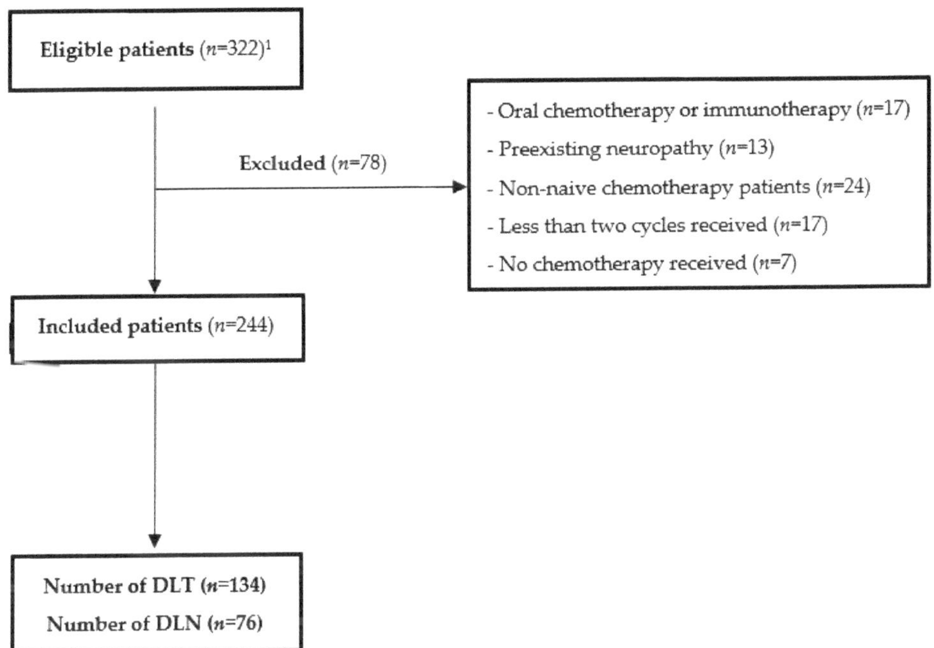

Figure 1. Flow chart of the FIGHTDIGOTOX study. Abbreviations: DLN: dose-limiting neurotoxicity; DLT: dose-limiting toxicity; UMA-CH, ambulatory cancer unit. [1] 69 patients were assessed in AgElOn.

The median age was 69 (IQR, 59.0–74.0) years and the sex ratio was balanced. Colorectal cancer was the most common primary tumor site (*n* = 105, 43.2%). Eighty-four patients (34.4%) were diagnosed with localized disease, whereas 103 (42.2%) were at a metastatic stage. Forty patients (16.4%) underwent a combination of chemotherapy and biotherapy. Most anticancer drugs were potentially neurotoxic (*n* = 189, 77.4%) and the most frequently received chemotherapy regimen was FOLFOX (infusional and bolus 5FU, leucovorin plus oxaliplatin) (*n* = 96, 39.3%).

3.2. Handgrip Strength (HGS)

The mean HGS value was 35.8 ± 8.5 kg for men and 22.8 ± 6.3 kg for women. According to the original EWGSOP2 criteria, 23 patients (9.4%) were defined as dynapenic, including 13 men and 10 women. The median HGS value, defining our exploratory threshold was 34 kg (IQR: 30–41.5) for men and 22 kg (IQR: 19–28) for women. According to our exploratory definition, 107 patients (43.8%) were considered dynapenic, including 57 men and 50 women.

Table 1. Overall population characteristics and according to dynapenia (EWGSOP2 criteria).

Characteristics of Patients	Level	Overall	Dynapenia [1]	Normal HGS [1]	*p*-Value
Total, *n* (%)		244	23 (9.4)	221 (90.6)	
Sex, *n* (%)	Female	109 (44.7)	10 (43.5)	99 (44.8)	1.000
	Male	135 (55.3)	13 (56.5)	122 (55.2)	
Age, median (IQR)		69.0 (59.0–74.0)	73.0 (69.0–81.5)	68.0 (58.0–73.0)	<0.001
BMI, median (IQR)		24.6 (21.5–28.6)	24.6 (21.1–29.4)	24.6 (21.6–28.6)	0.862
ECOG PS, *n* (%)	0	66 (27.0)	0 (0.0)	66 (29.9)	<0.001
	1	150 (61.5)	14 (60.9)	136 (61.5)	
	2	26 (10.7)	8 (34.8)	18 (8.1)	
	3	2 (0.8)	1 (4.3)	1 (0.5)	
Serum albumin level, median (IQR)		39.0 (36.0–42.0)	36.0 (33.0–40.0)	39.0 (37.0–42.0)	0.012
CRP, median (IQR)		9.0 (4.0–33.2)	19.0 (4.5–40.5)	8.0 (4.0–33.0)	0.114
mGPS, *n* (%)	0	118 (48.4)	5 (21.7)	113 (51.1)	0.006
	1	87 (35.7)	10 (43.5)	77 (34.8)	
	2	39 (16.0)	8 (34.8)	31 (14.0)	
Lymphopenia, *n* (%)	No	226 (92.6)	21 (91.3)	205 (92.8)	0.681
	Yes	18 (7.4)	2 (8.7)	16 (7.2)	
G8 score, median [2] (IQR)		12.0 (11.0–15.0)	10.0 (8.8–12.0)	13.0 (11.0–15.0)	0.002
Primary tumor location, *n* (%)	Colon and rectum	105 (43.2)	16 (69.6)	89 (40.5)	0.339
	Stomach	26 (10.7)	2 (8.7)	24 (10.9)	
	Esophagus	18 (7.4)	2 (8.7)	16 (7.3)	
	Pancreas	69 (28.4)	2 (8.7)	67 (30.5)	
	Others [3]	25 (10.2)		25 (11.1)	
Stage, *n* (%)	Localized	84 (34.4)	9 (39.1)	75 (33.9)	0.784
	Locally advanced	57 (23.4)	4 (17.4)	53 (24.0)	
	Metastatic	103 (42.2)	10 (43.5)	93 (42.1)	
Number of metastatic sites, *n* (%)	1	68 (65.4)	7 (70.0)	61 (64.9)	1.000
	≥2	36 (34.6)	3 (30.0)	33 (35.1)	
Chemotherapy regimen, *n* (%)	5FU + Oxaliplatin	96 (39.3)	12 (52.2)	84 (38.0)	0.170
	5FU + Irinotecan + Oxaliplatin	69 (28.3)	3 (13.0)	66 (29.9)	
	5FU alone	24 (9.8)	6 (26.1)	18 (8.1)	
	Gemcitabine	18 (7.4)		18 (8.1)	
	Others [4]	37 (15.1)	2 (8.6)	35 (16.2)	
Biotherapy, *n* (%)	None	204 (83.6)	18 (78.3)	186 (84.2)	0.305
	Bevacizumab	26 (10.7)	5 (21.7)	21 (9.5)	
	Others [5]	14 (5.7)		14 (6.3)	
Concomitant radiotherapy, *n* (%)	No	222 (91.0)	22 (95.7)	200 (90.5)	0.704
	Yes	22 (9.0)	1 (4.3)	21 (9.5)	

Abbreviations: 5FU: 5 Fluorouracil; BMI: Body Mass Index; CRP: C-reactive protein; ECOG PS: Eastern Cooperative Oncology Group Criteria Performance Status; HGS: handgrip strength; IQR: interquartile range; mGPS: modified Glasgow prognosis score. [1] According to the EWGSOP2 definition; [2] Data available for 82 patients; [3] Other localizations: biliary tract (*n* = 8), small intestine (*n* = 7), ampulla of Vater (*n* = 3), neuroendocrine tumor (*n* = 4), appendix (*n* = 1), anal (*n* = 1), unknown primary (*n* = 1); [4] Other chemotherapy: 5FU + Irinotecan (*n* = 8), 5FU + Oxaliplatin + Docetaxel (*n* = 9), 5FU + Cisplatin (*n* = 1), 5FU + Dacarbazine (*n* = 3), Carboplatin-Etoposide (*n* = 1), Gemcitabine + Cisplatin (*n* = 4), Gemcitabine + Oxaliplatin (*n* = 3), Capecitabine + Oxaliplatin (*n* = 7), Capecitabine + Mitomycin (*n* = 1); [5] Other biotherapy: Panitumumab (*n* = 10), Trastuzumab (*n* = 4).

3.3. Chemotherapy-Induced DLT

A total of 134 patients (54.9%) experienced chemotherapy-induced DLT. The most frequent DLT was neurotoxicity (*n* = 76, 41.3%). Patients with dynapenia according to the original EWGSOP2 definition were significantly older (*p* < 0.001), with worse ECOG PS (*p* < 0.001) and G8 score (*p* = 0.002), and lower serum albumin levels (*p* = 0.012).

The repartition of DLT according to dynapenia as defined by the original EWGSOP2 criteria is shown in Table 2.

Table 2. Association between dynapenia (original EWGSOP2 criteria) and chemotherapy-induced dose-limiting toxicity (DLT).

Dose Limiting Toxicity	Overall (n = 244)	Dynapenia [1] (n = 23)	Normal HGS [1] (n = 221)	p Value
All Type (%)	134 (54.9)	13 (56.5)	121 (54.8)	1.000
Neuropathy [2]	76 (41.3)	7 (46.7)	69 (40.8)	0.786
Asthenia (%)	24 (9.8)	5 (21.7)	19 (8.6)	0.059
Diarrhea (%)	20 (8.2)	2 (8.7)	18 (8.1)	1.000
Nausea [3] (%)	4 (2.0)	1 (5.9)	3 (1.6)	0.298
Vomiting [3] (%)	4 (2.0)	0 (0)	4 (2.2)	1.000
Neutropenia (%)	28 (11.5)	0 (0)	28 (12.7)	0.086
Anemia (%)	6 (2.5)	1 (4.3)	5 (2.3)	0.451
Thrombopenia (%)	13 (5.3)	2 (8.7)	11 (5.0)	0.352
Hand foot syndrome [4] (%)	4 (1.9)	2 (9.1)	2 (1.1)	0.075
Oral mucositis [4] (%)	3 (1.2)	0 (0)	3 (1.4)	1.000

Abbreviations: DLT; dose-limiting toxicity; HGS: handgrip strength. [1] According to the EWGSOP2 definition; [2] Only patients receiving neurotoxic chemotherapy (n = 184); [3] Patients receiving 5FU and gemcitabine alone were not analyzed for this adverse effect (n = 202); [4] Only patients receiving 5FU- or capecitabine-based chemotherapy regimen (n = 210).

There was no significant association between dynapenia and overall type of DLT. Asthenia (21.7% versus 8.6%, $p = 0.059$) and hand-foot syndrome (HFS) (9.1% versus 1.1%, $p = 0.075$) tended to be a more frequent cause of DLT in patients with dynapenia than in patients without. No association was found between dynapenia and DLN ($p = 0.786$). No additional multivariate analysis was performed for the original HGS threshold due to the limited number of patients diagnosed with dynapenia.

The repartition of DLT according to our exploratory HGS median-based threshold is described in Table 3. Using this definition, patients with exploratory low HGS were significantly older ($p < 0.001$) and had a worse ECOG PS ($p = 0.006$), mGPS score ($p = 0.020$), and G8 score ($p = 0.050$). A significantly higher rate of dose-limiting asthenia was observed in patients with below median-based HGS threshold (15% versus 5.8%, $p = 0.029$). The planned multivariate analysis for the median-based threshold, adjusted on age over 65 years, G8 score, ECOG PS, and mGPS did not show any significant relationship with asthenia ($p = 0.78$) or all DLT combined ($p = 0.2$).

3.4. All-Grade Toxicity (Dose-Limiting or Not)

The observed all-grade toxicities according to dynapenia as defined by the original EWGSOP2 criteria are shown in Table 4. Patients with dynapenia experienced more HFS (18.2% versus 3.2%, $p = 0.007$) and tended to experience more grade 3–4 diarrhea (25% versus 10%, $p = 0.071$) (Supplementary Table S3).

The observed all-grade toxicities according to our exploratory median-based HGS threshold are shown in Table 5. Patients with exploratory low HGS experienced more anemia (77.6% versus 59.9%, $p = 0.006$), more asthenia (97.2% versus 87.6% $p = 0.014$), and less vomiting (18.1% versus 30.3%, $p = 0.047$).

Table 3. Association between low handgrip strength on median-based analysis (exploratory low HGS) and chemotherapy-induced dose-limiting toxicity (DLT).

Dose Limiting Toxicity	Overall (*n* = 244)	Exploratory Low HGS [1] (*n* = 107)	Normal HGS [1] (*n* = 137)	*p*-Value for Univariate Analysis	*p*-Value for Multivariate Analysis *
All Type (%)	134 (54.9)	32 (29.9)	40 (29.2)	1.000	0.2
Neuropathy [2]	76 (41.3)	26 (36.1)	50 (44.6)	0.285	-
Asthenia (%)	24 (9.8)	16 (15.0)	8 (5.8)	0.029	0.78
Diarrhea (%)	20 (8.2)	9 (8.4)	11 (8.0)	1.000	-
Nausea [3] (%)	4 (2.0)	2 (2.4)	2 (1.7)	1.000	-
Vomiting [3] (%)	4 (2.0)	1 (1.2)	3 (2.5)	0.645	-
Neutropenia (%)	28 (11.5)	10 (9.3)	18 (13.1)	0.421	-
Anemia (%)	6 (2.5)	4 (3.7)	2 (1.5)	0.409	-
Thrombopenia (%)	13 (5.3)	5 (4.7)	8 (5.8)	0.779	-
Hand foot syndrome [4] (%)	4 (1.9)	2 (2.2)	2 (1.7)	1.000	-

Abbreviations: DLT: dose-limiting toxicity; HGS: handgrip strength; mGPS: modified Glasgow prognosis score. [1] HGS cut-off based on the median in the population as HGS <34 kg for men and <22 kg for women; [2] Only patients receiving neurotoxic chemotherapy (*n* = 184); [3] Patients receiving 5FU and gemcitabine alone were not analyzed for this adverse effect (*n* = 202); [4] Only patients receiving 5FU- or capecitabine-based chemotherapy regimen (*n* = 210); * Multivariate analysis was adjusted on age over 65 years, performance status, G8 score, and mGPS; -: statistical analysis not performed due to futility.

Table 4. Association between dynapenia (original EWGSOP2 criteria) and all-grade chemotherapy-induced toxicities (dose-limiting and not).

Toxicity (All Grade)	Overall (*n* = 244)	Dynapenia [1] (*n* = 23)	Normal HGS [1] (*n* = 221)	*p* Value
Neuropathy [2] (%)	174 (94.6)	14 (93.3)	160 (94.7)	0. 582
Asthenia (%)	224 (91.8)	23 (100.0)	201 (91.0)	0.303
Diarrhea (%)	139 (57.0)	12 (52.2)	127 (57.5)	0.693
Nausea [3] (%)	115 (56.9)	9 (52.9)	106 (57.3)	0.801
Vomiting [3] (%)	51 (25.2)	3 (17.6)	48 (25.9)	0.605
Neutropenia (%)	60 (24.6)	3 (13.0)	57 (25.8)	0.286
Anemia (%)	165 (67.6)	19 (82.6)	146 (66.1)	0.238
Thrombopenia (%)	73 (29.9)	10 (43.5)	63 (28.5)	0.235
Hand foot syndrome [4] (%)	10 (4.8)	4 (18.2)	6 (3.2)	0.007
Oral mucositis [4] (%)	29 (11.9)	2 (8.7)	27 (12.2)	0.836

Abbreviations: DLT: dose-limiting toxicity; HGS: handgrip strength. [1] According to the EWGSOP2 definition; [2] Only patients receiving neurotoxic chemotherapy (*n* = 184); [3] Patients receiving 5FU and gemcitabine alone were not analyzed for this adverse effect (*n* = 202); [4] Only patients receiving 5FU- or capecitabine-based chemotherapy regimen (*n* = 210).

Table 5. Association between low handgrip strength on median-based analysis (exploratory low HGS) and all-grade chemotherapy-induced toxicities (dose-limiting and not).

Toxicity (All Grade)	Overall (*n* = 244)	Exploratory Low HGS [1] (*n* = 107)	Normal HGS [1] (*n* = 137)	*p* Value
Neuropathy [2] (%)	174 (94.6)	66 (91.7)	108 (96.4)	0.193
Asthenia (%)	224 (91.8)	104 (97.2)	120 (87.6)	0.014
Diarrhea (%)	139 (57.0)	56 (52.3)	83 (60.6)	0.214

Table 5. *Cont.*

Toxicity (All Grade)	Overall (*n* = 244)	Exploratory Low HGS [1] (*n* = 107)	Normal HGS [1] (*n* = 137)	*p* Value
Nausea [3] (%)	115 (56.9)	41 (49.4)	74 (62.2)	0.084
Vomiting [3] (%)	51 (25.2)	15 (18.1)	36 (30.3)	0.047
Neutropenia (%)	60 (24.6)	26 (24.3)	34 (24.8)	1.000
Anemia (%)	165 (67.6)	83 (77.6)	82 (59.9)	0.006
Thrombopenia (%)	73 (29.9)	32 (29.9)	41 (29.9)	1.000
Hand foot syndrome [4] (%)	10 (4.8)	5 (5.4)	5 (4.3)	0.629
Oral mucositis [4] (%)	29 (11.9)	15 (14.0)	14 (10.2)	0.455

Abbreviations: DLT: dose-limiting toxicity; HGS: handgrip strength. [1] HGS cut-off based on the median in the population as HGS < 34 kg for men and <22 kg for women; [2] Only patients receiving neurotoxic chemotherapy (*n* = 184); [3] Patients receiving 5FU and gemcitabine alone were not analyzed for this adverse effect (*n* = 202); [4] Only patients receiving 5FU- or capecitabine-based chemotherapy regimen (*n* = 210).

4. Discussion

In the present study, pre-therapeutic dynapenia was not associated with chemotherapy-induced DLT in digestive cancer patients receiving first-line chemotherapy. Patients with dynapenia, as defined by EWGSOP2, seemed to experience more HFS and serious diarrhea. However, the current threshold used to define dynapenia is not consensual, especially in cancer patients [11]. Consequently, we performed an exploratory analysis based on the median HGS of our population. Using this new threshold to define dynapenia, we observed more dose-limiting asthenia and anemia.

The prevalence of DLT was 54.9% in this study. Previous studies observed similar DLT rates [20,23]. Botsen et al. reported 49% of chemotherapy-induced DLT in digestive cancer patients [20]. Celik et al. observed 52% of DLT in digestive cancer patients receiving platinum-based chemotherapy [23]. He also described higher rates (78.9%) in patients with sarcopenia (defined as a low muscle mass measured on a computed tomography), suggesting its potential role in predicting chemotherapy-induced DLT. Nonetheless, HGS alone seemed to be insufficient to predict the occurrence of DLT [23]. In another study, Lakenman et al. showed an association between low HGS and DLT during neoadjuvant chemoradiation in patients with esophageal cancer [24]. In this study, dynapenia was defined below the tenth percentile (HGS < 37.6 kg for men and <23.6 kg for women) [24].

Conversely to the FIGHTDIGO study, and despite a larger sample without a selection bias of non-neurotoxic treatment, we did not find any association between pre-therapeutic dynapenia and DLN [20]. Because dynapenic patients were less exposed to major neuropathic-providing chemotherapies such as docetaxel and cisplatin, they could experience less DLN.

Indeed, patients with exploratory low HGS were significantly older and had a worse ECOG PS, mGPS score, and geriatric G8 score. Our results are in line with the known association of low HGS with markers of functional and nutritional status [18,25], age [26], and geriatric G8 score [27]. These findings support the usefulness of HGS measurement as an interesting additional tool to identify frailty in cancer patients.

In the exploratory analysis, in which low HGS was defined by a HGS value below the sex-based median, patients experienced more dose-limiting asthenia in univariate analysis. Kilgour et al. have already observed more fatigue in patients with weaker muscle strength [14]. Additionally, asthenia could be a part of the cachexia syndrome [28], which is defined as a metabolic syndrome associated with an underlying chronic disease and characterized by a loss of skeletal muscle mass [6]. Cachexia is generally associated with chronic inflammation [29–31]. Sarcopenia, cachexia, and asthenia share common and overlapping characteristics. Our findings might be an additional item for the interaction between these different clinical entities. HGS could be used for the assessment of sarcopenia in daily practice as a part of the spectrum of cachexia [5]. However, due to low statistical power, the higher risk of a false positive association should be taken into account. Despite the absence of a strong statistical association, dynapenia seems to be a part of a larger frailty syndrome.

In the same analysis, patients with exploratory low HGS tend to experience more frequent all-grade anemia but not DLT. Previous studies have shown higher hematological toxicity in patients with lower HGS and sarcopenia [32,33]. However, this toxicity is commonly managed by blood transfusion support and/or erythropoiesis-stimulating agents before treatment's dose adaptation or delay [34,35]. Cancer-induced inflammation inhibits hematopoiesis by the interaction of interleukin 6 and tumor necrosis factor-alpha (TNF-α) [36]. However, the role of cachexia remains unclear. In a prospective study, Rocha et al. described an increased risk of grade 3–4 toxicities in the first three cycles of chemotherapy in patients with cachexia treated for gastrointestinal cancer [37]. However, their follow-up was shorter in comparison with our study, and potential consequences of cachexia on erythropoiesis have not been detected due to the lifespan of blood cells. In non-cancer populations, the association between anemia and low HGS has already been described [38]. Our results suggest that HGS could be interesting to improve the management of cancer-related anemia and be a part of the adaptation of supportive care.

Patients with dynapenia also experienced more HFS ($p = 0.007$). This result relies on a very low number of patients and should be interpreted with caution. Risk factors of HFS have been previously described, including age, sex, and genetic susceptibilities [39,40]. To our knowledge, no association between muscle strength and HFS has been previously observed. Gökyer et al. described higher DLT and HFS rates in the sarcopenic population with colorectal cancer receiving regorafenib [41]. Although, HGS was not measured [41].

Patients with dynapenia as defined by our exploratory thresholds experienced less vomiting without DLT. This group had fewer chemotherapy combinations (such as FOLFIRI-NOX or TFOX) which are associated with more vomiting [42]. However, the population with exploratory low HGS was older, whereas younger patients are at greater risk of chemotherapy-induced nausea and vomiting [43].

Sarcopenia had already shown promising results in predicting chemotherapy-related toxicity but adapted criteria for assessment are needed [11]. Low lean body mass was associated with the increased occurrence of chemotherapy-related toxicities [44]. The relationship between body composition and pharmacokinetics of chemotherapy is also well established [45]. Recently, Cereda et al. demonstrated that muscle weakness was a better predictor of survival than skeletal muscle mass estimated by bioelectrical impedance analysis [46].

Heterogeneous cut-off points have been previously described to define dynapenia [9,10,17,24,47]. Nowadays, the gold standard is the EWGSOP2 definition, based on Dodds et al. study [9]. However, most of these thresholds (including the current gold standard) have been established in non-oncologic geriatric populations. Thus, this lack of consensus is an issue that still needs to be addressed as it hampers comparisons between studies and the emergence of new guidelines for daily oncology practice. Indeed, the consequences of cancer on muscle strength are not included in the used definitions. Therefore, an exploratory cut-off point was assessed in our study, but further prospective studies are needed to validate its relevance. A recent study based on 6182 patients found that HGS cut-offs of <36 kg for men and <23 kg for women were the best ones to predict mortality in the elderly [48]. Their new thresholds are similar to those defined by Lakenman et al. in oesophageal cancer patients and in our exploratory analysis [24,48].

The present study had several limitations. First, the study was based on a retrospective examination of medical records which limits the exhaustive collection of every toxicity. However, our oncology unit has a strong culture of grading and tracing every toxicity. Moreover, chemotherapy was prescribed on a unique software limiting selection bias. Second, some comorbidities influencing DLT were not recorded, such as heart failure or renal insufficiency. The use of the Charlson Comorbidity Index could have been useful to prevent a possible confusion bias. Third, the included population was heterogeneous with various types of digestive cancers and chemotherapy regimens. Fourth, we observed a very limited number of DLT, hindering the pre-planned multivariate analysis that could have helped us better understand the interaction between potential confusion factors. This study also presented several strengths including the analysis of a large cohort of outpatients with digestive cancer and providing real-life daily-practice data.

HGS measurement with the JAMAR dynamometer is an easy-to-use, portable and economical way to screen for dynapenia in daily clinical practice [19]. Further studies could focus on the prevention of potential toxicities, the evolution of HGS throughout the anticancer treatment program, or the usefulness of Adapted Physical Activity (APA) programs in sarcopenic patients. Muscle strength follow-up could be used in daily practice during APA programs. Recently, APA has been reported as feasible in cancer outpatients beginning medical anticancer treatment [49]. Indeed, recent studies have shown improvement in quality of life and a reduction in fatigue through physical activity [50].

5. Conclusions

In conclusion, digestive cancer outpatients with pre-therapeutic dynapenia, according to EWGSOP2 criteria, do not seem to have more chemotherapy-induced DLT. Based on an exploratory higher cut-off point, low HGS could be a predictive factor of chemotherapy-induced anemia and asthenia. There is a growing need to better define the HGS cut-off points of dynapenia in cancer patients. The HGS measurement is easily use in daily practice, non-invasive and inexpensive. The diagnosis of dynapenia could help the care provider to better assess patients' frailty, and to adjust nutritional care and APA before the appearance of chemotherapy-induced toxicities.

Supplementary Materials: The following supporting information can be downloaded at: https://www.mdpi.com/article/10.3390/nu14214448/s1, Table S1: Population characteristics according to low and normal hand grip strength based on the median (exploratory low HGS); Table S2: Association between low hand grip strength based on the median (exploratory low HGS) and chemotherapy-induced toxicities and DLT (detailed grades and therapeutic modifications). Table S3: Association between dynapenia based on EWSGOP2 criteria and chemotherapy-induced toxicities and DLT (detailed grades and therapeutic modifications).

Author Contributions: Conceptualization, P.M., D.B., M.B., E.B., C.C., R.M., M.P. and O.B.; methodology P.M., D.B., M.B., C.C., M.P. and O.B.; software, M.B.; validation, P.M., D.B., M.B., E.B., C.C, R.M., F.S., M.P. and O.B.; formal analysis, P.M., D.B., M.B. and O.B.; investigation, P.M., D.B., M.B., C.C, R.M., F.S., M.P. and O.B.; resources, P.M., D.B., M.B., E.B., C.C., R.M., F.S., M.P. and O.B.; data curation, M.B. and O.B.; writing—original draft preparation, P.M., D.B., M.B., E.B., C.C., R.M., F.S., M.P. and O.B.; writing—review and editing, P.M., D.B., M.B., E.B., C.C., R.M., F.S., M.P. and O.B.; visualization, P.M., D.B., M.B., E.B., C.C., R.M., M.P. and O.B.; supervision, D.B., M.B., E.B., C.C., R.M., M.P. and O.B.; project administration, D.B. and O.B.; funding acquisition, R.M. and O.B. All authors have read and agreed to the published version of the manuscript.

Funding: AgElOn study was supported by the Champagne-Ardenne Region and Reims University Hospital.

Institutional Review Board Statement: The study was conducted in accordance with the Helsinki Declaration. Informed written consent was obtained from each patient enrolled in the AgElOn trial. This trial was approved by the ethics committee (Committee for the Protection of Person EST I DIJON, 25 March 2016) and was registered on Clinicaltrials.gov (NCT02807129). Patients' records were anonymized prior to analysis. The database was constituted in accordance with the reference methodology MR004 of the French National Commission on Informatics and Liberty (CNIL). As per French regulations concerning the retrospective study, no informed consent or additional ethical committee review was required.

Informed Consent Statement: As per French regulations concerning the retrospective study, no informed consent or additional ethical committee review was required.

Data Availability Statement: The data presented in this study are available on request from the corresponding author.

Acknowledgments: We would like to thank the patients and their families.

Conflicts of Interest: D.B. reports personal fees as a speaker and/or in an advisory role from Accord Healthcare, Amgen, Sanofi, Servier, and Pierre Fabre, outside the submitted work. C.C. reports personal fees as a speaker from Brystol Myers Squibb, outside the submitted work. F.S. reports personal fees as a speaker and/or in an advisory role from Astra Zeneca, outside the submitted work. O.B. reports personal fees as a speaker and/or in an advisory role from Merck KGaA, Apmomia Therapeutics Bayer, Grunenthal, MSD, Amgen, Servier, and Pierre Fabre, outside the submitted work. All other authors have no conflict of interest.

References

1. Sung, H.; Ferlay, J.; Siegel, R.L.; Laversanne, M.; Soerjomataram, I.; Jemal, A.; Bray, F. Global Cancer Statistics 2020: GLOBOCAN Estimates of Incidence and Mortality Worldwide for 36 Cancers in 185 Countries. *CA A Cancer J. Clin.* **2021**, *71*, 209–249. [CrossRef] [PubMed]
2. Hilmi, M.; Jouinot, A.; Burns, R.; Pigneur, F.; Mounier, R.; Gondin, J.; Neuzillet, C.; Goldwasser, F. Body Composition and Sarcopenia: The next-Generation of Personalized Oncology and Pharmacology? *Pharmacol. Ther.* **2019**, *196*, 135–159. [CrossRef] [PubMed]
3. Kurk, S.; Peeters, P.; Stellato, R.; Dorresteijn, B.; de Jong, P.; Jourdan, M.; Creemers, G.-J.; Erdkamp, F.; de Jongh, F.; Kint, P.; et al. Skeletal Muscle Mass Loss and Dose-Limiting Toxicities in Metastatic Colorectal Cancer Patients. *J. Cachexia Sarcopenia Muscle* **2019**, *10*, 803–813. [CrossRef] [PubMed]
4. Aleixo, G.F.P.; Williams, G.R.; Nyrop, K.A.; Muss, H.B.; Shachar, S.S. Muscle Composition and Outcomes in Patients with Breast Cancer: Meta-Analysis and Systematic Review. *Breast Cancer Res. Treat.* **2019**, *177*, 569–579. [CrossRef] [PubMed]
5. Cruz-Jentoft, A.J.; Bahat, G.; Bauer, J.; Boirie, Y.; Bruyère, O.; Cederholm, T.; Cooper, C.; Landi, F.; Rolland, Y.; Sayer, A.A.; et al. Sarcopenia: Revised European Consensus on Definition and Diagnosis. *Age Ageing* **2019**, *48*, 16–31. [CrossRef] [PubMed]
6. Cruz-Jentoft, A.J.; Baeyens, J.P.; Bauer, J.M.; Boirie, Y.; Cederholm, T.; Landi, F.; Martin, F.C.; Michel, J.-P.; Rolland, Y.; Schneider, S.M.; et al. Sarcopenia: European Consensus on Definition and Diagnosis: Report of the European Working Group on Sarcopenia in Older People. *Age Ageing* **2010**, *39*, 412–423. [CrossRef] [PubMed]
7. Neuzillet, C.; Anota, A.; Foucaut, A.-M.; Védie, A.-L.; Antoun, S.; Barnoud, D.; Bouleuc, C.; Chorin, F.; Cottet, V.; Fontaine, E.; et al. Nutrition and Physical Activity: French Intergroup Clinical Practice Guidelines for Diagnosis, Treatment and Follow-up (SNFGE, FFCD, GERCOR, UNICANCER, SFCD, SFED, SFRO, ACHBT, AFC, SFP-APA, SFNCM, AFSOS). *BMJ Support. Palliat. Care* **2021**, *11*, 381–395. [CrossRef]
8. Clark, B.C.; Manini, T.M. Sarcopenia =/= Dynapenia. *J. Gerontol. A Biol. Sci. Med. Sci.* **2008**, *63*, 829–834. [CrossRef]
9. Dodds, R.M.; Syddall, H.E.; Cooper, R.; Benzeval, M.; Deary, I.J.; Dennison, E.M.; Der, G.; Gale, C.R.; Inskip, H.M.; Jagger, C.; et al. Grip Strength across the Life Course: Normative Data from Twelve British Studies. *PLoS ONE* **2014**, *9*, e113637. [CrossRef]
10. Fried, L.P.; Tangen, C.M.; Walston, J.; Newman, A.B.; Hirsch, C.; Gottdiener, J.; Seeman, T.; Tracy, R.; Kop, W.J.; Burke, G.; et al. Frailty in Older Adults: Evidence for a Phenotype. *J. Gerontol. A Biol. Sci. Med. Sci.* **2001**, *56*, M146–M156. [CrossRef]
11. Findlay, M.; White, K.; Stapleton, N.; Bauer, J. Response to Comment: Evaluating Sarcopenia in Cancer Patients: The Role of Muscle Strength. *Clin. Nutr.* **2022**, *41*, 780–781. [CrossRef] [PubMed]
12. Norman, K.; Stobäus, N.; Gonzalez, M.C.; Schulzke, J.-D.; Pirlich, M. Hand Grip Strength: Outcome Predictor and Marker of Nutritional Status. *Clin. Nutr.* **2011**, *30*, 135–142. [CrossRef] [PubMed]
13. Olguín, T.; Bunout, D.; de la Maza, M.P.; Barrera, G.; Hirsch, S. Admission Handgrip Strength Predicts Functional Decline in Hospitalized Patients. *Clin. Nutr. ESPEN* **2017**, *17*, 28–32. [CrossRef] [PubMed]
14. Kilgour, R.D.; Vigano, A.; Trutschnigg, B.; Hornby, L.; Lucar, E.; Bacon, S.L.; Morais, J.A. Cancer-Related Fatigue: The Impact of Skeletal Muscle Mass and Strength in Patients with Advanced Cancer. *J. Cachexia Sarcopenia Muscle* **2010**, *1*, 177–185. [CrossRef] [PubMed]
15. Norman, K.; Stobäus, N.; Smoliner, C.; Zocher, D.; Scheufele, R.; Valentini, L.; Lochs, H.; Pirlich, M. Determinants of Hand Grip Strength, Knee Extension Strength and Functional Status in Cancer Patients. *Clin. Nutr.* **2010**, *29*, 586–591. [CrossRef] [PubMed]
16. Chen, C.-H.; Huang, Y.-Z.; Hung, T.-T. Hand-Grip Strength Is a Simple and Effective Outcome Predictor in Esophageal Cancer Following Esophagectomy with Reconstruction: A Prospective Study. *J. Cardiothorac. Surg.* **2011**, *6*, 98. [CrossRef]
17. Contreras-Bolívar, V.; Sánchez-Torralvo, F.J.; Ruiz-Vico, M.; González-Almendros, I.; Barrios, M.; Padín, S.; Alba, E.; Olveira, G. GLIM Criteria Using Hand Grip Strength Adequately Predict Six-Month Mortality in Cancer Inpatients. *Nutrients* **2019**, *11*, 2043. [CrossRef]
18. Perrier, M.; Ordan, M.-A.; Barbe, C.; Mazza, C.; Botsen, D.; Moreau, J.; Renard, Y.; Brasseur, M.; Tailliere, B.; Regnault, P.; et al. Dynapenia in Digestive Cancer Outpatients: Association with Markers of Functional and Nutritional Status (the FIGHTDIGO Study). *Support. Care Cancer* **2022**, *30*, 207–215. [CrossRef]

19. Ordan, M.-A.; Mazza, C.; Barbe, C.; Perrier, M.; Botsen, D.; Renard, Y.; Moreau, J.; Brasseur, M.; Taillière, B.; Bertin, É.; et al. Feasibility of Systematic Handgrip Strength Testing in Digestive Cancer Patients Treated with Chemotherapy: The FIGHTDIGO Study. *Cancer* **2018**, *124*, 1501–1506. [CrossRef]
20. Botsen, D.; Ordan, M.-A.; Barbe, C.; Mazza, C.; Perrier, M.; Moreau, J.; Brasseur, M.; Renard, Y.; Taillière, B.; Slimano, F.; et al. Dynapenia Could Predict Chemotherapy-Induced Dose-Limiting Neurotoxicity in Digestive Cancer Patients. *BMC Cancer* **2018**, *18*, 955. [CrossRef]
21. Proctor, M.J.; Morrison, D.S.; Talwar, D.; Balmer, S.M.; O'Reilly, D.S.J.; Foulis, A.K.; Horgan, P.G.; McMillan, D.C. An Inflammation-Based Prognostic Score (MGPS) Predicts Cancer Survival Independent of Tumour Site: A Glasgow Inflammation Outcome Study. *Br. J. Cancer* **2011**, *104*, 726–734. [CrossRef]
22. Roberts, H.C.; Denison, H.J.; Martin, H.J.; Patel, H.P.; Syddall, H.; Cooper, C.; Sayer, A.A. A Review of the Measurement of Grip Strength in Clinical and Epidemiological Studies: Towards a Standardised Approach. *Age Ageing* **2011**, *40*, 423–429. [CrossRef] [PubMed]
23. Celik, E.; Suzan, V.; Samanci, N.S.; Suzan, A.A.; Karadag, M.; Sahin, S.; Aslan, M.S.; Yavuzer, H.; Demirci, N.S.; Doventas, A.; et al. Sarcopenia Assessment by New EWGSOP2 Criteria for Predicting Chemotherapy Dose-Limiting Toxicity in Patients with Gastrointestinal Tract Tumors. *Eur. Geriatr. Med.* **2022**, *13*, 267–274. [CrossRef]
24. Lakenman, P.; Ottens-Oussoren, K.; Witvliet-van Nierop, J.; van der Peet, D.; de van der Schueren, M. Handgrip Strength Is Associated With Treatment Modifications During Neoadjuvant Chemoradiation in Patients With Esophageal Cancer. *Nutr. Clin. Pract.* **2017**, *32*, 652–657. [CrossRef] [PubMed]
25. de Almeida Marques, R.; de Souza, V.F.; do Rosario, T.C.; da Silva Garcia, M.R.P.; Pereira, T.S.S.; Marques-Rocha, J.L.; Guandalini, V.R. Agreement between Maximum and Mean Handgrip Strength Measurements in Cancer Patients. *PLoS ONE* **2022**, *17*, e0270631. [CrossRef] [PubMed]
26. Clark, B.C.; Manini, T.M. What Is Dynapenia? *Nutrition* **2012**, *28*, 495–503. [CrossRef]
27. Cavusoglu, C.; Tahtaci, G.; Dogrul, R.T.; Ileri, I.; Yildirim, F.; Candemir, B.; Kizilarslanoglu, M.C.; Uner, A.; Goker, B. Predictive Ability of the G8 Screening Test to Determine Probable Sarcopenia and Abnormal Comprehensive Geriatric Assessment in Older Patients with Solid Malignancies. *BMC Geriatr.* **2021**, *21*, 574. [CrossRef]
28. Peixoto da Silva, S.; Santos, J.M.O.; Costa e Silva, M.P.; Gil da Costa, R.M.; Medeiros, R. Cancer Cachexia and Its Pathophysiology: Links with Sarcopenia, Anorexia and Asthenia. *J. Cachexia Sarcopenia Muscle* **2020**, *11*, 619–635. [CrossRef]
29. Durham, W.J.; Dillon, E.L.; Sheffield-Moore, M. Inflammatory Burden and Amino Acid Metabolism in Cancer Cachexia. *Curr. Opin. Clin. Nutr. Metab. Care* **2009**, *12*, 72–77. [CrossRef]
30. Morley, J.E.; Anker, S.D.; Evans, W.J. Cachexia and Aging: An Update Based on the Fourth International Cachexia Meeting. *J. Nutr. Health Aging* **2009**, *13*, 47–55. [CrossRef]
31. McGovern, J.; Dolan, R.D.; Skipworth, R.J.; Laird, B.J.; McMillan, D.C. Cancer Cachexia: A Nutritional or a Systemic Inflammatory Syndrome? *Br. J. Cancer* **2022**, *127*, 379–382. [CrossRef] [PubMed]
32. Kurita, Y.; Kobayashi, N.; Tokuhisa, M.; Goto, A.; Kubota, K.; Endo, I.; Nakajima, A.; Ichikawa, Y. Sarcopenia Is a Reliable Prognostic Factor in Patients with Advanced Pancreatic Cancer Receiving FOLFIRINOX Chemotherapy. *Pancreatology* **2019**, *19*, 127–135. [CrossRef] [PubMed]
33. Huang, X.; Lv, L.-N.; Zhao, Y.; Li, L.; Zhu, X.-D. Is Skeletal Muscle Loss Associated with Chemoradiotherapy Toxicity in Nasopharyngeal Carcinoma Patients? A Prospective Study. *Clin. Nutr.* **2021**, *40*, 295–302. [CrossRef] [PubMed]
34. Bohlius, J.; Bohlke, K.; Castelli, R.; Djulbegovic, B.; Lustberg, M.B.; Martino, M.; Mountzios, G.; Peswani, N.; Porter, L.; Tanaka, T.N.; et al. Management of Cancer-Associated Anemia with Erythropoiesis-Stimulating Agents: ASCO/ASH Clinical Practice Guideline Update. *Blood Adv.* **2019**, *3*, 1197–1210. [CrossRef] [PubMed]
35. Gilreath, J.A.; Rodgers, G.M. How I Treat Cancer-Associated Anemia. *Blood* **2020**, *136*, 801–813. [CrossRef]
36. Dicato, M.; Plawny, L.; Diederich, M. Anemia in Cancer. *Ann. Oncol.* **2010**, *21*, vii167–vii172. [CrossRef]
37. da Rocha, I.M.G.; Marcadenti, A.; de Medeiros, G.O.C.; Bezerra, R.A.; Rego, J.F.M.; Gonzalez, M.C.; Fayh, A.P.T. Is Cachexia Associated with Chemotherapy Toxicities in Gastrointestinal Cancer Patients? A Prospective Study. *J. Cachexia Sarcopenia Muscle* **2019**, *10*, 445–454. [CrossRef]
38. Gi, Y.-M.; Jung, B.; Kim, K.-W.; Cho, J.-H.; Ha, I.-H. Low Handgrip Strength Is Closely Associated with Anemia among Adults: A Cross-Sectional Study Using Korea National Health and Nutrition Examination Survey (KNHANES). *PLoS ONE* **2020**, *15*, e0218058. [CrossRef]
39. Nikolaou, V.; Syrigos, K.; Saif, M.W. Incidence and Implications of Chemotherapy Related Hand-Foot Syndrome. *Expert Opin. Drug Saf.* **2016**, *15*, 1625–1633. [CrossRef]
40. Ruiz-Pinto, S.; Pita, G.; Martín, M.; Nuñez-Torres, R.; Cuadrado, A.; Shahbazi, M.N.; Caronia, D.; Kojic, A.; Moreno, L.T.; de la Torre-Montero, J.C.; et al. Regulatory CDH4 Genetic Variants Associate With Risk to Develop Capecitabine-Induced Hand-Foot Syndrome. *Clin. Pharmacol. Ther.* **2021**, *109*, 462–470. [CrossRef]
41. Gökyer, A.; Küçükarda, A.; Köstek, O.; Hacıoğlu, M.B.; Sunal, B.S.; Demircan, N.C.; Uzunoğlu, S.; Solak, S.; İşsever, K.; Çiçin, I.; et al. Relation between Sarcopenia and Dose-Limiting Toxicity in Patients with Metastatic Colorectal Cancer Who Received Regorafenib. *Clin. Transl. Oncol.* **2019**, *21*, 1518–1523. [CrossRef] [PubMed]
42. Adel, N. Overview of Chemotherapy-Induced Nausea and Vomiting and Evidence-Based Therapies. *Am. J. Manag. Care* **2017**, *23*, S259–S265. [PubMed]

43. Dranitsaris, G.; Molassiotis, A.; Clemons, M.; Roeland, E.; Schwartzberg, L.; Dielenseger, P.; Jordan, K.; Young, A.; Aapro, M. The Development of a Prediction Tool to Identify Cancer Patients at High Risk for Chemotherapy-Induced Nausea and Vomiting. *Ann. Oncol.* **2017**, *28*, 1260–1267. [CrossRef]
44. Prado, C.M.M.; Maia, Y.L.M.; Ormsbee, M.; Sawyer, M.B.; Baracos, V.E. Assessment of Nutritional Status in Cancer–The Relationship Between Body Composition and Pharmacokinetics. *Anti-Cancer Agents Med. Chem.* **2013**, *13*, 1197–1203. [CrossRef]
45. Pérez-Pitarch, A.; Guglieri-López, B.; Nacher, A.; Merino, V.; Merino-Sanjuán, M. Impact of Undernutrition on the Pharmacokinetics and Pharmacodynamics of Anticancer Drugs: A Literature Review. *Nutr. Cancer* **2017**, *69*, 555–563. [CrossRef]
46. Cereda, E.; Tancredi, R.; Klersy, C.; Lobascio, F.; Crotti, S.; Masi, S.; Cappello, S.; Stobäus, N.; Tank, M.; Cutti, S.; et al. Muscle Weakness as an Additional Criterion for Grading Sarcopenia-Related Prognosis in Patients with Cancer. *Cancer Med.* **2022**, *11*, 308–316. [CrossRef]
47. Retornaz, F.; Guillem, O.; Rousseau, F.; Morvan, F.; Rinaldi, Y.; Nahon, S.; Castagna, C.; Boulahssass, R.; Grino, M.; Gholam, D. Predicting Chemotherapy Toxicity and Death in Older Adults with Colon Cancer: Results of MOST Study. *Oncologist* **2020**, *25*, e85–e93. [CrossRef]
48. Spexoto, M.C.B.; Ramírez, P.C.; de Oliveira Máximo, R.; Steptoe, A.; de Oliveira, C.; Alexandre, T.D.S. European Working Group on Sarcopenia in Older People 2010 (EWGSOP1) and 2019 (EWGSOP2) Criteria or Slowness: Which Is the Best Predictor of Mortality Risk in Older Adults? *Age Ageing* **2022**, *51*, afac164. [CrossRef] [PubMed]
49. Lemoine, A.; Perrier, M.; Mazza, C.; Quinquenel, A.; Brasseur, M.; Delmer, A.; Vallerand, H.; Dewolf, M.; Bertin, E.; Barbe, C.; et al. Feasibility and Impact of Adapted Physical Activity (APA) in Cancer Outpatients Beginning Medical Anti-Tumoral Treatment: The UMA-CHAPA Study. *Cancers* **2022**, *14*, 1993. [CrossRef]
50. Chen, Y.-J.; Li, X.-X.; Ma, H.-K.; Zhang, X.; Wang, B.-W.; Guo, T.-T.; Xiao, Y.; Bing, Z.-T.; Ge, L.; Yang, K.-H.; et al. Exercise Training for Improving Patient-Reported Outcomes in Patients With Advanced-Stage Cancer: A Systematic Review and Meta-Analysis. *J. Pain Symptom Manag.* **2020**, *59*, 734–749.e10. [CrossRef] [PubMed]

Article

Medium-Chain Fatty Acids and Breast Cancer Risk by Receptor and Pathological Subtypes

Padmanabha Ganeshkodi Roopashree [1,†], Shilpa S. Shetty [1,*,†], Vijith Vittal Shetty [2] and Suchetha Kumari Nalilu [3,*]

1   Central Research Laboratory, KS Hegde Medical Academy, Nitte (Deemed to be University), Mangalore 575018, India
2   Department of Oncology, KS Hegde Medical Academy, Nitte (Deemed to be University), Mangalore 575018, India
3   Department of Biochemistry, KS Hegde Medical Academy, Nitte (Deemed to Be University), Mangalore 575018, India
*   Correspondence: shilpajshetty@nitte.edu.in (S.S.S.); kumarin@nitte.edu.in (S.K.N.)
†   These authors contributed equally to this work.

Abstract: Introduction: Medium-chain fatty acids contain 6–12 carbon atoms and are absorbed directly into the blood vessels, proceeding to the portal vein and, finally, to the liver, where they are immediately utilized for energy. We aimed to determine the medium-chain fatty acid levels in women with and without breast cancer. **Materials and Methods:** A total of 200 women (100 breast cancer subjects and 100 control subjects) were recruited for the study as per the inclusion and exclusion criteria. Blood samples were collected for biochemical estimations. Fatty acid methyl esters were isolated, and medium-chain fatty acid levels in plasma were analyzed using gas chromatography (GC-FID). Statistical analysis was performed using SPSS 20.0 software; $p \leq 0.05$ was considered statistically significant. **Results:** The fatty acid analysis revealed a significant decrease in the levels of caprylic acid (C:8) and lauric acid (C:12) and a significant increase in the level of capric acid (C:10) in the breast cancer subjects when compared to the control group. The level of caproic acid (C:6) was not significantly increased in the breast cancer subjects. In particular, the HER2- and ER-positive breast cancer subjects showed a decrease in their caprylic acid and lauric acid levels compared to other receptors. **Conclusions:** The results of the current study imply that lower levels of caprylic and lauric acid may be associated with a higher risk of breast cancer. The relevance of medium-chain fatty acids for preventive and therapeutic interventions will be amplified by further research on the possibility that alteration in a patient's medium-chain fatty acid composition may mechanistically contribute to disease progression or breast cancer risk.

Keywords: breast cancer; medium-chain fatty acids; lauric acid; caprylic acid

Citation: Roopashree, P.G.; Shetty, S.S.; Shetty, V.V.; Nalilu, S.K. Medium-Chain Fatty Acids and Breast Cancer Risk by Receptor and Pathological Subtypes. *Nutrients* 2022, 14, 5351. https://doi.org/10.3390/nu14245351

Academic Editors: Ashley J. Snider, Nuno Borges, Fernando Mendes and Diana Martins

Received: 3 November 2022
Accepted: 14 December 2022
Published: 16 December 2022

Publisher's Note: MDPI stays neutral with regard to jurisdictional claims in published maps and institutional affiliations.

1. Introduction

Breast cancer is the major cause of cancer-related death in women. In India, in the year 2018, 162,468 women were newly diagnosed with breast cancer and 87,090 women died of breast cancer [1]. It is indisputable that the existing chemotherapy is effective in cancer treatment, though the presence of undesirable adverse effects has activated a demand for new therapeutic agents. Breast cancer risk factors have been recognized, including hereditary inheritance, environmental exposure, infection, reproductive characteristics, and diet. The nutritional risk factors of breast cancer include dietary fats [2], even though the epidemiological evidence is still ambiguous. This controversy exists partially because body fat is usually studied according to the total fat content, the type of fatty acid (saturated, monounsaturated, or polyunsaturated), or in terms of origin. Therefore, even though certain fatty acids are thought to have a significant impact on a variety of biological processes, including tumor development and progression, little attention has been paid to them.

Medium-chain fatty acids (MCFAs) are saturated fatty acids with 6 to 12 carbon atoms. Coconut oil (CO) and palm kernels are some natural sources. Increased levels of long-chain fatty acids (LCFA) are also related to an increased risk of breast cancer [3]. Contrary to LCFAs, MCFAs must be combined into chylomicrons before entering the liver. However, dietary medium-chain fatty acids are quickly absorbed in the gastrointestinal tract and are transported into the bloodstream through direct contact with albumin [4,5] via the portal vein; once there, they reach the liver and are metabolized through β-oxidation in mitochondria [6]. Caproic acid (C:6), caprylic acid (C:8), capric acid (C:10), and lauric acid (C:12) are the most common medium-chain fatty acids. They have proven anticancer effects on human breast, skin, and colorectal cancer cells invitro [6]. Coconut oil, a source of MCFA's is a credible nutraceutical for cancer prevention [7]. A cohort study has revealed that MCFAs are valuable early diagnostic biomarkers of colorectal cancer [8]. To potentiate our understanding of the possible relationship between MCFAs and breast cancer, this case–control study was undertaken on a relatively stable and homogeneous population to determine the level of specific MCFAs in breast cancer subjects and controls and to assess the possible association between specific MCFAs and breast cancer risk.

2. Materials and Methods

The study was approved by the Central Ethics Committee of Nitte (Deemed to be a University). A total of 100 histopathologically proven breast cancer subjects and 100 control subjects attending the OPD for general health check-ups between the age of 25–60 were recruited for the study after obtaining informed consent. Demographic data such as age, BMI (Body Mass Index), diet, menopausal status, and first-degree family history with biochemical parameters like Hb (Hemoglobin), RBS (Random blood sugar), platelet count, blood urea, creatinine concentration, ALP (Alkaline phosphatase), AST (Aspartate amino-transferase), ALT (Alanine transaminase), Na^+, K^+, Cl^-, total bilirubin, albumin, globulins, A/G (Albumin/globulin) ratio, and total protein were noted. Clinical characteristics such as TNM (Tumor Node Metastases) stage, tumor size, receptor status, grade, and histological type were noted in breast cancer subjects. A total of 3 mL of blood was collected for from each subjects.

2.1. Selection of Subjects

The study group included 100 women as per the inclusion and exclusion criteria. Subjects diagnosed with breast cancer proven through pathology reports without any treatment were included in the study. Subjects who had breast implants or any malignancies were excluded from the study. We recruited 100 age-matched control women who attended the oncology OPD and not undergone breast-conserving surgery/not reported breast cancer at the time of enrollment. Control subjects, pregnant women, and any other benign proliferations were excluded from the study. A total of 3 mL of the blood was collected from the recruited subjects, plasma samples and stored at $-20\,^\circ\text{C}$ and until further analysis.

2.2. Estimation of Lipid Profiles

TC (Total Cholesterol), TG (Triglycerides), and HDL-C (High-density lipoprotein choles-terol) levels were analyzed using commercially available kits (Liqui CHEKTM AGAPPE). Friedewald formula was used for LDL-C (Low-density lipoprotein cholesterol) and VLDL-C (Very low-density lipoprotein cholesterol) calculation [9]. The formation of malondialdehyde (MDA) was estimated using a standardized protocol by Buege, J. A et al. [10].

2.3. Medium-Chain Fatty Acid Analysis

Lipid transesterification of stored plasma to fatty acid methyl esters (FAMEs) was performed according to the modified protocol of Metcalfe et al. [11]. Fatty acid levels were determined in the presence of internal standard (1 mg/mL methyl heptadecanoate-C17:0,

Sigma Aldrich). Extraction of the total plasma medium-chain fatty acid was performed via the hydrolysis of esters and then derivatization of esters under alkaline conditions in 14% boron trifluoride-methanol for 5 min at 100 °C to form FAMEs. Then, these fatty acids were measured using gas chromatography (GC-FID). Extracted FAMEs were analyzed on a 7820A Agilent GC-FID (flame ionization detector) with J and W DB-23 high-quality columns. Individual medium-chain fatty acids were identified by comparing their elution times with relative medium-chain fatty acid standards. Fatty acids were calculated according to their comparative abundance with respect to the internal standard added. The quantity of individual medium-chain fatty acid was calculated as the percentage of the total medium-chain fatty acid concentration within each sample.

3. Statistical Analysis

Statistical analysis of the obtained results was performed using GraphPad Prism, Version 8.0.2 (GraphPad Software, Inc., La Jolla, CA, USA). Categorical variables were analysed using a chi-square test. Student's *t*-test was used to compare the two groups. The Mann–Whitney U test was used for non-parametric variables. To correlate the non-parametric variables with each other for all subjects, the Spearman's correlation coefficient test was used. All statistical tests were two-sided, and $p < 0.05$ was considered significant.

4. Results

4.1. General Characteristics of the Study Population

A total of 200 participants were included in this study. Of them, 100 were control subjects and 100 were breast cancer subjects. The demographic characteristics are summarized in Table 1. The mean age ($\pm$SD) was 46.731 ± 10.846 years for the control subjects and 50.04 ± 10.611 years for the breast cancer subjects, and the age distribution of the control subjects and breast cancer subjects was similar. Out of 100 control subjects, 52% were pre-menopausal and 48% were post-menopausal, whereas in the 100 breast cancer subjects, 31% were pre-menopausal and 69% were post-menopausal. Out of the 100 control subjects, 3% had a BMI of <18.5, 70% had BMI of 18.5–24.9, and 27% had a BMI of $\geq$25. Whereas in the breast cancer subjects, 2% had a BMI of <18.5, 58% had a BMI of 18.5–24.9, and 40% had a BMI of $\geq$25. The mean BMI and age showed a significant difference between the breast cancer subjects and the control subjects. The family history of the breast cancer in the first-degree relatives among the cases and controls indicated that the cases were likely to have a higher proportion of first-degree relatives (mother, sisters, and daughters) with breast cancer (<0.040). History regarding menopausal status and diet did not exhibit a significant case–control difference in the present study.

Among the 100 breast cancer subjects, 63 had invasive ductal carcinoma and 37 had invasive lobular carcinoma. Concerning the clinicopathological differences among the individuals in the two types of breast cancer groups, there was no significant difference observed with TNM stage, tumor size, lymph node status, and receptor status of breast cancer, but a significant difference was observed concerning grade ($p < 0.027$). The clinicopathological characteristics are summarized in Table 2.

Table 1. Demographic characteristics of the study population.

	Variables	Control ($n = 100$)	Case ($n = 100$)	*p* Value
Age				
	Total	46.731 ± 10.846	50.04 ± 10.611	<0.04 *
	<45	59	35	
	45–54	24	33	0.072
	55–64	9	24	
	$\geq$65	8	8	

Table 1. *Cont.*

Variables	Control (*n* = 100)	Case (*n* = 100)	*p* Value
BMI			
Total	21.232 ± 2.870	22.117 ± 3.448	<0.001 **
<18.5	3	2	
18.5–24.9	70	58	0.050
≥25	27	40	
Menopausal status (%)			
Pre-menopausal	52	31	
Post-menopausal	48	69	0.072
First-degree family history (%)			
Yes	2	11	
No	98	89	<0.040 *
Diet (%)			
Mixed	87	85	
Vegetarian	13	15	0.833

Age and BMI is represented as mean ± SD. Subgroups of Age and BMI, Menopausal status, First-degree family history, and Diet variables are given in percentages. *p* value was calculated using the Student's *t*-test for parametric variables. ** *p* < 0.001; * *p* < 0.05. Categorical variables was tested using the chi-square test.

Table 2. Clinical characteristics of women with breast cancer.

Tumor Characteristics	Invasive Ductal Carcinoma *n* (%) (*n* = 63)	Invasive Lobular Carcinoma *n* (%) (*n* = 37)	*p* Value
TNM Stage			
I	11 (18)	9 (24)	
II	24 (38)	10 (27)	0.477
III/IV	28 (44)	18 (49)	
Tumor size (cm)			
<2.0	16 (25)	16 (43)	
2.0–4.9	20 (32)	6 (16)	0.105
≥5.0	27 (43)	15 (41)	
Lymph Node status			
Positive	42 (67)	25 (68)	
Negative	21 (33)	12 (32)	0.553
Receptor status			
ER + Ve	6 (10)	11 (30)	
HER2 + Ve	14 (22)	4 (11)	
ER/PR + Ve	11 (18)	5 (13)	
ER/HER2 + Ve	11 (17)	7 (19)	0.144
ER/PR/HER2 + Ve	11 (17)	4 (11)	
TNBC	10 (16)	6 (16)	
Grade			
2	41 (65)	16 (43)	
4	22 (35)	21 (57)	<0.027 *

The variables—TNM stage, tumor size, lymph node status, receptor status, and grade—are presented as percentages. For categorical variables, the statistical significance of the two groups was tested using a chi-square test. * *p* < 0.05 was considered statistically significant. Abbreviations: TNM—Tumor Node Metastases; ER—Estrogen Receptor; HER2—Human Epidermal Growth Factor Receptor 2; PR—Progesterone Receptor; TNBC—Triple-Negative Breast Cancer.

4.2. Comparison of Biochemical Parameters in the Study Population

The random blood glucose level was significantly higher in the breast cancer subjects than in the control group. The mean Hb level was significantly lower in the breast cancer subjects than in the control group. The levels of TC, TG, HDL-C, LDL-C, VLDL-C, and MDA were also examined and are shown in Table 3. The levels of TC, TG, VLDL-C, and MDA in the breast cancer subjects were significantly increased when compared to the control group. Whereas the HDL-C levels were significantly decreased in the breast cancer subjects. The mean values regarding the ALT, blood urea, creatinine, albumin, and globulin levels were significantly increased in the breast cancer group compared to the control group.

Table 3. Comparison of the biochemical parameters in subjects with and without breast cancer.

Parameters	Control (*n* = 100)	Breast Cancer (*n* = 100)	*p* Value
RBS (mg/dL)	105.332 ± 7.695	136.600 ± 72.074	<0.001 **
Hematological Parameters			
Hb (g/dL)	13.273 ± 0.849	10.894 ± 1.375	<0.001 **
Platelet count (10^3/µL)	336.549 ± 107.376	252.360 ± 93.006	0.061
Lipid Profiles			
TC (mg/dL)	157.430 ± 27.185	197.180 ± 33.065	<0.021 *
TG (mg/dL)	123.191 ± 34.028	152.228 ± 53.672	<0.001 **
HDL-C (mg/dL)	59.075 ± 19.592	53.766 ± 15.669	<0.017 *
LDL-C (mg/dL)	60.279 ± 19.865	65.568 ± 19.777	0.888
VLDL-C (mg/dL)	24.546 ± 7.046	29.977 ± 11.191	<0.001 **
Lipid Peroxidation			
MDA (µM/L)	3.867 ± 1.882	4.137 ± 0.441	<0.04 *
Liver Function tests			
ALP (IU/L)	107.381 ± 30.061	107.520 ± 29.892	0.586
AST (IU/L)	24.448 ± 16.435	26.686 ± 9.913	0.571
ALT (IU/L)	17.553 ± 9.814	33.087 ± 15.573	<0.001 **
Total Bilirubin (mg/dL)	0.464 ± 0.166	1.257 ± 0.520	0.137
Kidney Function tests			
Blood Urea (mg/dL)	18.314 ± 6.224	14.539 ± 4.752	<0.001 **
Serum Creatinine (mg/dL)	0.707 ± 0.140	1.607 ± 8.526	<0.032 *
Albumin (g/dL)	3.958 ± 0.856	4.290 ± 0.635	<0.011 *
Globulin (g/dL)	2.840 ± 0.280	3.167 ± 0.434	<0.001 **
A/G Ratio	1.574 ± 0.680	1.676 ± 0.733	0.306
Total Protein (mg/dL)	7.256 ± 0.946	7.394 ± 0.860	0.086
Na^+ (mmol/L)	137.442 ± 15.349	139.442 ± 6.509	0.796
K^+ (mmol/L)	4.180 ± 0.485	4.333 ± 0.694	<0.013 *
Cl^- (mmol/L)	99.604 ± 11.432	102.323 ± 4.176	0.937
Medium-chain fatty acids			
Caproic Acid (%, C:6)	3.594 (2.684–5.309)	4.516 (2.671–6.529)	0.302
Caprylic Acid (%, C:8)	2.256 (1.794–3.318)	0.902 (0.624–1.547)	<0.001 #
Capric Acid (%, C:10)	10.709 (9.198–12.544)	12.559 (10.393–14.956)	<0.002 #
Lauric Acid (%, C:12)	3.882 (3.470–4.761)	3.083 (2.366–3.700)	<0.001 #

p value ≤ 0.05 was considered statistically significant. * Student's *t*-test was used for parametric variables; the data are represented as mean ± SD. # Mann–Whitney U test was used for non-parametric variables; data are shown as medians (interquartile range). ** *p* < 0.001 and * *p* < 0.05. Abbreviations: RBS—Random blood sugar, Hb—Hemoglobin, MDA—Malondialdehyde, TC—Total Cholesterol, TG—Triglycerides, HDL-C—High-density lipoprotein cholesterol, LDL—Low-density lipoprotein cholesterol, VLDL-C—Very low-density lipoprotein cholesterol, ALP—Alkaline phosphatase, AST—Aspartate aminotransferase, ALT—Alanine transaminase, and A/G (Albumin/globulin) ratio.

In addition, the percentages of medium-chain fatty acids, i.e., caproic acid, caprylic acid, capric acid, and lauric acid, in each group are represented in a box plot (Table 3). The results showed the levels of MCFAs in the control and breast cancer subjects, while the most statistically significant decrease in caprylic acid and lauric acid was seen in the breast cancer subjects. In particular, the levels of capric acid were statistically significantly higher in the breast cancer subjects than in the control group. The levels of caproic acid were higher in the breast cancer subjects than in the control group, but no significant difference was observed.

4.3. Age-Wise Distribution of MCFA Levels in the Study Population

In the 100 breast cancer subjects, 35% of the subjects were <45 years old, 33% of the subjects were 45–54 years old, 24% of the subjects were 55–64 years old, and 8% of the subjects were ≥65 years old. Whereas in the 100 control subjects, 59% of the subjects were <45 years old, 24% of the subjects were 45–54 years old, 9% of the subjects were 55–64 years old, and 8% of the subjects were ≥65 years old. The box plot shows lower levels of caprylic and lauric acids in the different age groups of the breast cancer subjects than in the control group, which were statistically significant. The distribution of MCFAs in each age group of the study subjects is reported in the box plot (Figure 1). In the entire <45-year-old demographic, the level of capric acid was significantly higher in the breast cancer subjects compared to the control group. However, there were no significant differences observed in the other age groups of the study subjects.

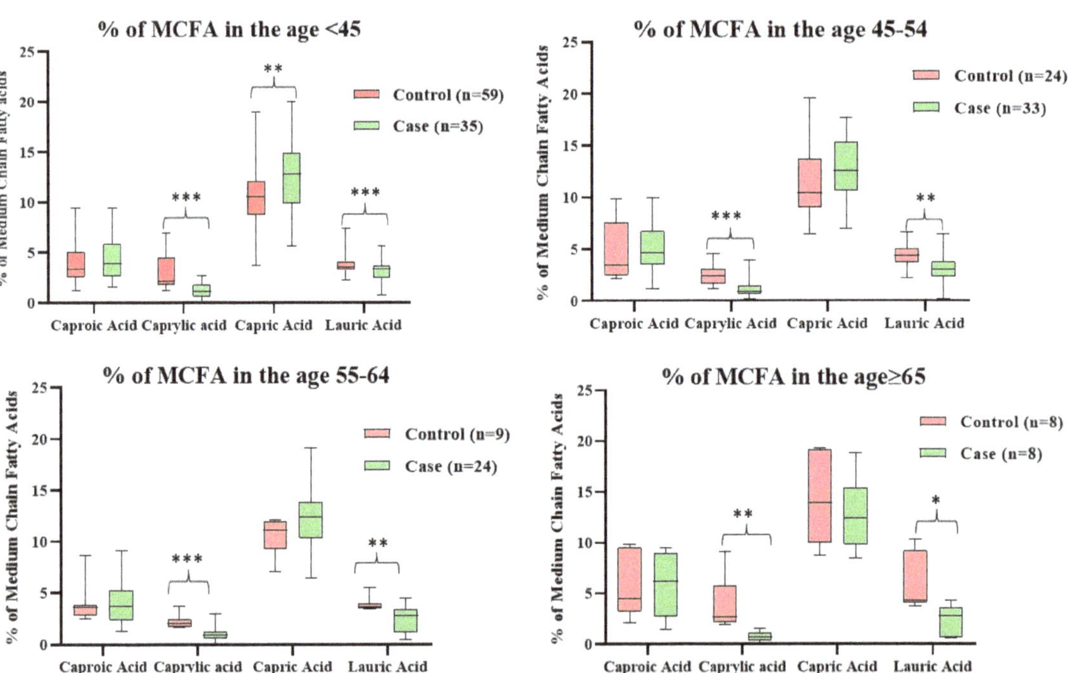

Figure 1. Age-wise distribution of MCFA levels in control and breast cancer subjects. The amount of each MCFA was calculated as a % of total MCFAs. *p*-value was calculated using Mann–Whitney U test for non-parametric variables. Data are shown as median (interquartile range). *** $p < 0.0001$, ** $p < 0.001$, and * $p < 0.05$. Abbreviations: MCFA—Medium-chain fatty acid.

4.4. Comparison of MCFA Levels Regarding Histological Types, TNM Stage, and Grade of Breast Cancer Subjects

A comparison of the MCFAs with respect to the histological types, TNM stage, and grade of the breast cancer subjects is shown in Figure 2. Regarding tumor histology, 63%

of the breast cancer subjects had invasive ductal carcinoma and 37% of the breast cancer subjects had invasive lobular carcinoma. There were no significant differences observed regarding the histological types of the breast cancer subjects. Regarding the TNM stage, 20% of the breast cancer subjects corresponded to stage I, 34% of the breast cancer subjects corresponded to stage II, and 46% of the breast cancer subjects corresponded to stage III/IV. The level of capric acid was significantly increased in the stage II tumor compared to the stage I tumor breast cancer subjects, whereas the level of capric acid was significantly decreased in the stage III/IV tumor compared to the stage II tumor subjects. However, there was no significant difference observed with respect to the other MCFA levels. Concerning grade, 57% of the breast cancer subjects corresponded to grade I, and 43% of the breast cancer subjects corresponded to grade II. The level of capric acid was significantly decreased in grade III compared to grade II, but no significant difference was observed in the other MCFA levels.

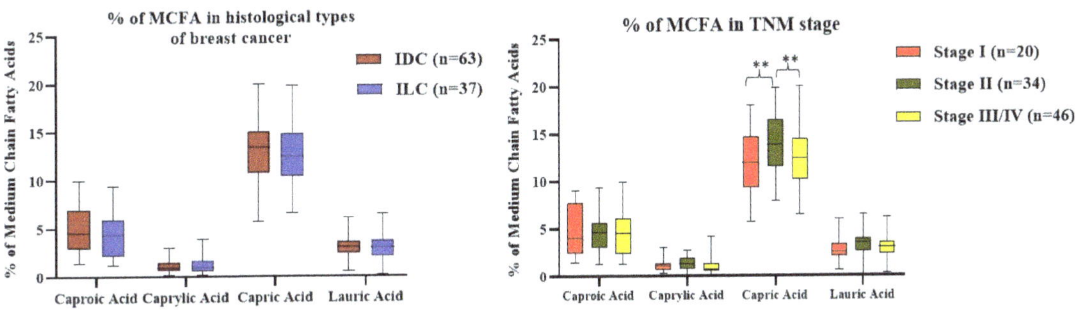

Figure 2. Comparison of MCFA levels in terms of histological types, TNM stage, and grades of breast cancer subjects. The amount of each MCFA was calculated as a % of total MCFAs. *p*-value was calculated using the Mann–Whitney U test for non-parametric variables. Data are shown as median (interquartile range). ** $p < 0.001$. Abbreviations: TNM—Tumor Node Metastases; Tumor IDC—Invasive ductal carcinoma; ILC—Invasive lobular carcinoma; MCFA—Medium-chain fatty acid.

4.5. Distribution of MCFAs Levels with Respect to the Receptor Status of the Breast Cancer Subjects

According to receptor status, the differences in the MCFA levels in the breast cancer subjects became more evident (Figure 3). Of the 100 breast cancer subjects, 17% of the breast cancer subjects were ER-positive, 18% of the breast cancer subjects were HER2-positive, 16% of the breast cancer subjects were ER/PR-positive, 18% of the breast cancer subjects were ER/HER2-positive, 15% of the breast cancer subjects were ER/PR/HER2-positive, and 16% of the breast cancer subjects had TNBC. Both HER2-positive and ER/HER2-positive breast cancer were frequent findings in our study compared to other receptor statuses.

The corresponding box plot shows significant decreases in the levels of caprylic and lauric acids at various receptor statuses compared to the control group. In particular, the level of caprylic acid was significantly lower in the HER2-positive breast cancer subjects than in the other receptor statuses of the breast cancer subjects. The level of lauric acid was lower and statistically significant in the ER/HER2-positive breast cancer subjects compared to the other receptor status. Caproic and capric acids did not show any statistically significant differences with receptor status in breast cancer subjects.

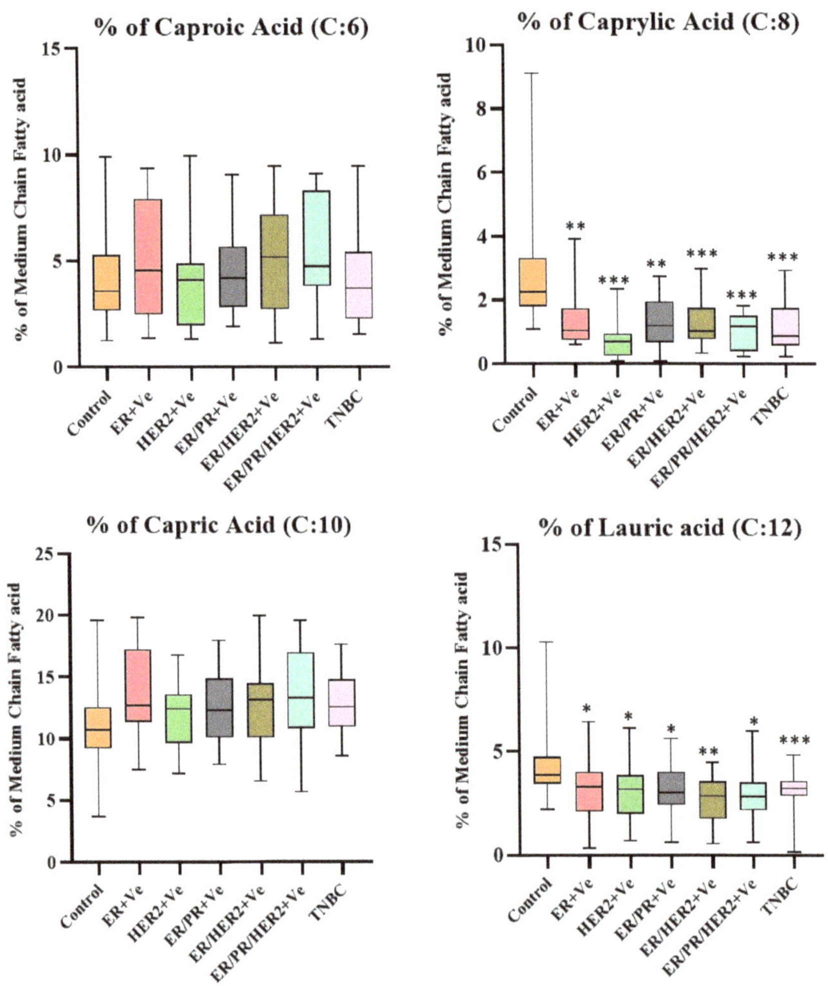

Figure 3. Tukey boxplots showing the distributions of the levels of each MCFA corresponding to various receptor statuses of breast cancer subjects. The quantity of each MCFA was calculated as a % of total MCFAs. *p* value was calculated using Mann–Whitney U test for non-parametric variables. ER + Ve (*n* = 17), HER2 + Ve (*n* = 18), ER/PR + Ve (*n* = 16), ER/HER2 + Ve (*n* = 18), ER/PR/HER2 + Ve (*n* = 15), and TNBC (*n* = 16). *** $p < 0.0001$, ** $p < 0.001$, and * $p < 0.05$. Abbreviations: ER—Estrogen Receptor; HER2—Human Epidermal Growth Factor Receptor 2; PR—Progesterone Receptor; TNBC—Triple-Negative Breast Cancer; MCFA—Medium-chain fatty acid.

4.6. Correlation between Hematological Parameters, Lipid Profile, Kidney Function Test, Liver Function Test, and Medium-Chain Fatty Acids in Breast Cancer

A significant positive correlation was observed between caproic acid and TC, whereas a significant negative correlation was seen between ALT, the total bilirubin content, and the A/G ratio. Caprylic acid showed a significant negative correlation with TC, AST, and globulin compared to the other parameters. Capric acid showed a significant negative correlation with ALT and AST. Lauric acid showed a significantly negative correlation with TC, LDL-C, VLDL-C, AST, and globulin (Supplementary Table S1).

5. Discussion

The complex variations in the metabolism of carbohydrates, proteins, and lipids are essential for tumor cells' growth and proliferation. In many epidemiological studies, increased dietary fat consumption is positively associated with breast cancer. In the present study, significantly high levels of TC, TG, and VLDL-C were observed, but the level of HDL-C was significantly decreased in breast cancer subjects. More lipids are needed to improve signaling and the resistance to apoptosis in rapidly multiplying cancer cells [12]. An increased plasma LDL-C concentration increases exposure to oxidation, causing higher lipid peroxidation in breast cancer subjects [13]. In this study, the level of VLDL-C was significantly increased in the breast cancer subjects (Table 3).

High levels of SGOT (Serum glutamic-oxaloacetic transaminase) and SGPT (Serum glutamic-pyruvic transaminase) suggest that liver and kidney function impairment might be triggered by tumor invasion [14]. This study observed a significantly decreased albumin level in breast cancer subjects. This reduced albumin level corresponds to the poor survival of breast cancer subjects [15–17]. An elevated creatinine level causes kidney impairment [18]. This study found a statistically significant realtion to creatinine level. Glucose plays a vital role in breast cancer therapy. Some studies have shown that hematological and solid tumor hyperglycemia is linked with increased toxicity [19]. In this study, significant changes were noted concerning random blood glucose levels in breast cancer subjects. Based on the clinical findings concerning the subjects diagnosed with breast cancer, our investigations showed that the levels of urea were significantly lower compared to the control group. Decreased urea levels might suggest a link between the dysregulation of protein catabolic processes and the aggressive behavior of cancer cells [20]. A common complication in breast cancer subjects is anemia. This study observed significantly low hemoglobin levels in breast cancer subjects (Table 3). Higher levels of inflammatory markers, IL-6, leptin, hepcidin, ferritin, and ROS all contribute to anemia in cancer subjects.

The traditional sources of fatty acids include diet, circulation from adipose tissue, and surplus carbohydrates the liver that turn into fat. Most human diets contain various saturated fatty acids of different carbon chain lengths. MCFAs are saturated fatty acids with 6–12 carbon atoms that are more quickly taken from the intestine to the liver via the portal vein and immediately used for energy. Naturally, medium-chain fatty acids are found in coconut oil, palm kernel oil, and in milk fat [21–23]. Though no association between medium-chain fatty acids and gut microbes has been documented, Caprylate (C8), one of the MCFAs, has been reported to be produced by specific yeast strains [24]. Clostridium kluyveri [25–27], a bacterial strain found in the rumen intestine [28,29], can produce MCFAs for industrial uses. However, there are few findings on MCFAs generated from gut bacteria in non-rumen animals. It is interesting to note that MCFAs have anti-bacterial and anti-fungal effects on specific bacterial strains [30,31]. Less than 2% of dietary energy is typically contributed by MCFAs in the modern human diet [32]. While PUFAs are generally known as anticancer dietary components, MCFAs have also been described to have a therapeutic role [33,34]. In the present study, we analyzed the plasma medium-chain fatty acid levels in control and breast cancer subjects. Our results showed a significant decrease in caprylic acid and lauric acid levels in the breast cancer subjects compared to the control group, except for caproic and capric acid (Table 3). In this study, the plasma medium-chain fatty acid levels were measured because plasma fatty acid levels depend on dietary intake, and these sources can provide a more objective measure of fatty acid levels than estimates based on dietary intake [35].

Among the saturated medium-chain fatty acids, the caprylic acid concentration was higher in the control group and lower in the breast cancer subjects. Caprylic acid is enriched in coconut and goat's milk. Narayanan et al. [6] reported that caprylic acid inhibited the viability of skin, colorectal, and breast cancer cells and downregulated the expression of genes such as CDK2, CDK4, CCNA2, and CCND1, which are mainly involved in progression and cell cycle division in colon cancer cells. A mechanism connected to ABCA1 and the p-JAK2/pSTAT3 signaling pathway suggests that caprylic acid may be crucial for lipid metabolism and the inflammatory response [36]. Yamasaki et al. [37] suggested that octanoic acid inhibits bladder cancer cell proliferation but does not reduce cell migration and invasion. Studies on the association between the levels of caprylic acid and the prognosis of breast cancer are limited. Jansen et al. [38] indicated that consuming full-fat products including saturated fatty acids such as octanoic acid increases the risk of pancreatic cancer dose-dependently. According to Cuizhe Wang et al. [39], caprylic acid (C8:0) enhances COX2 and PGE2 expression in the bone marrow cavity, increases adipocyte growth and proliferation, and causes bone metastases of prostate cancer. According to a study by Iemoto et al. [40], people with colorectal cancer who had lower levels of serum caprylic acid (C8:0) had a better prognosis than those who had greater levels of caprylic acid. However, to the best of our knowledge, no studies assessing the effect of caprylic acid on the risk of breast cancer have been detailed; hence, more research is required.

In this study, we found that the level of lauric acid in the breast cancer subjects was significantly lower than in the control group. The major source of lauric acid is coconut oil and palm kernel oil. Lauric acid promotes cell death, which is assisted through the activation of EGFR and the Rho-associated kinase pathway, according to Lappano et al. [41]. According to research by Sheela et al. [7], lauric acid significantly inhibits the growth of human hepatocellular carcinoma and murine macrophage cells. Lauric acid and increased intracellular reactive oxygen species with a corresponding decrease in the intracellular reduced glutathione levels, have been demonstrated to cause apoptotic alterations and cell cycle arrest in the G0/G1 and G2/M phases.

According to numerous studies, postmenopausal older women in industrialized countries are more likely to acquire breast cancer than younger premenopausal women. In our study, the age-wise distribution of caprylic and lauric acids was significantly lower in the breast cancer subjects than in the control group (Figure 1). The younger population and unique demographics of developing countries may have a significant impact on these findings [1,42].

In this study, Figure 2 depicted the levels of medium-chain fatty acids by grade. Out of the 100 breast cancer individuals, 43% of those with grade III and 57% of those with grade II had the disease. When compared to other saturated medium-chain fatty acids, grade III had much less capric acid than grade II. In the stage II breast cancer patients, compared to stage I, the capric acid levels were higher; however, in the stage III/IV breast cancer patients, compared to stage II, the capric acid levels were lower. The most prevalent finding in our investigation related to grade 2 and stage 2 or 3 tumors, which was comparable to the results from a study of symptomatic cases conducted in the UK [43].

In the present study, we found that the caprylic and lauric acid levels were significantly lower in all breast cancer subjects receptor subtypes compared to control subjects (Figure 3). Based on the receptor status of the breast cancer subjects, caprylic acid was lower in the HER2-positive breast tumors, whereas lauric acid was lower in the ER/HER2-positive breast tumors compared to the other receptor groups. These findings suggested that out of the 100 breast cancer subjects, 18% had HER2-positive and ER/HER2-positive tumors. The ER status in the breast tumor cells was assessed in a study by Mirtavoos-Mahyari et al. (2014) to determine how it affected the activation of the tyrosine kinase human epidermal growth factor receptor 2 (HER2). Consequently, 31% of breast cancer participants and 67% overall had HER2 + tumors, according to the study [44]. Similarly, Rodrigue et al. (2014) found that ER + tumors were present in 61% of their study's breast cancer participants [45]. These studies linked increased steroid hormone responsiveness, a higher BMI, and increased body fat to hormone receptor-positive breast cancer.

The caproic acid levels showed a weakly positive link with total cholesterol and a weakly negative correlation with ALT, total bilirubin, and the A/G ratio in the breast cancer individuals. Between total cholesterol, AST, and globulin, there was a weakly negative association with the caprylic acid levels. While the lauric acid levels showed a weakly negative association with triglycerides, LDL-C, VLDL-C, AST, and globulin, capric acid levels showed a weakly negative correlation with AST and ALT (Supplementary Table S1).

There is a knowledge gap concerning the mechanism of action of medium-chain fatty acids on the oncogenic signal transduction pathway, and also clinical studies are limited. Identifying how medium-chain fatty acids are associated with the oncogenic signal transduction regulation and clinical presentation of breast cancer will be key in elucidating the mechanism behind this disease. Our study is vital in demonstrating the importance of medium chain fatty acid in cancer management and prevention.

6. Conclusions

There are considerable barriers o improving the prognosis of breast cancer subjects due to the lack of diagnostic technologies that are characterized by substantial patient compliance and good clinical applicability. Here, we have described the levels of different MCFAs in human samples. Among the medium-chain fatty acids investigated, the levels of caprylic and lauric acid were decreased in the subjects with breast cancer. Our findings imply that increasing the intake of caprylic and lauric acid while lowering the intake of caproic and capric acids may be an effective strategy for preventing breast cancer. To fully comprehend the impact of these medium-chain fatty acids on the development of breast cancer, more in vitro and in vivo investigations are required. The possible mechanisms of MCFA's effects must be determined, to further understand and enhance the use of MCFAs as a complementary breast cancer treatment and their effectiveness across breast cancer receptor and pathological subtypes. Further investigation would increase the knowledge and understanding of fatty acids and breast cancer, including the impact of MCFAs on the overexpressing receptor subtypes. Future research in this field should focus on the effect of early MCFA exposure on long-term breast cancer risk.

Supplementary Materials: The following supporting information can be downloaded at: https://www.mdpi.com/article/10.3390/nu14245351/s1, Table S1. Correlation between lipid profiles, liver function test, hematological parameters, kidney function test, and medium chain fatty acids in breast cancer.

Author Contributions: S.K.N.: Conceptualization, Visualization, Supervision, Reviewing. S.S.S.: Conceptualization, Visualization, Supervision, Methodology, Writing—Reviewing and Editing, P.G.R.: Data curation, Writing—Original draft preparation, Writing—Reviewing and Editing, V.V.S.: Supervision and Validation. All authors have read and agreed to the published version of the manuscript.

Funding: This research received no external funding. The APC was funded by Nitte (Deemed to be University).

Institutional Review Board Statement: The study was conducted in accordance with the Declaration of Helsinki, and approved by the Central Ethics Committee of Nitte (Deemed to be University) (NU/CEC/2020/0334 and 09/09/2020; NU/CEC/2022/327 and 15/11/2022) and Institutional Biosafety Committee (IBSC) of Nitte (Deemed to be University) (NU/IBSC/2020-21/15F and 24/08/2020).

Informed Consent Statement: Informed consent was obtained from all subjects involved in the study.

Data Availability Statement: The data are not publicly available.

Conflicts of Interest: The authors declare no conflict of interest.

References

1. Bray, F.; Ferlay, J.; Soerjo Mataram, I.; Siegel, R.L.; Torre, L.A.; Jemal, A. Global cancer statistics 2018: GLOBOCAN estimates of incidence and mortality worldwide for 36 cancers in 185 countries. *CA Cancer J. Clin.* **2018**, *68*, 394–424. [CrossRef] [PubMed]
2. Greenwald, P. Role of dietary fat in the causation of breast cancer: Point. *Cancer Epidemiol. Biomark. Prev.* **1999**, *8*, 3–7.
3. Cottet, V.; Vaysse, C.; Scherrer, M.L.; Ortega-Deballon, P.; Lakkis, Z.; Delhorme, J.B.; Deguelte-Lardiere, S.; Combe, N.; Bonithon-Kopp, C. Fatty acid composition of adipose tissue and colorectal cancer: A case-control study. *Am. J. Clin. Nutr.* **2015**, *101*, 192–201. [CrossRef] [PubMed]
4. Kenyon, M.A.; Hamilton, J.A. 13C NMR studies of the binding of medium-chain fatty acids to human serum albumin. *J. Lipid Res.* **1994**, *35*, 458–467. [CrossRef] [PubMed]
5. Vorum, H.; Pedersen, A.; Honoré, B. Fatty acid and drug binding to a low-affinity component of human serum albumin, purified by affinity chromatography. *Int. J. Pept. Protein Res.* **1992**, *40*, 415–422. [CrossRef]
6. Nagao, K.; Yanagita, T. Medium-chain fatty acids: Functional lipids for the prevention and treatment of the metabolic syndrome. *Pharmacol. Res.* **2010**, *61*, 208–212. [CrossRef]
7. Sheela, D.L.; Narayanankutty, A.; Nazeem, P.A.; Raghavamenon, A.C.; Muthangaparambil, S.R. Lauric acid induce cell death in colon cancer cells mediated by the epidermal growth factor receptor downregulation: An in silico and in vitro study. *Hum. Exp. Toxicol.* **2019**, *38*, 753–761. [CrossRef]
8. Crotti, S.; Agnoletto, E.; Cancemi, G.; di Marco, V.; Traldi, P.; Pucciarelli, S.; Nitti, D.; Agostini, M. Altered plasma levels of decanoic acid in colorectal cancer as a new diagnostic biomarker. *Anal. Bioanal. Chem.* **2016**, *408*, 6321–6328. [CrossRef]
9. Krauss-Etschmann, S.; Shadid, R.; Campoy, C.; Hoster, E.; Demmelmair, H.; Jiménez, M.; Gil, A.; Rivero, M.; Veszprémi, B.; Decsi, T.; et al. Effects of fish-oil and folate supplementation of pregnant women on maternal and fetal plasma concentrations of docosahexaenoic acid and eicosapentaenoic acid: A European randomized multicenter trial. *Am. J. Clin. Nutr.* **2007**, *85*, 1392–1400.
10. Buege, J.A.; Aust, S.D. Microsomal lipid peroxidation. *Methods Enzymol.* **1978**, *52*, 302–310.
11. Metcalfe, L.D.; Schmitz, A.A.; Pelka, J.R. Preparation of fatty acid esters from lipids for gas chromatography. *Anal. Chem.* **1966**, *38*, 514–515. [CrossRef]
12. Ray, G.; Husain, S.A. Role of lipids, lipoproteins and vitamins in women with breast cancer. *Clin. Biochem.* **2001**, *34*, 71–76. [CrossRef] [PubMed]
13. Kumar, K.; Sachdanandam, P.; Arivazhagan, R. Studies on the changes in plasma lipids and lipoproteins in patients with benign and malignant breast cancer. *Biochem. Int.* **1991**, *23*, 581–589. [PubMed]
14. Thangaraju, M.; Rameshbabu, J.; Vasavi, H.; Ilanchezhian, S.; Vinitha, S.; Sachdanandam, P. The salubrious effect of tamaxifen on serum marker enzymes, glycoproteins, and lysosomal enzymes level in breast cancer women. *Mol. Cell. Biochem.* **1998**, *185*, 85–94. [CrossRef] [PubMed]
15. Fatima, T.; Roohi, N.; Abid, R. Circulatory Proteins in Women with Breast Cancer and their Chemotherapeutic Responses. *Pak. J. Zool.* **2013**, *45*, 1207–1213.
16. Lis, C.G.; Grutsch, J.F.; Vashi, P.G.; Lammersfeld, C.A. Is serum albumin an independent predictor of survival in patients with breast cancer? *J. Parenter. Enter. Nutr.* **2003**, *27*, 10–15. [CrossRef]
17. Wheler, J.; Tsimberidou, A.M.; Moulder, S.; Cristofanilli Mong, D.; Naing, A.; Pathak, R.; Liu, S.; Feng, L.; Kurzrock, R. Clinical outcomes of patients with breast cancer in a phase I clinic: The MD Anderson cancer center experience. *Clin. Breast Cancer* **2010**, *10*, 46–51. [CrossRef]
18. Devi, L.I.; Ralte, L.; Ali, M.A. Serum Biochemical Profile of Breast cancer patients. *Eur. J. Pharm. Med. Res.* **2015**, *2*, 210–214.
19. Brunello, A.; Kapoor, R.; Extermann, M. Hyperglycemia during chemotherapy for hematologic and solid tumors is correlated with increased toxicity. *Am. J. Clin. Oncol. Cancer Clin. Trials* **2011**, *34*, 292–296. [CrossRef]
20. Fares, M.S.M.; Karuppannan, M.; Abdulrahman, E.; Suriyon Uitrakul, B.A.H.R.; Mohammed, A.H. Prevalence and associated factors of anemia among breast cancer patients undergoing chemotherapy: A prospective study. *Adv. Pharmacol. Pharm. Sci.* **2022**, *2022*, 7611733.
21. Guillot, E.; Vaugelade, P.; Lemarchali, P.; Rat, A.R. Intestinal absorption, and liver uptake of medium-chain fatty acids in non-anaesthetized pigs. *Br. J. Nutr.* **1993**, *69*, 431–442. [CrossRef] [PubMed]
22. Papamandjaris, A.A.; Macdougall, D.E.; Jones, P.J. Medium chain fatty acid metabolism and energy expenditure: Obesity treatment implications. *Life Sci.* **1998**, *62*, 1203–1215. [CrossRef]
23. Lieber, C.S.; Lefèvre, A.; Spritz, N.; Feinman, L.; DeCarli, L.M. Difference in Hepatic Metabolism of Long- and Medium-Chain Fatty Acids: The Role of Fatty Acid Chain Length in the Production of the Alcoholic Fatty Liver. *J. Clin. Investig.* **1967**, *46*, 1451–1460. [CrossRef]
24. Saerens, S.M.G.; Verstrepen, K.; Van Laere, S.D.M.; Voet, A.; Van Dijck, P.; Delvaux, F.R.; Thevelein, J. The Saccharomyces cerevisiae EHT1 and EEB1 Genes Encode Novel Enzymes with Medium-chain Fatty Acid Ethyl Ester Synthesis and Hydrolysis Capacity. *J. Biol. Chem.* **2006**, *281*, 4446–4456. [CrossRef] [PubMed]
25. Reddy, M.V.; Mohan, S.V.; Chang, Y.-C. Sustainable production of medium chain fatty acids (MCFA) with an enriched mixed bacterial culture: Microbial characterization using molecular methods. *Sustain. Energy Fuels* **2018**, *2*, 372–380. [CrossRef]
26. Diender, M.; Stams, A.J.M.; Sousa, D.Z. Production of medium-chain fatty acids and higher alcohols by a synthetic co-culture grown on carbon monoxide or syngas. *Biotechnol. Biofuels* **2016**, *9*, 82. [CrossRef] [PubMed]

27. Smith, G.M.; Kim, B.W.; Franke, A.A.; Roberts, J.D. 13C NMR studies of butyric fermentation in Clostridium kluyveri. *J. Biol. Chem.* **1985**, *260*, 13509–13512. [CrossRef]
28. Weimer, P.J.; Stevenson, D.M. Isolation, characterization, and quantification of Clostridium kluyveri from the bovine rumen. *Appl. Microbiol. Biotechnol.* **2012**, *94*, 461–466. [CrossRef]
29. Kenealy, W.R.; Cao, Y.; Weimer, P.J. Production of caproic acid by cocultures of ruminal cellulolytic bacteria a and Clostridium kluyveri grown on cellulose and ethanol. *Appl. Microbiol. Biotechnol.* **1995**, *44*, 507–513. [CrossRef]
30. Amalaradjou, M.A.R.; Annamalai, T.; Marek, P.; Rezamand, P.; Schreiber, D.; Hoagland, T.; Venkitanarayanan, K. Inactivation of Escherichia coli O157:H7 in Cattle Drinking Water by Sodium Caprylate. *J. Food* **2006**, *69*, 2248–2252. [CrossRef]
31. Valipe, S.R.; Nadeau, J.A.; Annamali, T.; Venkitanarayanan, K.; Hoagland, T. In vitro antimicrobial properties of caprylic acid, monocaprylin, and sodium caprylate against Dermatophilus congolensis. *Am. J. Vet. Res.* **2011**, *72*, 331–335. [CrossRef] [PubMed]
32. Lemarié, F.; Beauchamp, E.; Legrand, P.; Rioux, V. Revisiting the metabolism and physiological functions of caprylic acid (C8:0) with special focus on ghrelin octanoylation. *Biochimie* **2016**, *120*, 40–48. [CrossRef] [PubMed]
33. Roopashree, P.G.; Shilpa Shetty, S.; Suchetha Kumari, N. Effect of medium chain fatty acid in human health and disease. *J. Funct. Foods* **2021**, *87*, 104724. [CrossRef]
34. Kadochi, Y.; Mori, S.; Fujiwara-Tani, R.; Luo, Y.; Nishiguchi, Y.; Kishi, S.; Fujii, K.; Ohmori, H.; Kuniyasu, H. Remodeling of energy metabolism by a ketone body and medium-chain fatty acid suppressed the proliferation of CT26 mouse colon cancer cells. *Oncol. Lett.* **2017**, *14*, 673–680. [CrossRef]
35. Tiuca, I.D.; Nagy, K.; Oprean, R. Development and optimization of a gaschromatographic separation method of fatty acids in human serum. *World J. Pharm. Sci.* **2015**, *3*, 1713–1719.
36. Zhang, X.; Zhang, P.; Liu, Y.; Xu, Q.; Zhang, Y.; Li, H.; Liu, L.; Liu, Y.; Yang, X.; Xue, C. Caprylic Acid Improves Lipid Metabolism, Suppresses the Inflammatory Response and Activates the ABCA1/p-JAK2/pSTAT3 Signaling Pathway in C57BL/6J Mice and RAW264.7 Cells. *Biomed. Environ. Sci.* **2022**, *35*, 95–106.
37. Yamasaki, M.; Soda, S.; Sakakibara, Y.; Suiko, M.; Nishiyama, K. The importance of 1,2-dithiolane structure in α-lipoic acid for the downregulation of cell surface β1-integrin expression of human bladder cancer cells. *Biosci. Biotechnol. Biochem.* **2014**, *78*, 1939–1942. [CrossRef]
38. Jansen, R.J.; Robinson, D.P.; Frank, R.D.; Anderson, K.E.; Bamlet, W.R.; Oberg, A.L.; Rabe, K.G.; Olson, J.E.; Sinha, R.; Petersen, G.M.; et al. Fatty acids found in dairy, protein and unsaturated fatty acids are associated with risk of pancreatic cancer in a case–control study. *Int. J. Cancer* **2014**, *134*, 1935–1946. [CrossRef]
39. Wang, C.; Wang, J.; Chen, K.; Pang, H.; Li, X.; Zhu, J.; Ma, Y.; Qiu, T.; Li, W.; Xie, J.; et al. Caprylic acid (C8:0) promotes bone metastasis of prostate cancer by dysregulated adi-po-osteogenic balance in bone marrow. *Cancer Sci.* **2020**, *111*, 3600–3612. [CrossRef]
40. Iemoto, T.; Nishiumi, S.; Kobayashi, T.; Fujigaki, S.; Hamaguchi, T.; Kato, K.; Shoji, H.; Matsumura, Y.; Honda, K.; Yoshida, M. Serum level of octanoic acid predicts the efficacy of chemotherapy for colorectal cancer. *Oncol. Lett.* **2019**, *17*, 831–842. [CrossRef]
41. Lappano, R.; Sebastiani, A.; Cirillo, F.; Rigiracciolo, D.C.; Galli, G.R.; Curcio, R.; Malaguarnera, R.; Belfiore, A.; Cappello, A.R.; Maggiolini, M. The lauric acid-activated signaling prompts apoptosis in cancer cells. *Cell Death Discov.* **2017**, *3*, 1–9. [CrossRef] [PubMed]
42. Gukas, I.D.; Jennings, B.A.; Mandong, B.M.; Manasseh, A.N.; Harvey, I.; Leinster, S.J. A comparison of the pattern of occurrence of breast cancer in Nigerian and British women. *Breast* **2006**, *15*, 90–95. [CrossRef] [PubMed]
43. Robinson, D.; Bell, J.; Møller, H.; Salman, A. A 13-year follow-up of patients with breast cancer presenting to a District General Hospital breast unit in southeast England. *Breast* **2006**, *15*, 173–180. [CrossRef]
44. Mirtavoos-Mahyari, H.; Khosravi, A.; Esfahani-Monfared, Z. Human epidermal growth factor receptor 2 and estrogen receptor status in respect to tumor characteristics in non-metastatic breast cancer. *Tanaffos* **2014**, *13*, 26–34. [PubMed]
45. Hasiniatsy, N.R.; Vololonantenaina, C.R.; Rabarikoto, H.F.; Razafimanjato, N.; Ranoharison, H.D.; Rakotoson, J.L.; Samison, L.H.; Rafaramino, F. First results of hormone receptors' status in Malagasy women with invasive breast cancer. *Pan Afr. Med. J.* **2014**, *17*, 153. [PubMed]

nutrients

Article

Malnutrition, Cancer Stage and Gastrostomy Timing as Markers of Poor Outcomes in Gastrostomy-Fed Head and Neck Cancer Patients

Diogo Sousa-Catita [1,2,3,*], Cláudia Ferreira-Santos [4], Paulo Mascarenhas [1], Cátia Oliveira [2], Raquel Madeira [2], Carla Adriana Santos [2], Carla André [4], Catarina Godinho [1], Luís Antunes [4] and Jorge Fonseca [1,2]

[1] Grupo de Patologia Médica, Nutrição e Exercício Clínico (PaMNEC) do Centro de Investigação Interdisciplinar Egas Moniz (CiiEM), 2829-511 Almada, Portugal
[2] GENE—Artificial Feeding Team, Gastroenterology Department, Hospital Garcia de Orta, 2805-267 Almada, Portugal
[3] Residências Montepio—Serviços de Saúde, SA—Rua Julieta Ferrão Nº 10–5º, 1600-131 Lisboa, Portugal
[4] Otorhinolaryngology Department, Hospital Garcia de Orta, 2805-267 Almada, Portugal
* Correspondence: diogo.rsc2@gmail.com

Abstract: For percutaneous endoscopic gastrostomy (PEG)-fed head and neck cancer (HNC) patients, risk markers of poor outcomes may identify those needing more intensive support. This retrospective study aimed to evaluate markers of poor outcomes using TNM-defined stages, initial anthropometry [body mass index (BMI), mid-upper arm circumference (MUAC), tricipital skinfold (TSF), mid-arm muscle circumference (MAMC)] and laboratory data (albumin, transferrin, cholesterol), with 138 patients, 42–94 years old, enrolled. The patients had cancer, most frequently in the larynx ($n = 52$), predominantly stage IV ($n = 109$). Stage IVc presented a four times greater death risk than stage I (OR 3.998). Most patients presented low parameters: low BMI ($n = 76$), MUAC ($n = 114$), TSF ($n = 58$), MAMC ($n = 81$), albumin ($n = 47$), transferrin ($n = 93$), and cholesterol ($n = 53$). In stages I, III, IVa, and IVb, MAMC and PEG-timing were major survival determinants. Each MAMC unit increase resulted in 16% death risk decrease. Additional 10 PEG-feeding days resulted in 1% mortality decrease. Comparing IVa/IVb vs. IVc, albumin and transferrin presented significant differences ($p = 0.042$; $p = 0.008$). All parameters decreased as severity of stages increased. HNC patients were malnourished before PEG, with advanced cancer stages, and poor outcomes. Initial MAMC, reflecting lean tissue, significantly increases survival time, highlighting the importance of preserving muscle mass. PEG duration correlated positively with increased survival, lowering death risk by 1% for every additional 10 PEG-feeding days, signaling the need for early gastrostomy.

Keywords: head and neck cancer; nutritional status; percutaneous endoscopic gastrostomy

Citation: Sousa-Catita, D.; Ferreira-Santos, C.; Mascarenhas, P.; Oliveira, C.; Madeira, R.; Santos, C.A.; André, C.; Godinho, C.; Antunes, L.; Fonseca, J. Malnutrition, Cancer Stage and Gastrostomy Timing as Markers of Poor Outcomes in Gastrostomy-Fed Head and Neck Cancer Patients. *Nutrients* **2023**, *15*, 662. https://doi.org/10.3390/nu15030662

Academic Editor: Ashley J. Snider

Received: 19 December 2022
Revised: 21 January 2023
Accepted: 25 January 2023
Published: 28 January 2023

1. Introduction

Head and neck cancer (HNC) include cancers in the lips, mouth, nasal cavity, paranasal sinus, pharynx, larynx, and proximal esophagus, that share some common features. Most of them (90%) are squamous cell carcinomas related to tobacco smoking, heavy alcohol consumption, or human papillomavirus infections, and tend to affect swallowing and oral feeding. HNC patients present a very high risk of developing malnutrition for several reasons. First, tobacco smoking and heavy alcohol consumption are associated with malnutrition [1]. Moreover, heavy alcohol consumption frequently results in social disruption, which may lead to a further decline in nutritional status. HNC patients may be malnourished before cancer development due to these unhealthy habits.

The wasting effects of cancer have a major impact on nutritional status. Cancer malnutrition is considered as malnutrition associated with mild to moderate inflammation [2]. It is much more catabolic than simple starvation, with greater consumption of body lean mass and muscle proteins. HNC patients frequently present a reduced oral intake due to mechanical obstacles, causing dysphagia or odynophagia [3]. The mass position and cancer therapy may affect these patients' chewing and swallowing. Therapeutic procedures like surgery, chemotherapy or radiotherapy could significantly impact nutritional status [3–5] by reducing food intake, contributing to malnutrition [6,7].

Malnutrition is very frequent (>70%) in HNC patients with severe weight loss and impaired immune function, leading to incomplete or postponed treatment cycles, and decreased quality of life [8]. Malnourished HNC patients present an increase in the number and severity of complications, and decreased survival [9,10]. Maintaining an optimal nutritional status is mandatory for improving treatment tolerance, outcome, and survival for all patients receiving cancer-directed treatment [11]. These patients suffering from malnutrition need specialized nutritional support. When oral intake is insufficient, and there is no other digestive tract disturbance, tube feeding is the obvious option. Most of these patients need it for some period during the evolution of the disease [12]. If tube feeding is required for more than 3 weeks, percutaneous endoscopic gastrostomy (PEG) is the gold standard. This is associated with fewer treatment failures and provides better nutritional support than long-standing nasogastric feeding tubes [13–15]. In HNC patients, PEG feeding improves clinical outcomes and survival [16].

HNC dysphagic patients frequently present speech deficiency that evolves parallelly with the swollen impairment. Speech difficulties may inhibit the use of several nutrition evaluation tools, and artificial feeding teams must often rely on objective data (e.g., anthropometric and laboratory data) for the nutritional status follow-up of PEG patients. Serum albumin, transferrin, and total cholesterol levels are non-specific, but may be used as serum markers of malnutrition, inflammation, and/or prognosis [17]. In fact, albumin and transferrin are negative acute-phase proteins and their production may be impaired in a long-term inflammatory stage, as well as in starvation. Although anthropometric and laboratory evaluation may reflect other influences that are diverse from nutritional issues, taken together they may become very useful tools for teams following PEG patients [18,19]. These tools are frequently used to assess the patient status, as they are low-cost, easy to obtain, and widely available [20–24].

Although the guidelines recommend an early gastrostomy of HNC patients [25], many of them present evident malnutrition when referred to the PEG procedure. We previously built a predictive model that helps us to identify patients with a probable life expectancy shorter than 3 weeks [18]. Additionally, we have previously identified nutritional and laboratory factors associated with poor outcomes for HNC patients after PEG [26]. These previous studies focused only on nutritional issues to identify prognostic factors and produce predictive models. In the present study, the classification of malignant tumors (TNM) was added to include the cancer severity and evolution. We remain interested in analyzing whether:

1. According to guidelines, is the patient a suitable candidate for gastrostomy, with a life expectancy longer than 3 weeks?
2. How to use nutritional and laboratory data to identify HNC patients with severely impaired nutritional status, and unfavorable outcomes months after the gastrostomy, requiring more powerful nutritional support with larger protein energy intake?

In the present study, we aim to answer the question: can the cancer staging severity, anthropometric and laboratory data help us identify severely compromised patients requiring special attention?

Specifically, we aim to:

1. Evaluate the clinical and nutritional status of HNC patients when referred to endoscopic gastrostomy for long-term enteral nutrition, using anthropometry, laboratory data and accessible tools, even with patients who cannot speak.
2. Evaluate the clinical outcome of PEG-fed HNC patients.
3. Evaluate the relations between survival, severity, and nutritional status:

- Compare the nutritional status with the different TNM-defined stages.
- Compare the nutritional status with the different grades of stage IV, grouped as having metastases and no distant metastases (IVa and IVb, against IVc).
- Evaluate the impact of clinical and nutritional status on the survival of PEG-fed HNC patients, using TNM-defined stages, anthropometry, and laboratory data.

2. Materials and Methods

2.1. Patients

We studied consecutive adult HNC patients who underwent endoscopic gastrostomy to have PEG nutritional support, between January 2006 and December 2019 (the last year before the COVID-19 pandemic period). We included patients with cancers in the oral cavity, pharyngeal, laryngeal, esophageal proximal, and neck regions, arising from other organs or tissues. All patients were routinely evaluated in Garcia de Orta hospital Outpatient Artificial Nutrition Clinic, before the gastrostomy procedure, and one week, one month, and three months after the gastrostomy. After the third month, stable patients were followed every 4 to 6 months. Patients who experienced difficulties adjusting to PEG feeding were evaluated more often, until the patient and the caregiver achieved complete adaptation.

2.2. The Clinic, Anthropometric and Laboratory Data

All clinic, anthropometric and laboratory data are part of the routine evaluation of PEG patients, and were collected from the clinical files of the Artificial Feeding Team (GENE–Grupo de Estudo de Nutrição Entérica/parentérica). We recorded data on the day of the endoscopic gastrostomy or the day before. A blood sample was obtained in the endoscopy room, just before the gastrostomy procedure. Incomplete patient data was an exclusion criterion.

2.3. Head and Neck Cancer TNM Classification of Malignant Tumors

We searched data in electronic clinical files and the otorhinolaryngology oncology multidisciplinary reunion database for each patient by process number. Exclusion criteria were applied: no data for neoplasia location, incomplete cancer staging and advanced liver or kidney disorders.

Each patient's cancer staging was obtained using the manual American Joint Committee on Cancer (AJCC) eighth edition to standardize the data. Each patient was classified within a TNM-defined stage: I, II, III, IVa, IVb, or IVc.

2.4. Clinical Outcome

According to the outcome, we divided patients into three categories: dead, lost to follow-up and alive. The time span from the gastrostomy procedure until death or until December 2019 was expressed in months.

2.5. Anthropometric Evaluation

The anthropometric evaluation was performed according to the International Society for the Advancement of Kinanthropometry manual on the day of the gastrostomy procedure or the day before. We obtained three consecutive measurements; the clinical file record represents the mean of those three measurements.

1. Body mass index (BMI) was obtained in most patients using the equation Weight (Kg)/Height (m)^2. If patients were bedridden and could not stand up for weight and height evaluation, BMI was estimated using the mid-upper arm circumference (MUAC) and regression equations described by Powell-Tuck and Hennessy [27]; this method has been previously used and proved to provide a reliable BMI estimation in PEG patients [28,29]. Each patient was classified by the WHO classification according to their age as having low BMI if was <18.5 kg/m^2 or <22 kg/m^2, normal BMI if 18.5–25 kg/m^2 or 22–27 kg/m^2, and high BMI if >25 kg/m^2 or >27 kg/m^2, for patients under 65 years or 65 years old or older, respectively [30] (Table 1).

Table 1. Body mass index (BMI) classification according to age.

	Low	Normal	High
<65 Years	<18.5 kg/m^2	>18.5–<25 kg/m^2	>25 kg/m^2
≥65 Years	<22 kg/m^2	≥22–<27 kg/m^2	≥27 kg/m^2

2. Mid-upper arm circumference (MUAC) was evaluated using an inextensible measuring tape, with a 1 mm resolution. MUAC results were obtained from evaluating several tissues representing fat and lean mass.
3. Tricipital skinfold (TSF), was measured using a Lange skinfold caliper with a 1 mm resolution. TSF evaluates the subcutaneous adipose tissue and estimates adipose reserves.
4. The mid-arm muscle circumference (MAMC) was calculated according to the equation: MAMC = MUAC (cm) − 0.314 × TSF (mm). The MAMC allows us to estimate lean and muscle mass.

For each patient, MUAC, MAMC, and TSF were compared with reference values of the National Health and Nutrition Examination Survey (NHANES), through the comparison with the Frisancho reference tables [31–34].

2.6. Laboratory Evaluation

A blood sample was obtained from these patients, minutes before the endoscopic gastrostomy procedure. Blood samples were obtained between 8:00 and 10:00 A.M. following at least 12 h of fasting. Serum albumin <3.5 g/dL, serum transferrin <200 mg/dL, and total serum cholesterol <160 mg/dL were considered low values, suggestive of poor prognosis and/or malnutrition [18,35–38].

2.7. Statistics

All statistical analyses were computed by SPSS software version 26. Survival analysis (Kaplan Meier/Cox regression) provided all results evaluating the impact of covariates on PEG patient survival time. Linear regression analysis allowed us to estimate the impact of TMN-defined stages on HNC patient nutritional status biomarkers before PEG, by Z-testing the obtained marginal estimates. Statistical significance for each model and associated parameters were set at $p < 0.05$.

3. Results

3.1. Subjects

We enrolled 138 HNC patients (129 males, and 9 females), who underwent endoscopic gastrostomy to be PEG fed. Participants ages ranged from 42 to 94 years (mean: 61.3 years; median: 60.0 years). The characteristics of the study population, including the demographic data (age and gender), are presented in Table 2. HNCs arise from several organs: oral cavity (mouth), pharynx, larynx, and other organs and tissues. Patients presented HNCs at stages I to IVc.

3.2. Head or Neck Cancers

Cancer Location

For 129 males, the primary tumor was located in the pharynx ($n = 41$) or the larynx ($n = 52$). For the nine females, the primary tumor was mainly in the mouth ($n = 5$) (Table 3).

Table 2. Subject characteristics.

Subject Characteristics		*n* (%)
Age (years)		
	42–94 (mean 61.3)	
Gender		
	Female	9 (6.5%)
	Male	129 (93.5%)
Cancer site		
	Mouth	39 (28.3%)
	Pharyngeal	42 (30.4%)
	Laryngeal	55 (39.9%)
	Others	2 (1.4%)

Table 3. Characterization of subjects by primary tumor location, TNM classification, anthropometry and laboratory serum data.

		Total (*n* = 138)	Total Mean
Primary Tumor located			
	Mouth	39	
	Pharynx	42	
	Larynx	55	
	Other HNC location	2	
Classification of Malignant Tumors (TNM) Classification			
	Stage I	6	
	Stage II	1	
	Stage III	22	
	Stage Iva	81	
	Stage IVb	12	
	Stage IVc	16	
Anthropometry Results			
	BMI	76 Low BMI 46 Normal BMI 16 High BMI	
	MUAC	114 Low 24 Normal	
	TSF	58 Low 80 Normal	
	MAMC	81 Low 57 Normal	
Laboratory serum data			
	Albumin	47 Low 91 Normal	3.7 g/dL
	Transferrin	93 Low 45 Normal	182.0 mg/dL
	Total Cholesterol	53 Low 85 Normal	173.2 mg/dL

(BMI)—Body mass index; BMI classification according to age, <65 y, low BMI is <18.5 Kg/m^2, normal BMI is between 18.5 Kg/m^2 and <25 Kg/m^2, and high BMI is $\geq$25 Kg/m^2, $\geq$65 y, low BMI is <22 Kg/m^2, a normal BMI is between 22 Kg/m^2 and <27 Kg/m^2, and high BMI is $\geq$ 27 Kg/m^2; (MUAC)—mid-upper arm circumference <90% low, $\geq$90–110% normal; (TSF)—tricipital skinfold results, <90% low, $\geq$90–110% normal and (MAMC)—mid-arm muscle circumference <90% low, $\geq$90–110% normal; albumin < 3.5 g/dL (low), transferrin < 200 mg/dL (low), total cholesterol < 160 mg/dL (low).

All patients were classified according to the TNM classification, from data searched in the clinical files of the otorhinolaryngology oncology multidisciplinary reunion, and an otorhinolaryngology specialist validated each datum (Table 3). The most frequent tumor stage was IV, present in 102 male and seven female patients.

3.3. Anthropometry

3.3.1. Body Mass Index (BMI)

For eight patients, BMI was estimated using the Powell-Tuck and Hennessy regression equations. BMI ranged from 14 Kg/m^2 to 48 Kg/m^2 (mean: 20.64 Kg/m^2; median: 19.4 Kg/m^2). Classification was used according to age. Following this classification, 76 (55%) patients displayed a low BMI. The results are summarized in Table 3.

3.3.2. Mid-Upper Arm Circumference (MUAC)

Compared with Frisancho criteria [31], 114 (83%) patients showed MUAC in the low range (Table 3).

3.3.3. Tricipital Skinfold (TSF)

In this anthropometric parameter, 58 (42%) patients displayed low TSF (Table 3).

3.3.4. Mid-Arm Muscle Circumference (MAMC)

In this anthropometric parameter, 81 (59%) patients showed MAMC in the low range (Table 3).

3.4. Laboratory Assessment

3.4.1. Serum albumin

In 92 patients, albumin was in the normal range, and 47 displayed low serum albumin.

3.4.2. Serum Transferrin

In 46 patients, transferrin was in the normal range, and 93 showed low serum transferrin.

3.4.3. Total Serum Cholesterol

In 86 patients, cholesterol was in the normal range, while 53 patients displayed low serum total cholesterol. Of 130 males, 49 displayed a low serum total cholesterol. Of nine females, three displayed a low serum total cholesterol.

Laboratory data are summarized in Table 3.

3.5. Clinical Outcome

At the end of December 2019, out of the 138 patients, six (4.3%) were lost to follow-up, 111 (80.5%) patients were deceased, six (4.3%) were still PEG-fed and followed by the Artificial Nutrition Outpatients Clinic, and 15 (10.9%) resumed oral feeding with the tube removed and gastrostomy closed. Comparing all patients, the ones who had a longer survival time were patients with cancer classification TNM defined as stage I, and with the location of the cancer in the pharynx.

3.6. Kaplan–Meier Survival Analysis

Stage I cancer was associated with increased survival in any type of cancer. Stage III and IVa showed a similar survival time, and the stage with the least survivability was type IVc (Figure 1).

3.7. Cox Regression Analysis

We applied a Cox regression to obtain a statistical model adjusted for HNC TNM-defined stage to evaluate the tumor site, age, gender, anthropometrics, biochemical and PEG covariates effect on a patient's survival time. Throughout the model fit process, the tumor site, age, and gender resulted in redundant variables and were removed from the reduced final model (Table 6).

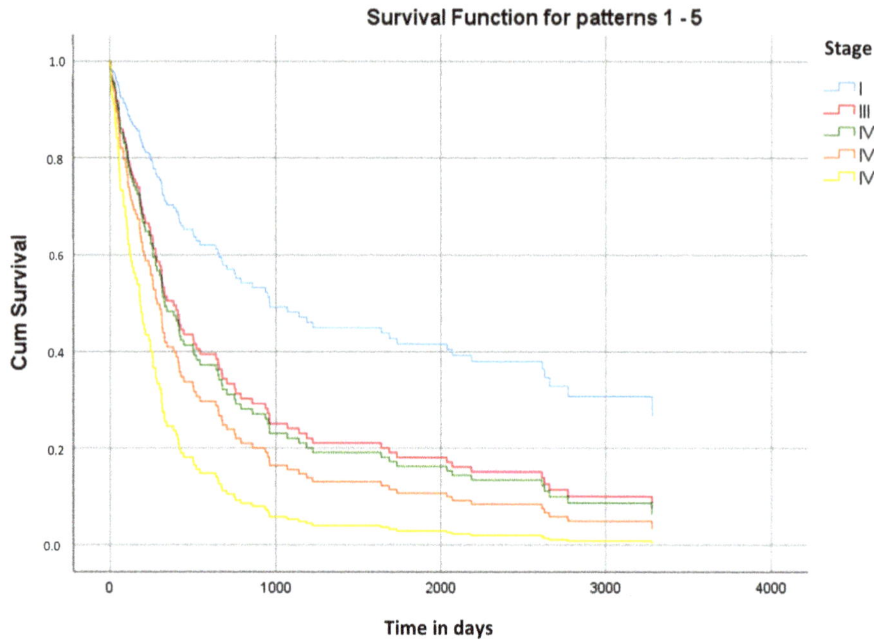

Figure 1. Kaplan–Meier curve of the cumulative survival for different cancer stages.

The mean survival time was 996 days (Table 4).

Table 4. Means and Medians for survival time (days) by stage of cancer.

Stage (N)	Mean				Median	
	Survival Time	Std. Error	95% Confidence Interval		Survival Time	Std. Error
			Lower Bound	Upper Bound		
I (6)	1135	496	164	2106	178	515
III (22)	1074	291	503	1645	275	268
IVa (81)	1054	149	762	1347	397	93
IVb (12)	1082	431	237	1927	233	202
IVc (16)	219	55	111	327	135	26
Overall (137)	996	117	767	1226	316	56

The pharynx appeared to be the type of cancer associated with the longest survival time, mainly in stages I and III. In stage IVc, any type of cancer had a much shorter life span (Table 5).

Stage IVc was the only stage that had significance for the impact on survival time regardless of MAMC and PEG time (CI = [0.775, 0.901], $p < 0.001$; CI = [0.999, 0.999], $p < 0.001$). In the earlier stages (I, III, Iva and IVb), PEG time and MAM seemed to be major determinants of survival. Stage II was withdrawn due to having a single patient in this stage.

Patients with stage IVc had a four-times higher risk of death than those with stage I (OR 3.998).

Table 5. Median survival time by cancer location and stage.

				Time in Days
Local	Mouth	Stage	I ($n = 1$)	555
			III ($n = 6$)	75
			IVa ($n = 27$)	450
			IVb ($n = 2$)	400
			IVc ($n = 3$)	175
	Pharynx	Stage	I ($n = 1$)	2700
			III ($n = 4$)	2000
			IVa ($n = 25$)	305
			IVb ($n = 6$)	1700
			IVc ($n = 5$)	120
	Larynx	Stage	I ($n = 3$)	950
			III ($n = 11$)	631
			IVa ($n = 29$)	452
			IVb ($n = 4$)	250
			IVc ($n = 8$)	200
	Other locations	Stage	I ($n = 1$)	150
			III ($n = 1$)	150

Table 6. Cox regression analysis.

		Coef	SE	*p*-Value	OR	95.0% CI for OR	
						Lower	Upper
TNM Stage	I			0.117			
	III	0.668	0.583	0.252	1.949	0.622	3.276
	IVa	0.728	0.529	0.169	2.071	0.734	3.408
	IVb	0.934	0.616	0.130	2.546	0.761	4.331
	IVc	1.386	0.592	0.019	3.998	1.252	6.744
MAMC		−0.177	0.040	0.000	0.838	0.775	0.901
Albumin		0.251	0.162	0.122	1.285	0.935	1.635
Time with PEG		−0.001	0.000	0.000	0.999	0.999	0.999

Mid-Arm Muscle Circumference (MAMC); percutaneous endoscopic gastrostomy (PEG); Coef variable or level coefficient; standard error (SE), Ref reference level; Odds ratio (OR).

Globally, for all stages and cancer locations, the 3000-day survival rate was less than 10% (Figure 2).

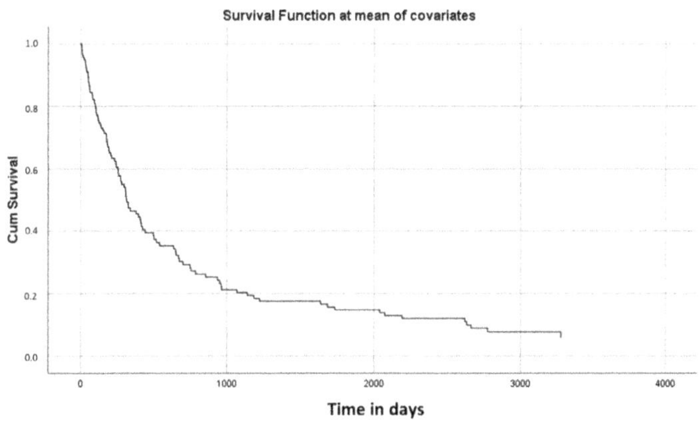

Figure 2. Kaplan–Meier survival curve in all stages and cancer location.

The MAMC had an average odds ratio of 0.838, and each unit increase of MAMC was associated with a 16% decrease in the risk of death. The PEG duration time increasing by one unit was associated with a 0.1% reduction in the risk of death. That is, for every 10 days more of PEG (limited by the study time-frame), the risk of death decreased by 1%.

3.8. Regression Analysis of Cancer Stage Impact on Nutrition Markers

We performed a linear regression analysis to estimate the impact of TMN-defined stages on HNC patient nutritional status biomarkers before PEG.

3.8.1. TNM-Defined Stages (I vs. II vs. III vs. IVa vs. IVb vs. IVc)

Model results showed significant differences among the BMI ($p = 0.039$), and TSF ($p = 0.007$) of the TNM-defined stages. The BMI and TSF tended to decrease as the severity of the TNM-defined stages increased.

Model results showed no significant differences among the MUAC ($p = 0.0231$); MAMC, ($p = 0.584$); albumin ($p = 0.165$); transferrin ($p = 0.074$); and cholesterol ($p = 0.035$) of the TNM-defined stages. Nevertheless, MUAC, MAMC, albumin and transferrin tended to decrease as the severity of the TNM-defined stage increased. Cholesterol presented non-linear changes in the different TNM-defined stages.

3.8.2. TNM-Defined Stages IVa and IVb vs. IVc

When comparing IVa, IVb vs. IVc, there were no significant differences in BMI ($p = 0.169$), MUAC ($p = 0.149$), MAMC ($p = 0.307$), TSF ($p = 0.068$) and cholesterol ($p = 0.135$).

Albumin and transferrin, when comparing IVa, IVb vs. IVc, showed significant differences ($p = 0.042$ and 0.008, respectively). Nevertheless, all parameters tended to decrease as the severity of the TNM-defined stage increased.

4. Discussion

Head and neck cancer (HNC) patients present a high frequency of malnutrition compared with other cancers, due to the direct effects of the disease, therapy side effects, and poor food intake [39]. HNC patients are often malnourished at diagnosis, having involuntary weight loss before starting treatment [40]. Good nutritional management is essential to the patient's ability to complete the prescribed treatment courses, minimize nutrition-related side effects, and foster healing [41].

Our results show that this type of cancer affects most of the male gender, probably, due to poor lifestyle habits, such as smoking and alcohol consumption. Comparing the genders, both presented a high percentage of malnutrition. In the male gender, we can see in our study that a significant percentage of obese patients are likely linked to poor eating habits. In this study, we have a higher percentage of patients in advanced cancer stages (TNM defined as stage III and IV), characterized by the worst nutritional status and poor prognosis.

Although nutritional evaluation could benefit from sophisticated devices for measuring body composition, such as bioelectrical impedance analysis (BIA) or CT scan analysis, those devices were not available for all patients. Although less precise, BMI and anthropometry are inexpensive and widespread nutritional evaluation tools, classically used as an approach to the evaluation of fat/lean mass [33,34] and available everywhere, even in institutions with scarce resources. Most of our anthropometric data display low values, related to a poor nutrition status due to an advanced cancer stage. Arm anthropometry (MUAC, TSF, and MAMC) data show malnutrition in over eighty per cent of the patients. Estimation of fat and fat-free reserves, also reveals a poor nutritional status. MAMC recognizes more malnourished patients than TSF, which suggests that lean tissue is depleted at the beginning of the disease, fat reserves are more preserved, and over time, they are slowly degraded. Also, MAMC is an independent outcome predictor, highlighting the importance of lean mass in patient survival. From another perspective, MAMC is strikingly reduced since the early stages, and the other anthropometric and laboratory data reduce gradually, as the disease stage and severity progress [42].

Serum proteins are negative acute-phase proteins, and, like cholesterol, they may be modified by various biological influences. However, the usefulness of biochemical data is well recognized in several nutritional studies [17,20,26]. In our study, most patients display laboratory markers in a normal range, but albumin and transferrin tend to decrease with increasing severity of the TNM-defined stage.

Globally, our anthropometric and biochemical results demonstrate the strong influence that HNC had on the lean tissue and, later, on the fat mass of these patients, leading to malnutrition. Other authors have addressed the problem of malnutrition in cancer patients and their outcomes, such as the impact of nutrition management and status of head and neck cancer patients, on the success of treatment and survival [9,16,21].

Regarding the impact of the cancer stage (by TNM-defined stage) on the different nutritional parameters evaluated, when compared to all stages, only the BMI and the TSF had a significant difference between stages, with a progressive decrease as the severity of the cancer stage increases. In contrast, MUAC and MAMC are reduced since the early stages. This suggests that lean tissue is consumed during the initial stages of the disease, as expected in cancer-related inflammation. In contrast, fat tissue suffers a progressive loss, unlike fat-free mass, which is severely depleted since the beginning of cancer progression. When we focus on the most severe cancer stages (Iva, IVb and IVc), only albumin and transferrin had a significant difference, decreasing as the severity of the cancer stage increases. Likely, fasting is more severe in this advanced cancer stage than in less advanced stages [43].

When we tried to create a model that evaluates the role of all parameters of this study against the clinical outcome as survival time, only the MAMC was statistically significant, except for the most advanced cancer stage (Stage IVc). Therefore, this anthropometric parameter seems to no longer influence survival, as the severity of the disease increases to stage IVc. On the other hand, this result demonstrates the importance of preserving lean tissue in the early cancer stages, to maintain a better nutritional status and outcome. In fact, lean tissue is also associated with better treatment response and, consequently, a better prognosis [44–46]. Moreover, this study suggests that the PEG duration time positively impacts survival time in HNC patients. This supports the importance of early PEG feeding for HNC patient prognostics, suggesting that PEG feeding is important for better patient outcomes.

Early PEG is generally recommended in the treatment of HNC patients. Nevertheless, our results suggest that special attention should be addressed to patients with lower lean mass, evaluated through anthropometry, as in our study, or any other method.

Our study has some limitations resulting in some missing data. We completed processing patient data in December of 2019 because it was the last year before the COVID-19 pandemic, and several patients did not continue their follow-up, (refusing the hospital), and their records were incomplete. Other missing data included the causes of death. More than half of our patients (58.7%) were in TNM-defined stage IVa. The only TNM-defined stage II patient was excluded from inferential statistics to improve the statistical model parameter estimation.

5. Conclusions

HNC patients are malnourished when referred to undergo endoscopic gastrostomy and have advanced cancer as defined by TNM-defined stage, a marker of poor outcomes.

MAMC, the anthropometric parameter reflecting the lean tissue, was the only one with statistical significance in survival time, highlighting the necessity to preserve the muscle mass of these patients. PEG duration time was shown to correlate with increased survival time, at a rate of 1% decrease in the risk of death for every 10 days of PEG extension, suggesting that gastrostomy should be performed in an early stage of the disease progression.

Author Contributions: D.S.-C.—data curation; formal analysis; investigation; and writing. C.F.-S.—data curation; and writing. P.M.—formal analysis; writing and reviewing. C.O.—formal analysis; investigation. R.M.—formal analysis; investigation. C.A.S.—data curation; formal analysis; investigation; writing and reviewing. C.A.—methodology and reviewing. C.G.—formal analysis; data curation; writing and reviewing. L.A.—methodology and reviewing. J.F.—conceptualization; data curation; formal analysis; methodology; writing and reviewing. All authors have read and agreed to the published version of the manuscript.

Funding: This research received no external funding.

Institutional Review Board Statement: The study was conducted in accordance with the Declaration of Helsinki and approved by the local health research ethics boards—Garcia de Orta Ethical Committee, Garcia de Orta Centre. All patients gave their informed consent to participate.

Informed Consent Statement: Informed consent was obtained from all subjects involved in the study.

Data Availability Statement: The data presented in this study are available on request from the first author.

Acknowledgments: This work is financed by national funds through the FCT—Foundation for Science and Technology, I.P., under the project UIDB/04585/2020. The researchers would like to thank the Centro de Investigação Interdisciplinar Egas Moniz (CiiEM) for the support provided for the publication of this article.

Conflicts of Interest: The authors declare no conflict of interest.

References

1. Ross, L.J.; Wilson, M.; Banks, M.; Rezannah, F.; Daglish, M. Prevalence of Malnutrition and Nutritional Risk Factors in Patients Undergoing Alcohol and Drug Treatment. *Nutrition* **2012**, *28*, 738–743. [CrossRef] [PubMed]
2. White, J.V.; Guenter, P.; Jensen, G.; Malone, A.; Schofield, M. Consensus Statement: Academy of Nutrition and Dietetics and American Society for Parenteral and Enteral Nutrition. *J. Parenter. Enter. Nutr.* **2012**, *36*, 275–283. [CrossRef] [PubMed]
3. O'Neill, J.P.; Shaha, A.R. Nutrition Management of Patients with Malignancies of the Head and Neck. *Surg. Clin. N. Am.* **2011**, *91*, 631–639. [CrossRef] [PubMed]
4. Silander, E.; Nyman, J.; Hammerlid, E. An Exploration of Factors Predicting Malnutrition in Patients with Advanced Head and Neck Cancer. *Laryngoscope* **2013**, *123*, 2428–2434. [CrossRef]
5. Alshadwi, A.; Nadershah, M.; Carlson, E.; Young, L.; Burke, P.; Daley, B. Nutritional Considerations for Head and Neck Cancer Patients: A Review of the Literature. *J. Oral Maxillofac. Surg.* **2013**, *71*, 1853–1860. [CrossRef]
6. Kubrak, C.; Olson, K.; Jha, N.; Jensen, L.; McCargar, L.; Seikaly, H.; Harris, J.; Scrimger, R.; Parliament, M.; Baracos, V.E. Nutrition Impact Symptoms: Key Determinants of Reduced Dietary Intake, Weight Loss, and Reduced Functional Capacity of Patients with Head and Neck Cancer before Treatment. *Head Neck* **2009**, *32*, 290–300. [CrossRef]
7. Paccagnella, A.; Morello, M.; Da Mosto, M.C.; Baruffi, C.; Marcon, M.L.; Gava, A.; Baggio, V.; Lamon, S.; Babare, R.; Rosti, G.; et al. Early Nutritional Intervention Improves Treatment Tolerance and Outcomes in Head and Neck Cancer Patients Undergoing Concurrent Chemoradiotherapy. *Support. Cancer Ther.* **2009**, *18*, 837–845. [CrossRef] [PubMed]
8. Gómez, C.C.; Olivar, R.J.; García, M.; Marín, M.; Madero, R.; Pérez-Portabella, C.; Planás, M.; Mokoroa, A.; Pereyra, F.; Martín Palmero, A. Assessment of a malnutrition screening tool in cancer patients. *Nutr. Hosp.* **2010**, *25*, 400–405.
9. Ravasco, P.; Monteiro-Grillo, I.; Vidal, P.M.; Camilo, M.E. Nutritional Deterioration in Cancer: The Role of Disease and Diet. *Clin. Oncol.* **2003**, *15*, 443–450. [CrossRef]
10. Arends, J.; Bodoky, G.; Bozzetti, F.; Fearon, K.; Muscaritoli, M.; Selga, G.; Von Meyenfeldt, M.; Zürcher, G.; Fietkau, R.; Aulbert, E.; et al. ESPEN Guidelines on Enteral Nutrition: Non-Surgical Oncology. *Clin. Nutr.* **2006**, *25*, 245–259. [CrossRef]
11. Sandmæl, J.A.; Sand, K.; Bye, A.; Solheim, T.S.; Oldervoll, L.; Helvik, A. Nutritional Experiences in Head and Neck Cancer Patients. *Eur. J. Cancer Care* **2019**, *28*, e13168. [CrossRef] [PubMed]
12. Wang, J.; Liu, M.; Liu, C.; Ye, Y.; Huang, G. Percutaneous Endoscopic Gastrostomy versus Nasogastric Tube Feeding for Patients with Head and Neck Cancer: A Systematic Review. *J. Radiat. Res.* **2014**, *55*, 559–567. [CrossRef] [PubMed]
13. Yagishita, A.; Kakushima, N.; Tanaka, M.; Takizawa, K.; Yamaguchi, Y.; Matsubayashi, H.; Ono, H. Percutaneous Endoscopic Gastrostomy Using the Direct Method for Aerodigestive Cancer Patients. *Eur. J. Gastroenterol. Hepatol.* **2012**, *24*, 77–81. [CrossRef] [PubMed]
14. Nicholl, M.B.; Lyons, D.A.; Wheeler, A.A.; Jorgensen, J.B. Repeat PEG Placement Is Safe for Head and Neck Cancer Patients. *Am. J. Otolaryngol.* **2014**, *35*, 89–92. [CrossRef]
15. Rutter, C.E.; Yovino, S.; Taylor, R.; Wolf, J.; Cullen, K.J.; Ord, R.; Athas, M.; Zimrin, A.; Strome, S.; Suntharalingam, M. Impact of Early Percutaneous Endoscopic Gastrostomy Tube Placement on Nutritional Status and Hospitalization in Patients with Head and Neck Cancer Receiving Definitive Chemoradiation Therapy. *Head Neck* **2010**, *33*, 1441–1447. [CrossRef]

16. Müller-Richter, U.; Betz, C.; Hartmann, S.; Brands, R.C. Nutrition Management for Head and Neck Cancer Patients Improves Clinical Outcome and Survival. *Nutr. Res.* **2017**, *48*, 1–8. [CrossRef]
17. Reis, T.G.; da Silva, R.A.W.P.; Nascimento, E.D.S.; de Bessa, J.; Oliveira, M.C.; Fava, A.S.; Lehn, C.N. Early Postoperative Serum Albumin Levels as Predictors of Surgical Outcomes in Head and Neck Squamous Cell Carcinoma. *Braz. J. Otorhinolaryngol.* **2022**, *88* (Suppl. S1), S48–S56. [CrossRef]
18. Fonseca, J.; Adriana Santos, C.; Brito, J. Predicting Survival of Endoscopic Gastrostomy Candidates Using the Underlying Disease, Serum Cholesterol, Albumin and Transferrin Levels. *Nutr. Hosp.* **2013**, *28*, 1280–1285. [CrossRef]
19. Santos, C.A.; Fonseca, J.; Brito, J.; Fernandes, T.; Gonçalves, L.; Sousa Guerreiro, A. Serum Zn Levels in Dysphagic Patients Who Underwent Endoscopic Gastrostomy for Long Term Enteral Nutrition. *Nutr. Hosp.* **2014**, *29*, 359–364. [CrossRef]
20. Fuhrman, M.P.; Charney, P.; Mueller, C.M. Hepatic Proteins and Nutrition Assessment. *J. Am. Diet. Assoc.* **2004**, *104*, 1258–1264. [CrossRef]
21. Duguet, A.; Bachmann, P.; Lallemand, Y.; Blanc-Vincent, M.P. Summary Report of the Standards, Options and Recommendations for Malnutrition and Nutritional Assessment in Patients with Cancer (1999). *Br. J. Cancer* **2003**, *89* (Suppl. S1), S92–S97. [CrossRef]
22. Guerra, L.T.; Rosa, A.R.; Romani, R.F.; Gurski, R.R.; Schirmer, C.C.; Kruel, C.D.P. Serum Transferrin and Serum Prealbumin as Markers of Response to Nutritional Support in Patients with Esophageal Cancer. *Nutr. Hosp.* **2009**, *24*, 241–242.
23. Ellegård, L.H.; Bosaeus, I.G. Biochemical Indices to Evaluate Nutritional Support for Malignant Disease. *Clin. Chim. Acta* **2008**, *390*, 23–27. [CrossRef]
24. Valenzuela-Landaeta, K.; Rojas, P.; Basfi-fer, K. Nutritional assessment for cancer patient. *Nutr. Hosp.* **2012**, *27*, 516–523. [CrossRef]
25. Löser, C.; Aschl, G.; Hébuterne, X.; Mathus-Vliegen, E.M.H.; Muscaritoli, M.; Niv, Y.; Rollins, H.; Singer, P.; Skelly, R.H. ESPEN Guidelines on Artificial Enteral Nutrition–Percutaneous Endoscopic Gastrostomy (PEG). *Clin. Nutr.* **2005**, *24*, 848–861. [CrossRef]
26. Fonseca, J.; Santos, C.A.; Brito, J. Malnutrition and Clinical Outcome of 234 Head and Neck Cancer Patients Who Underwent Percutaneous Endoscopic Gastrostomy. *Nutr. Cancer* **2016**, *68*, 589–597. [CrossRef]
27. Powell-Tuck, J.; Hennessy, E.M. A Comparison of Mid Upper Arm Circumference, Body Mass Index and Weight Loss as Indices of Undernutrition in Acutely Hospitalized Patients. *Clin. Nutr.* **2003**, *22*, 307–312. [CrossRef] [PubMed]
28. Pereira, M.; Santos, C.; Fonseca, J. Body mass index estimation on gastrostomy patients using the mid-upper arm circumference. *J. Aging Res. Clin. Pract.* **2012**, *1*, 252–255.
29. Barosa, R.; Roque Ramos, L.; Santos, C.A.; Pereira, M.; Fonseca, J. Mid Upper Arm Circumference and Powell-Tuck and Hennessy's Equation Correlate with Body Mass Index and Can Be Used Sequentially in Gastrostomy Fed Patients. *Clin. Nutr.* **2018**, *37*, 1584–1588. [CrossRef] [PubMed]
30. World Health Organization—Europe. Nutrition—Body Mass Index—BMI. Available online: https://www.euro.who.int/en/healthtopics/disease-prevention/nutrition/a-healthylifestyle/body-mass-index-bmi~{} (accessed on 6 January 2020).
31. Frisancho, A.R. New Standards of Weight and Body Composition by Frame Size and Height for Assessment of Nutritional Status of Adults and the Elderly. *Am. J. Clin. Nutr.* **1984**, *40*, 808–819. [CrossRef]
32. Fonseca, J.; Santos, C.A. Clinical anatomy: Anthropometry for nutritional assessment of 367 adults who underwent endoscopic gastrostomy. *Acta Med. Port.* **2013**, *26*, 212–218. [PubMed]
33. McDowell, M.A.; Fryar, C.D.; Ogden, C.L.; Flegal, K.M. Anthropometric Reference Data for Children and Adults: United States, 2003–2006. *Natl. Health Stat. Rep.* **2008**, *10*, 1–48.
34. Sultan, S.; Nasir, K.; Qureshi, R.; Dhrolia, M.; Ahmad, A. Assessment of the Nutritional Status of the Hemodialysis Patients by Anthropometric Measurements. *Cureus* **2021**, *13*, e18605. [CrossRef] [PubMed]
35. Friedenberg, F.; Jensen, G.; Gujral, N.; Braitman, L.E.; Levine, G.M. Serum Albumin Is Predictive of 30-Day Survival after Percutaneous Endoscopic Gastrostomy. *JPEN J. Parenter. Enter. Nutr.* **1997**, *21*, 72–74. [CrossRef] [PubMed]
36. Kuzuya, M.; Izawa, S.; Enoki, H.; Okada, K.; Iguchi, A. Is Serum Albumin a Good Marker for Malnutrition in the Physically Impaired Elderly? *Clin. Nutr.* **2007**, *26*, 84–90. [CrossRef] [PubMed]
37. Badia-Tahull, M.B.; Llop-Talaveron, J.; Fort-Casamartina, E.; Farran-Teixidor, L.; Ramon-Torrel, J.M.; Jódar-Masanés, R. Preoperative Albumin as a Predictor of Outcome in Gastrointestinal Surgery. *e-SPEN Eur. J. Clin. Nutr. Metab.* **2009**, *4*, e248–e251. [CrossRef]
38. Vyroubal, P.; Chiarla, C.; Giovannini, I.; Hyspler, R.; Ticha, A.; Hrnciarikova, D.; Zadak, Z. Hypocholesterolemia in Clinically Serious Conditions—Review. *Biomed. Pap. Med. Fac. Univ. Palacky Olomouc Czech Repub.* **2008**, *152*, 181–189. [CrossRef]
39. Casas-Rodera, P.; Gómez-Candela, C.; Benítez, S.; Mateo, R.; Armero, M.; Castillo, R.; Culebras, J.M. Immunoenhanced Enteral Nutrition Formulas in Head and Neck Cancer Surgery: A Prospective, Randomized Clinical Trial. *Nutr. Hosp.* **2008**, *23*, 105–110.
40. van Wayenburg, C.A.M.; Rasmussen-Conrad, E.L.; van den Berg, M.G.A.; Merkx, M.A.W.; van Staveren, W.A.; van Weel, C.; van Binsbergen, J.J. Weight Loss in Head and Neck Cancer Patients Little Noticed in General Practice. *J. Prim. Health Care* **2010**, *2*, 16–21. [CrossRef]
41. Platek, M.E. The Role of Dietary Counseling and Nutrition Support in Head and Neck Cancer Patients. *Curr. Opin. Support. Palliat. Care* **2012**, *6*, 438–445. [CrossRef]
42. Cao, J.; Xu, H.; Li, W.; Guo, Z.; Lin, Y.; Shi, Y.; Hu, W.; Ba, Y.; Li, S.; Li, Z.; et al. Nutritional Assessment and Risk Factors Associated to Malnutrition in Patients with Esophageal Cancer. *Curr. Probl. Cancer* **2021**, *45*, 100638. [CrossRef] [PubMed]
43. Bossi, P.; Delrio, P.; Mascheroni, A.; Zanetti, M. The Spectrum of Malnutrition/Cachexia/Sarcopenia in Oncology According to Different Cancer Types and Settings: A Narrative Review. *Nutrients* **2021**, *13*, 1980. [CrossRef] [PubMed]

44. Ackerman, D.; Laszlo, M.; Provisor, A.; Yu, A. Nutrition Management for the Head and Neck Cancer Patient. *Cancer Treat. Res.* **2018**, *174*, 187–208. [CrossRef] [PubMed]
45. Meerkerk, C.D.A.; Chargi, N.; de Jong, P.A.; van den Bos, F.; de Bree, R. Low Skeletal Muscle Mass Predicts Frailty in Elderly Head and Neck Cancer Patients. *Eur. Arch. Otorhinolaryngol.* **2022**, *279*, 967–977. [CrossRef] [PubMed]
46. Lapornik, N.; Avramovič Brumen, B.; Plavc, G.; Strojan, P.; Rotovnik Kozjek, N. Influence of Fat-Free Mass Index on the Survival of Patients with Head and Neck Cancer. *Eur. Arch. Otorhinolaryngol.* **2022**, 1–9. [CrossRef]

Systematic Review

Nutritional Interventions during Chemotherapy for Pancreatic Cancer: A Systematic Review of Prospective Studies

Marco Cintoni [1], Futura Grassi [1], Marta Palombaro [1,*], Emanuele Rinninella [1,2], Gabriele Pulcini [1], Agnese Di Donato [1], Lisa Salvatore [3], Giuseppe Quero [4], Giampaolo Tortora [2,3], Sergio Alfieri [2,4], Antonio Gasbarrini [2,5] and Maria Cristina Mele [1,2]

[1] UOC di Nutrizione Clinica, Dipartimento di Scienze Mediche e Chirurgiche, Fondazione Policlinico Universitario A. Gemelli IRCCS, Largo A. Gemelli 8, 00168 Rome, Italy

[2] Dipartimento di Medicina e Chirurgia Traslazionale, Università Cattolica del Sacro Cuore, 00168 Rome, Italy

[3] Comprehensive Cancer Center, Dipartimento di Scienze Mediche e Chirurgiche, Fondazione Policlinico Universitario A. Gemelli IRCCS, Largo A. Gemelli 8, 00168 Rome, Italy

[4] Digestive Surgery Unit, Dipartimento di Scienze Mediche e Chirurgiche, Fondazione Policlinico Universitario A. Gemelli IRCCS, Largo A. Gemelli 8, 00168 Rome, Italy

[5] UOC Medicina Interna e Gastroenterologia, Dipartimento di Scienze Mediche e Chirurgiche, Fondazione Policlinico Universitario A. Gemelli IRCCS, Largo A. Gemelli 8, 00168 Rome, Italy

* Correspondence: marta.palombaro@guest.policlinicogemelli.it; Tel.: +39-06-3015-3410

Abstract: Background: Pancreatic cancer incidence is growing, but the prognosis for survival is still poor. Patients with pancreatic cancer often suffer from malnutrition and sarcopenia, two clinical conditions that negatively impact oncological clinical outcomes. The aim of this systematic review was to analyze the impact of different nutritional interventions on clinical outcomes in patients with pancreatic cancer during chemotherapy. Methods: A systematic review of MedLine, EMBASE, and Web of Science was carried out in December 2022, identifying 5704 articles. Titles and abstracts of all records were screened for eligibility based on inclusion criteria, and nine articles were included. Results: All nine articles included were prospective studies, but a meta-analysis could not be performed due to heterogenicity in nutritional intervention. This Systematic Review shows an improvement in Quality of Life, nutritional status, body composition, oral intake, and Karnofsky Performance Status, following nutritional interventions. Conclusions: This Systematic Review in pancreatic cancer patients during chemotherapies does not allow one to draw firm conclusions. However, nutritional support in pancreatic cancer patients is advisable to ameliorate oncological care. Further well-designed prospective studies are needed to identify nutritional support's real impact and to establish a reliable way to improve nutritional status of pancreatic cancer patients during chemotherapy.

Keywords: pancreatic cancer; nutritional support; Oral Nutritional Supplements; body composition; supportive care; Quality of Life

Citation: Cintoni, M.; Grassi, F.; Palombaro, M.; Rinninella, E.; Pulcini, G.; Di Donato, A.; Salvatore, L.; Quero, G.; Tortora, G.; Alfieri, S.; et al. Nutritional Interventions during Chemotherapy for Pancreatic Cancer: A Systematic Review of Prospective Studies. *Nutrients* **2023**, *15*, 727. https://doi.org/10.3390/nu15030727

Academic Editor: Yoichi Matsuo

Received: 31 December 2022
Revised: 19 January 2023
Accepted: 28 January 2023
Published: 1 February 2023

1. Introduction

In 2020, Pancreatic Cancer (PC), with 495,773 new cases, was the 12th most common tumor worldwide and the seventh leading cause of cancer mortality [1]. Incidence is higher in industrialized countries compared to developing countries, suggesting that environmental factors play a significant role as risk factors for the disease [2]. Cigarette smoking, alcohol drinking, physical inactivity, obesity, hypertension, chronic pancreatitis, diabetes, and high cholesterol are recognized as modifiable risk factors for PC development [2,3]. Other risk factors include age, gender, ethnicity, and inherited genetic syndromes [3].

The prognosis of PC patients is generally poor with a relative 5-year survival rate of 10.8%, because it is difficult to diagnose the disease at an early stage, since only 11%

of PC are at a local stage at the time of diagnosis, and only a few patients can benefit from surgical resection [4,5]. However, recent progress in diagnosis and chemotherapies give hope for better outcomes in PC patients, even though chemotherapy (CHT) is often burdened with important toxicity, and only fit patients can fully complete the planned treatments [6].

Malnutrition is a common feature in cancer patients due to both cancer itself and the related treatments and, when the neoplasm involves the gastrointestinal system, the maintenance of a proper nutritional balance can be very challenging [7,8]. When the pancreas is the site of cancer, both its exocrine and endocrine functions can be impaired [9]. The altered secretion of pancreatic enzymes determines a series of gastrointestinal symptoms with abdominal pain, bloating, gastric emptying delay, diarrhea, poor appetite, nausea, dyspepsia, malabsorption, and, consequently, weight loss [8,10]. Malnutrition's prevalence varies between 33.7% and 70.6% in PC patients, while the presence of sarcopenia has a great impact on this population, reaching 74% of PC patients according to some studies [8,11]. Malnutrition and sarcopenia are associated with an increased risk of chemotherapy-related toxicity (CIT), postoperative morbidity, poorer survival, and reduced Quality of Life (QoL) [12–14]. Thus, nutritional support during CHT may play a very special role in these patients [15,16].

Therefore, the aim of this systematic review was to analyze the impact of different nutritional interventions on clinical outcomes in PC patients during CHT.

2. Materials and Methods

This systematic review was performed according to the Cochrane Handbook for systematic reviews and to the Preferred Reporting Items for Systematic Reviews and Meta-Analyses (PRISMA) statement [17,18] It was registered in the International Prospective Register of Systematic Reviews PROSPERO 2020 CRD42020185706 [19]. The PRISMA checklist is detailed in Table S1.

2.1. Eligibility Criteria

We included studies with all the following PICOS Criteria:

- Population: eligible patients must (i) be at least 18 years old with any nutritional status (well-nourished, at risk of malnutrition, and malnourished), (ii) have a PC diagnosis, while (iii) undergoing CHT. Due to the limited number of studies which involve PC patients only, we decided to consider also papers with PC and other gastrointestinal tumors;
- Intervention: studies with nutritional interventions including nutritional counseling, supplementary food or drink, fortified foods, oral nutrition supplements, and enteral or parenteral nutrition during CHT were considered for inclusion in this review;
- Comparison: any types of comparison were considered as possible (i.e., no nutritional intervention, isocaloric diet without specific nutrients, etc.);
- Outcomes: the outcomes considered were CIT, changes in body composition, QoL, survival, and patient's functional capacity;
- Study designs: eligible study designs included randomized clinical trials (RCTs), prospective non-randomized studies, and other types of prospective studies.

2.2. Electronic Searches

The search was carried out on 2 December 2022 using three different electronic databases: Medical Literature Analysis and Retrieval System Online (MedLine) via PubMed, ISI Web of Science (WOS), and Excerpta Medica Database (EMBASE). Databases were screened for search terms in titles and abstract, limiting the search to English papers, without any restriction for date of publication. The comprehensive string search for each database is shown in Table S2.

2.3. Study Selection

The study selection process was independently carried out by three reviewers (M.C.; F.G.; M.P.) All articles generated from the electronic search were imported into Mendeley© (Elsevier, Amsterdam, The Netherlands), a reference management software, and duplicates were removed. Titles and abstracts of all records were screened for eligibility based on inclusion criteria, and all judged as ineligible were excluded. After the first title and abstract screening process, the three reviewers performed a second deeper title and/or abstract screening. A full text screening was performed on 43 studies, and 34 were excluded: 14 studies were not prospective ones, 17 were not performed during CHT, and 3 studies had no full text available.

Differences in judgment during the selection process between the three reviewers were settled by discussion and consensus.

2.4. Data Extraction

Information was collected using an Excel© (Microsoft Office, Redmond, WA, USA) spreadsheet specifically developed for this study. Each full-text article was retrieved, the ineligible articles were excluded, and the reasoning reported. Differences in judgment during the selection process between the three reviewers were settled by discussion and consensus.

2.5. Risk of Bias and Quality Assessment

The risk of bias instruments was used for randomized controlled trials and non-randomized comparative prospective studies. Risk of bias was independently assessed by two reviewers (M.C. and F.G.) and was further entered into the software «Review Manager 5.3.5» (The Nordic Cochrane Centre, Copenhagen, Denmark). According to the "Cochrane Handbook for Systematic Reviews of Interventions" [20], all the articles included were assessed as high, low, or unclear risk of bias.

In total, seven areas were assessed: (1) Random sequence generation; (2) Allocation concealment; (3) Blinding of participants and personnel; (4) Blinding of outcome assessment; (5) Incomplete outcome data; (6) Selective reporting; (7) Other bias (other any important concerns about bias not covered in the other domains (i.e., presence of data regarding diet during other nutritional treatment).

2.6. Data Synthesis

Given the high heterogeneity of the studies' measures, the variability of nutritional intervention, and the variety of the outcomes considered, a meta-analysis resulted unfeasible, and thus, a systematic review was performed. The main results of the review were displayed on a summary of findings table. For each study, a description of the population, type of intervention, outcome measures, and results were presented.

3. Results

3.1. Study Selection

The study selection process and the results of the literature search are shown in Figure 1. In particular, starting from the 5704 studies identified from the three different databases (1532 from PubMed, 1117 from Web of Science, and 3055 from EMBASE), nine were finally included into the systematic review process [21–29].

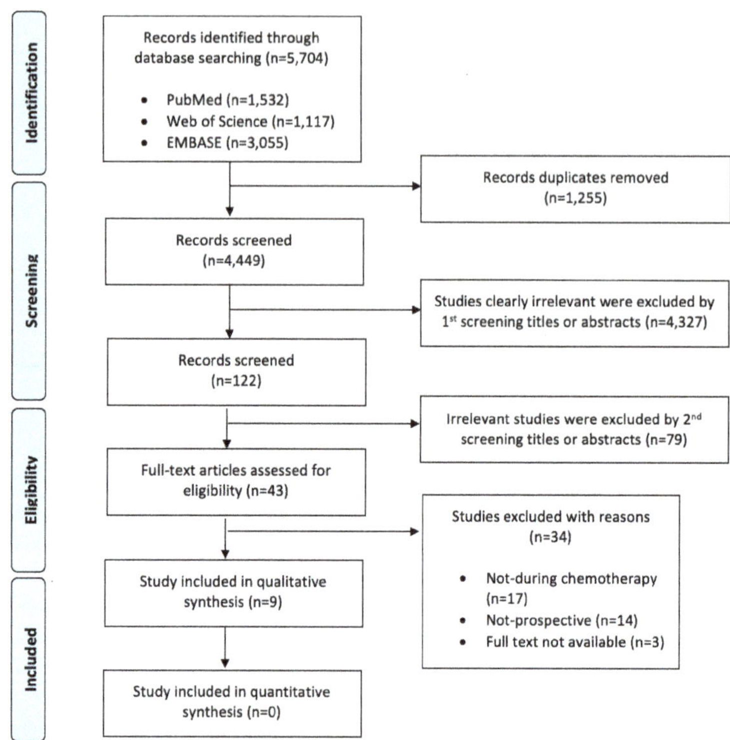

Figure 1. PRISMA Flow Diagram.

3.2. Study Characteristics

The most important characteristics of the nine studies are shown in Table 1.

The percentage of PC patients in all the included studies ranges from 7% [25] to 100%, with 6 studies enrolling only PC patients [22–24,26,27,29]. The Sample size varies from 7 [21] to 201 patients [25]. Three studies had no comparison groups [21,23,24], two studies had a placebo-controlled group [22,29], and three studies had normal or isocaloric diet as the controlled group [25,27,28]. Four papers considered the use of an Oral Nutritional Supplement (ONS) as nutritional intervention [21,25,27,28], two papers analyzed the role of a parenteral supplementation of fatty acids [23,24], one paper used oral carnitine supplementation [22], and the last one used oral supplement of sulforaphane and glucoraphanin or methylcellulose [29].

Table 2 shows the main characteristics of used ONS. In particular, three papers considered the use of a liquid premixed product [21,27,28], while one used a powder to be mixed with water prior to use [25]. ONS energy intake ranged from 310 [21] to 691 kcal per day [25], while protein intake was from 16 [21] to 45.75 g per day [25].

Table 1. Main characteristics of included studies.

Refs	Author and Year	Study Design	% Pancreatic Cancer	Sample Size (IG/CG)	Time of Intervention	Type of Nutritional Intervention	Comparison	Results
[21]	Bauer JD 2005	Single-arm trial	71.4	7 (7/0)	8 weeks	ONS	-	- ↑ protein ($p = 0.011$), energy ($p = 0.011$), and fiber ($p = 0.006$) intake ↑ nutritional status ($p = 0.019$) ↑ KPS ($p = 0.01$) ↑ QoL($p = 0.019$) Not statistically significant improvements in: - BW ($p = 0.368$) - LBM ($p = 0.225$)
[22]	Kraft M et al. 2012	Prospective, multi-center, placebo-controlled, randomized, and double-blinded trial	100	72 (38/34)	12 weeks	Oral liquid formulation of L-Carnitine	Placebo	in the IG group vs. CG: ↑ BCM after 6 weeks ($p = 0.013$) ↑ BF after 12 weeks ($p = 0.041$) ↑ BMI after 12 weeks ($p < 0.018$) ↑ cognitive function after 6 weeks ($p < 0.034$) ↑ global health status after 12 weeks ($p < 0.041$) ↓ gastrointestinal symptoms after 12 weeks ($p < 0.033$) No significant differences between the two groups in survival
[23]	Arshad A et al. 2013	Single-arm phase II clinical trial	100	32 (32/0)	Weekly for 3 weeks followed by a rest week during the CHT period	Parenteral supplement n-3FA-rich lipid emulsion	-	↓ OS in high expressors of IL-6 ($p = 0.009$) and IL-8 ($p = 0.02$) ↓ PFS in high expressors of IL-8 ($p = 0.002$)
[24]	Arshad A et al. 2014	Single-arm phase II clinical trial	100	21 (21/0)	Weekly for 3 weeks followed by a rest week for up to six months	Parenteral supplement n-3FA-rich lipid emulsion	-	Over the entire treatment course of up to six months: ↑ ECM pellet uptake of EPA ($p = 0.005$) and DHA ($p < 0.001$) ↓ n6:n3 ratio ($p < 0.001$)
[25]	Khemissa F et al. 2016	Double-blind, randomized, controlled, and multicenter trial	7	201 (99/102)	Five days before the start of each CHT cycle	ONS	Isocaloric ONS	No significant differences between the two groups in term of compliance and toxicities

Table 1. *Cont.*

Refs	Author and Year	Study Design	% Pancreatic Cancer	Sample Size (IG/CG)	Time of Intervention	Type of Nutritional Intervention	Comparison	Results
[26]	Werner K et al. 2017	Randomized, double-blind, controlled trial	100	60 (31/29)	6 weeks	FO capsules	MPL capsules	in both groups: BW stabilization ($p = 0.001$ in FO group; $p = 0.003$ in MPL group) ↑ meal portions ($p = 0.02$ in FO group; $p = 0.05$ in MPL group) No significant changes in both groups in QoL, and food intake.
[27]	Akita H et al. 2019	RCT	100	62 (31/31)	5 weeks	ONS	Normal diet	in CG group: ↓ Post/pre ratio of SMM ($p = 0.014$) in both groups: ↓ PMA (IG $p = 0.002$; CG $p < 0.001$) ↓ BMI (IG $p = 0.011$; CG $p = 0.001$) in IG group: ↑ Post/pre ratio of PMA ($p = 0.001$) ↑ Post/pre ratio of SMM ($p = 0.042$) ↑ Post/pre ratio of PMA ($p < 0.001$) No significant difference between the two groups in NACRT-related toxicity
[28]	Kim SH et al. 2019	Prospective randomized study	29.4	58 enrolled (36/22)	8 weeks	ONS	Nutritional care only	No significant difference between the two groups in BW, FFM, SMM, BCM, QoL, and biochemical tests (all patients) (dividing population based on CHT cycles) In IG vs. CG: ↑ dietary intake ↓ reduction of fatigue ($p = 0.041$) ↑ PG-SGA grade ratio ($p < 0.05$) ↑ BW ($p = 0.049$) ↑ FFM ($p = 0.034$) ↑ SMM ($p = 0.049$) ↑ BCM ($p = 0.049$)

Table 1. *Cont.*

Refs	Author and Year	Study Design	% Pancreatic Cancer	Sample Size (IG/CG)	Time of Intervention	Type of Nutritional Intervention	Comparison	Results
[29]	Lozanovski et al. 2020	Prospective, placebo-controlled trial	100	40 (29/11)	12 months	Daily intake of broccoli sprouts containing 90 mg sulforaphane and 180 mg glucoraphanin or methylcellulose a	Placebo	In IG: Drop out: 72% IG vs. CG: ↑ Survival at 180 days ($p = 0.291$)

BCM: Body Cell Mass; BF: Body Fat; BMI: Body Mass index; BW: Body Weight; CAF: pro-angiogenic cytokines and growth factors; CG: control group; DHA: Docosahexaenoic Acis; ECM: Erythrocyte Cell Membrane; EPA: Eicosapentaenoic Acid, FFM: Fat-Free Mass; FO: fish oil; KPS: Karnofsky Performance Status; IG: intervention Group; LBM: Lean Body Mass; NACRT: Neoadjuvant Chemoradiotherapy; MPL: Marine Phospholipids; OS: Overall Survival; PFS: Progression Free Survival; PG-SGA: Patient-Generated Subjective Global Assessment; PMA: Psoas Major Muscle Area; QoL: Quality of Life; SMM: Skeletal muscle mass.

Table 2. Main characteristics of Oral Nutritional Supplements used in the enrolled studies.

Refs	Author	ONS Type	ONS Quantity	Amount (per Day)	Energy (kcal per Day)	Protein (g per Day)	Other
[21]	Bauer JD et al.	L	Not reported	At least 1	310	16	1.1 g EPA
[25]	Khemissa F et al.	P	75 g	2	691	45.75	13.5 g glutamine + TGF-β2 20 mg
[27]	Akita H et al.	L	220 mL	2	560	29.3	1.98 g EPA
[28]	Kim SH et al.	L	150 mL	2	400	18	2.5 g fiber

Abbreviations: EPA: Eicosapentaenoic Acid; L: Liquid Formula; ONS: Oral Nutritional Supplements; P: Powder Formula; Refs: Bibliographic references; TGF-β2: Tumor Growth Factor.

3.3. Study Quality Assessment

The risk of bias was assessed in each included study. Figure 2 reports the different types of bias for each study, while Figure 3 shows the cumulative risk of bias expressed in percentage.

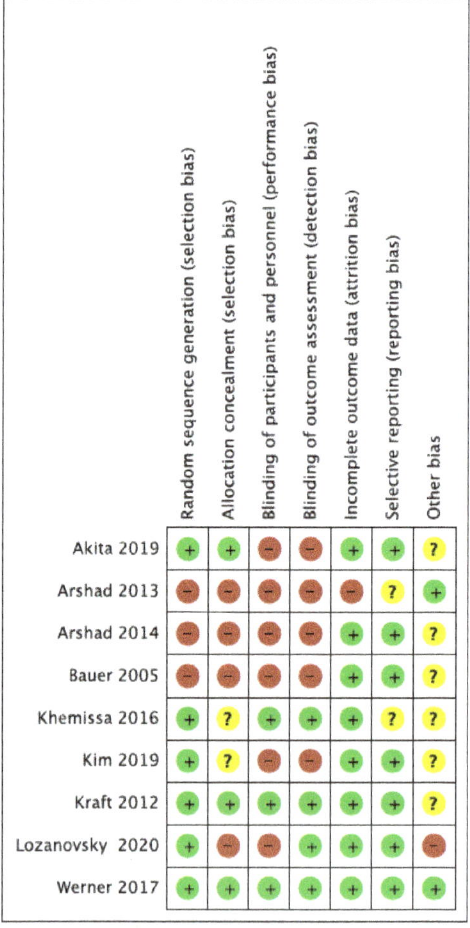

Figure 2. Risk of bias summary [21–29].

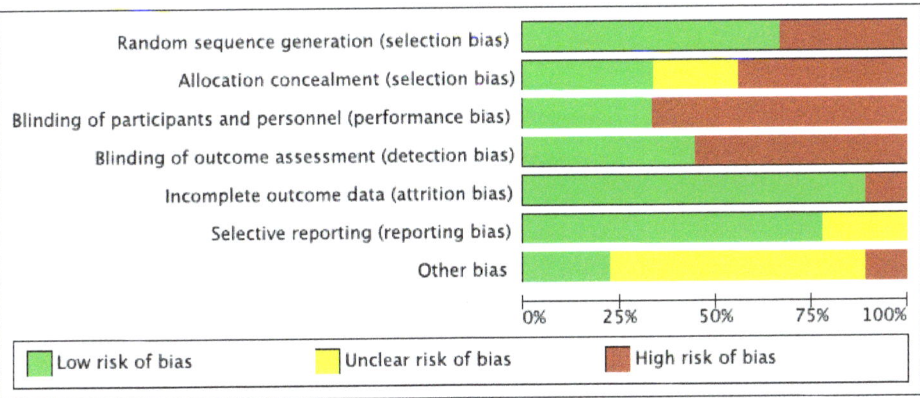

Figure 3. Risk of bias graph.

3.4. Summary of Results

3.4.1. Survival Analysis

Three papers analyzed the impact of nutritional support on survival in PC patients [22,23,29]. In particular, Kraft et al., using oral L-carnitine vs. placebo for 12 weeks, found only a non-statistically significative trend of increased overall survival (OS) (median 519 ± 50 vs. 399 ± 43 days) [22]. Lozanovski et al. showed longer survival in the intervention group (IG) during the first three months after the study (death raw rate of 25% in IG vs. 45% in control group (CG) [29]. Arshad et al. dosed the plasma cytokines at baseline and found a significant correlation between high expression of IL-6 and IL-8 and shorter OS [23]. Moreover, authors evidenced that platelet-derived growth factor (PDGF) and fibroblast growth factor (FGF) serum concentrations decreased at the end of the treatment period and FGF responders had a significantly improved progression free survival (PFS). In the case of PDGF reduction, a tendency toward improved OS was noticed [23].

3.4.2. Quality of Life

Four papers examined the impact of nutritional interventions on QoL [21,22,26,28]. All papers analyzed QoL using the European Organization for Research and Treatment of Cancer (EORTC) QLQ-C30 [21,22,26,28]; only two papers also added the PAN-26 analysis, the QLQ specific for PC [22,26].

In particular, Bauer et al. showed a stability in QLQ-C30 global scale at 4 weeks, with an increase at 8 weeks [21]. Kraft et al. reported an improvement in cognitive functions and global health status, a reduction in gastrointestinal symptoms, while a non-significative difference in fatigue was found [22]. Werner et al. did not find any differences in terms of QoL after 6 weeks of treatment, but only a non-significant slight increase in sub-scale physical, role, social, pain, appetite loss, and global health; moreover, the authors described a significant decrease in hepatic sub-scale of PAN-26 [26]. The study by Kim et al. reported a non-statistical increase in QoL in both IG and CG, while a decrease in subscale fatigue in IG and a pain reduction in CG was described [28].

3.4.3. Chemotherapy-Induced Toxicity

CIT was observed in two studies [25,27]. In the phase III study from Khemissa et al., the authors aimed to evaluate the possible role of oral supplementation with glutamine and TGF-β2 in the prevention of grade 3 and 4 non-hematological CIT. However, the results did not confirm this hypothesis, and no difference was evidenced between IG and CG for all kinds of CITs [25]. Akita et al. analyzed the incidence of adverse events during neoadjuvant CHT between patients on a normal diet and those who received hypercaloric,

Eicosapentaenoic Acid (EPA)-enriched oral supplements; even in this case, no significant difference between the two groups in terms of CIT was evidenced [27].

3.4.4. Nutritional Status

Only two studies evaluated PC patients' nutritional status with a Patient-Generated Subjective Global Assessment (PG-SGA) score [21,28]. According to this score, patients could be defined: well nourished (PG-SGA A), moderated or suspected of being malnourished (PG-SGA B), and severely malnourished (PG-SGA C). Bauer et al. showed a significant reduction in PG-SGA score from the baseline to 8 weeks (median 13 range 4.0–19.0 vs. median 4 range 1.0–16.0, $p = 0.019$); this improvement was significantly associated with a change in QoL ($p = 0.020$), Karnofsky Performance Status (KPS) ($p = 0.009$), and lean body mass ($p = 0.040$) [21]. Kim et al. revealed a reduction on PG-SGA score after 8 weeks of intervention (9.5 ± 0.9 vs. 5.6 ± 0.8, $p = 0.002$) [28].

3.4.5. Body Composition

Five studies evaluated the effect of different nutritional interventions on patients' body weight and/or body composition [21,22,26–28].

Akita et al. showed a higher skeletal muscle mass in those patients who consumed more than 50% of the prescribed ONS (p 0.042) and an improvement in psoas muscle ratio in the same population [27]. Another study evaluated the effect of n-3 fatty acid-enriched ONS reporting a clinical improvement in weight and lean body mass although not statistically significant [21]. Werner et al. showed a significant body weight stabilization, with a gain of body weight in half of the patients, but no significative change in fat mass, muscle mass, or body water was detected [26]. In addition, Kim et al. evaluated the effect of ONS administration in PC patients showing a significantly increase of fat mass from the baseline to 8 weeks of intervention, and a stabilization of fat-free mass, skeletal muscle mass, and body cell mass [28]. In the study by Kraft et al., who evaluated the effect of carnitine supplementation, an increase of body weight in the IG, as well as increase in body cell mass and body fat mass, was found [22].

3.4.6. Oral Intake

Three studies evaluated the effect of nutritional intervention on oral intake in PC patients undergoing CHT [21,26,28]. Bauer et al. showed no reduction of meal protein and energy intake with supplementation and observed 1.4 (1.2–2.2) g/kg/day of total protein intake and 33 (25–42) kcal/kg/day of total energy intake after 8 weeks of intervention [21]. Kim et al. revealed significant increases in dietary intakes of calories (1488.1 kcal vs. 1946.4 kcal, $p = 0.001$), proteins (64.1 g vs. 89.9 g, $p = 0.001$), carbohydrates (247.9 g vs. 289.2 g, $p = 0.015$), and lipids (38.6 g vs. 51.9 g, $p = 0.023$) in the ONS group from the baseline to 8 weeks of intervention. However, there was no significant difference between the change of values of dietary intake between baseline and 8 weeks in the ONS and non-ONS group [28]. Werner et al. evaluated appetite and meal portions in PC patients supplemented with n-3 fatty acids from MPL or FO and showed stabilization of appetite in both groups of patients. Moreover, meal portions increased significantly in the FO group ($p = 0.02$) and MPL group ($p = 0.05$) [26].

3.4.7. Karnofsky Performance Status

Two papers considered variation in patients' functional capacity, measured according to KPS [21,29]. Bauer et al. showed that nutritional intervention with high-protein, high-calorie nutritional supplement containing EPA, not only improved patients' nutritional status but equally increased their KPS after 8 weeks of treatment ($p = 0.01$) [21]. Similarly, Lozanovski et al. described a decrease in KPS in both groups (intervention and control group), stating that broccoli sprouts did not impact patient self-care and overall abilities severely [29].

4. Discussion

At our knowledge, this is the first Systematic Review on nutritional intervention in PC patients, during chemotherapies, enrolling only prospective studies. Even though nutritional interventions in PC patients should be a routine [7], the number of robust studies remains scarce.

Nutritional intervention strategy should include personalized nutritional counseling with a trained physician or dietitian specialized in oncological cures [30], with the evaluation of nutritional targets and intakes. According to the European Society of Nutrition and Metabolism (ESPEN) guidelines on nutrition in cancer patients, an energy intake of 25–30 kcal/kg/day and a protein intake of 1.0–1.5 g/kg/day should be guaranteed to all cancer patients [7,31,32]. However, only a few papers analyzed PC patients' nutritional requirements and performed a malnutrition risk assessment. To modify this situation, an integration in the cure pathway to include nutritional evaluation and intervention in clinical routine for oncological patients was proposed [33].

QoL represents a major concern in PC patients undergoing CHT treatments [34]. In addition to the more generic QLQ-C30 questionnaire, valid for all oncological patients, a specific module called PAN26 has been developed for the evaluation of the QoL in PC patient; however, QoL was analyzed in only four studies (less than half), and the specific module PAN26 was studied in only two of them. While data from the literature showed a general improvement in QoL in oncological patients who undergo nutritional interventions [35,36], our results are not conclusive; in fact, two papers showed an increase in QoL, one paper showed an increase only in subscale fatigue, while one other showed no differences.

KPS is a scale that tries to quantify the patient's well-being and their capacity to do all the daily-life activities [37]. In the two papers enrolled in this Systematic Review, nutritional supplementation showed an increase in patient function. In line with this, the same results were obtained also in other neoplasm-affected patients [38,39], demonstrating that better nutritional status is related to better functional capacity and physical resistance to therapies.

Different papers suggest an association between body composition and treatment related toxicity. In particular, sarcopenia is related with an increased incidence of severe adverse reactions and treatment interruption [40–43]. In patients with pancreatic neoplasm, results are still not conclusive [44]. In our systematic review, only two studies analyzed the effect of nutritional intervention on CIT, but no significant correlation was found [25,27].

Notably, malnutrition affects prognosis and survival in PC patients [45–47]. However, studies reporting the effects of high-energy ONS on survival outcomes are limited and heterogeneous, and there is no consensus [48–50]. None of the papers included in our review that supplemented PC patients with ONS evaluated their effect on survival. In our analysis, one study reported longer survival after three months with daily supplementation of broccoli sprouts [29]. Moreover, a trend of increase in OS was found with oral supplementation of L-carnitine (4 g/day) for 12 weeks, even if no statistical significance was achieved [22].

Recently a prospective cohort study showed that daily protein intake influenced the prognosis of patients with unresectable PC undergoing CHT. Interestingly, authors found that protein intake <1.1 g/kg/day was an independent poor prognostic factor in this setting [49]. Since the loss of appetite and the consequent reduction of calories and protein intake are common features of PC patients, the use of ONS can be an advisable, powerful strategy. Papers included in our analysis globally demonstrated an increase of dietary intake of all macronutrients with the use of ONS [21,26]. Oral supplementation with n3-fatty acids induced a stabilization of appetite and meal portions tended to increase [28].

Furthermore, the use of ONS seemed to have a direct impact on malnutrition. Indeed, two studies reported a significant reduction in PG-SGA score after 8 weeks of intervention with high-energy and high-protein oral supplementation [21,28].

Globally, studies focusing on nutritional interventions in gastrointestinal (GI) cancer showed heterogeneous results. A systematic review conducted on GI cancers (stomach, esophagus, pancreas) undergoing surgery presented scarce evidence of the effectiveness of using ONS in terms of body weight gain and increased energy intake both in pre- and post-operative period [51]. However, a meta-analysis based only on gastric cancer patients highlighted a positive association between the use of ONS and reduced weight loss, especially in the postoperative period [50]. Another recent meta-analysis on the role of oral supplementation with an amino acid-enriched formula containing glutamine, vitamins, and minerals during CHT and/or radiotherapy in 445 patients with GI and head-neck cancer showed that this type of nutritional intervention could be beneficial in preventing CIT and, in particular, oral mucositis [52]. In other malignancies, a proper nutritional intervention is associated with benefits in terms of body weight and body composition [53–55]. However, data on the correct timing and the proper type of nutritional intervention are still unconclusive. Results from our study are in line with these findings [21,22,26–28].

Moreover, in PC, particular attention should be paid to pancreatic exocrine insufficiency (PEI). Notably, the reduction of pancreatic secretions leads to maldigestion and malabsorption and remarkably contributes to the development of malnutrition. Thus, when considering nutritional intervention in PC, pancreatic enzyme replacement therapy (PERT) must always be taken into account. PEI can be caused by local tumor-induced changes (i.e., Warburg effect, production of tumor-specific factors, tumor location, etc.) or can be the consequence of surgery [56]. It can be present even before the onset of clinical symptoms, and the estimated prevalence in patients with advanced PC is 72% [57,58]. A few studies evidence a positive association between PERT prescription with survival and QoL [59,60]. However, PERT is not always adequate in common practice and frequently enzyme dosages are lower than needed [58]. According to this observation, none of the studies collected in our review considered PERT. Due to the complex etiology of malnutrition in PC patients, we believe that close attention should be given to any aspect that can improve nutritional status and that PERT must be part of nutritional intervention.

The present Systematic Review has some limitations: (i) the small number of included studies (only nine papers); (ii) the necessity to include papers which enrolled PC patients during CHT together with other gastrointestinal cancers; (iii) the large variability in term of nutritional intervention, population, and outcomes.

5. Conclusions

Pancreatic cancer remains one of the most challenging cancers for oncologists and surgeons. Due to the paucity of studies, the scarcity of sample size, the heterogeneity of the studies, and the lack of robust randomized clinical trials, it is not feasible to draw strong conclusions on the role of nutritional support during CHT for PC patients. The main results of this Systematic Review are an improvement in QoL, nutritional status, body composition, oral intake, and KPS when nutritional support is provided in PC patients. Nonetheless, nutritional intervention in PC patients remains advisable, particularly during CHT, to contribute to the oncological care.

Nevertheless, further well-designed prospective studies are needed to identify the real impact of nutritional support during oncological pathway in PC patients and to establish the most effective strategy aiming to reduce the burden of malnutrition in this population.

Supplementary Materials: The following supporting information can be downloaded at: https://www.mdpi.com/article/10.3390/nu15030727/s1; Table S1: PRISMA Checklist; Table S2: Full search strategies for electronic databases.

Author Contributions: Conceptualization, M.C. and F.G.; validation, G.Q. and G.T.; formal analysis, M.C.; resources, A.D.D.; data curation, G.P.; writing—original draft preparation, M.C. and F.G.; writing—review and editing, M.P. and E.R.; visualization, L.S.; supervision, S.A., A.G. and M.C.M. All authors have read and agreed to the published version of the manuscript.

Funding: This research received no external funding.

Institutional Review Board Statement: Not applicable.

Informed Consent Statement: Not applicable.

Data Availability Statement: No new data were created or analyzed in this study. Data sharing is not applicable to this article.

Conflicts of Interest: The authors declare no conflict of interest.

References

1. Sung, H.; Ferlay, J.; Siegel, R.L.; Laversanne, M.; Soerjomataram, I.; Jemal, A.; Bray, F. Global cancer statistics 2020: GLOBOCAN estimates of incidence and mortality worldwide for 36 cancers in 185 countries. *CA Cancer J. Clin.* **2021**, *71*, 209–249. [CrossRef]
2. Huang, J.; Lok, V.; Ngai, C.H.; Zhang, L.; Yuan, J.; Lao, X.Q.; Ng, K.; Chong, C.; Zheng, Z.J.; Wong, M.C.S. Worldwide Burden of, Risk Factors for, and Trends in Pancreatic Cancer. *Gastroenterology* **2021**, *160*, 744–754. [CrossRef]
3. McGuigan, A.; Kelly, P.; Turkington, R.C.; Jones, C.; Coleman, H.G.; McCain, R.S. Pancreatic cancer: A review of clinical diagnosis, epidemiology, treatment and outcomes. *World J. Gastroenterol.* **2018**, *24*, 4846–4861. [CrossRef]
4. National Cancer Institute. In Pancreatic Cancer Cancer Stat Facts. 2022. Available online: https://seer.cancer.gov/ (accessed on 13 March 2022).
5. Ghaneh, P.; Kleeff, J.; Halloran, C.M.; Raraty, M.; Jackson, R.; Melling, J.; Jones, O.; Palmer, D.H.; Cox, T.F.; Smith, C.J.; et al. The Impact of Positive Resection Margins on Survival and Recurrence Following Resection and Adjuvant Chemotherapy for Pancreatic Ductal Adenocarcinoma. *Ann. Surg.* **2019**, *269*, 520–529. [CrossRef]
6. Tempero, M.A. NCCN Guidelines Updates: Pancreatic Cancer. *J. Natl. Compr. Cancer Netw.* **2019**, *17*, 603–605. [CrossRef]
7. Muscaritoli, M.; Arends, J.; Bachmann, P.; Baracos, V.; Barthelemy, N.; Bertz, H.; Bozzetti, F.; Hütterer, E.; Isenring, E.; Kaasa, S.; et al. ESPEN practical guideline: Clinical Nutrition in cancer. *Clin. Nutr.* **2021**, *40*, 2898–2913. [CrossRef]
8. Bossi, P.; Delrio, P.; Mascheroni, A.; Zanetti, M. The Spectrum of Malnutrition/Cachexia/Sarcopenia in Oncology According to Different Cancer Types and Settings: A Narrative Review. *Nutrients* **2021**, *13*, 1980. [CrossRef]
9. Kordes, M.; Larsson, L.; Engstrand, L.; Löhr, J.M. Pancreatic cancer cachexia: Three dimensions of a complex syndrome. *Br. J. Cancer* **2021**, *124*, 1623–1636. [CrossRef]
10. Hendifar, A.E.; Chang, J.I.; Huang, B.Z.; Tuli, R.; Wu, B.U. Cachexia, and not obesity, prior to pancreatic cancer diagnosis worsens survival and is negated by chemotherapy. *J. Gastrointest. Oncol.* **2018**, *9*, 17–23. [CrossRef]
11. Griffin, O.M.; Bashir, Y.; O'Connor, D.; Peakin, J.; McMahon, J.; Duggan, S.N.; Geoghegan, J.; Conlon, K.C. Measurement of body composition in pancreatic cancer: A systematic review, meta-analysis and recommendations for future study design. *Dig. Surg.* **2022**, *39*, 141–152. [CrossRef]
12. Bundred, J.; Kamarajah, S.K.; Roberts, K.J. Body composition assessment and sarcopenia in patients with pancreatic cancer: A systematic review and meta-analysis. *HPB* **2019**, *21*, 1603–1612. [CrossRef]
13. Zhang, Y.X.; Yang, Y.F.; Han, P.; Ye, P.C.; Kong, H. Protein-energy malnutrition worsens hospitalization outcomes of patients with pancreatic cancer undergoing open pancreaticoduodenectomy. *Updates Surg.* **2022**, *74*, 1627–1636. [CrossRef]
14. Poulia, K.A.; Antoniadou, D.; Sarantis, P.; Karamouzis, M.V. Pancreatic Cancer Prognosis, Malnutrition Risk, and Quality of Life: A Cross-Sectional Study. *Nutrients* **2022**, *14*, 442. [CrossRef]
15. Trestini, I.; Carbognin, L.; Sperduti, I.; Bonaiuto, C.; Auriemma, A.; Melisi, D.; Salvatore, L.; Bria, E.; Tortora, G. Prognostic impact of early nutritional support in patients affected by locally advanced and metastatic pancreatic ductal adenocarcinoma undergoing chemotherapy. *Eur. J. Clin. Nutr.* **2018**, *72*, 772–779. [CrossRef]
16. Trestini, I.; Cintoni, M.; Rinninella, E.; Grassi, F.; Paiella, S.; Salvia, R.; Bria, E.; Pozzo, C.; Alfieri, S.; Gasbarrini, A.; et al. Neoadjuvant treatment: A window of opportunity for nutritional prehabilitation in patients with pancreatic ductal adenocarcinoma. *World J. Gastrointest. Surg.* **2021**, *13*, 885–903. [CrossRef]
17. Cochrane Handbook for Systematic Reviews of Interventions Version 6.3 [Updated February 2022]. Available online: www.cochrane-handbook.org (accessed on 12 January 2023).
18. Moher, D.; Liberati, A.; Tetzlaff, J.; Altman, D.G.; PRISMA Group. Preferred reporting items for systematic reviews and meta-analyses: The PRISMA statement. *BMJ* **2009**, *339*, b2535. [CrossRef]
19. Mele, M.C.; Rinninella, E.; Grassi, F. Nutritional Interventions during Chemotherapy for Pancreatic Cancer: A Systematic Review of Prospective Studies. PROSPERO 2020 CRD42020185706. Available online: https://www.crd.york.ac.uk/prospero/display_record.php?ID=CRD42020185706 (accessed on 12 January 2023).
20. Higgins, J.P.; Altman, D.G.; Gøtzsche, P.C.; Jüni, P.; Moher, D.; Oxman, A.D.; Savovic, J.; Schulz, K.F.; Weeks, L.; Sterne, J.A.; et al. The Cochrane Collaboration's tool for assessing risk of bias in randomised trials. *BMJ* **2011**, *343*, d5928. [CrossRef]
21. Bauer, J.D.; Capra, S. Nutrition intervention improves outcomes in patients with cancer cachexia receiving chemotherapy–a pilot study. *Support Care Cancer* **2005**, *13*, 270–274. [CrossRef]
22. Kraft, M.; Kraft, K.; Gärtner, S.; Mayerle, J.; Simon, P.; Weber, E.; Schütte, K.; Stieler, J.; Koula-Jenik, H.; Holzhauer, P.; et al. L-Carnitine-supplementation in advanced pancreatic cancer (CARPAN)—A randomized multicentre trial. *Nutr. J.* **2012**, *11*, 52. [CrossRef]

23. Arshad, A.; Chung, W.Y.; Steward, W.; Metcalfe, M.S.; Dennison, A.R. Reduction in circulating pro-angiogenic and pro-inflammatory factors is related to improved outcomes in patients with advanced pancreatic cancer treated with gemcitabine and intravenous omega-3 fish oil. *HPB* **2013**, *15*, 428–432. [CrossRef]
24. Arshad, A.; Chung, W.Y.; Isherwood, J.; Mann, C.D.; Al-Leswas, D.; Steward, W.P.; Metcalfe, M.S.; Dennison, A.R. Cellular and plasma uptake of parenteral omega-3 rich lipid emulsion fatty acids in patients with advanced pancreatic cancer. *Clin. Nutr.* **2014**, *33*, 895–899. [CrossRef] [PubMed]
25. Khemissa, F.; Mineur, L.; Amsellem, C.; Assenat, E.; Ramdani, M.; Bachmann, P.; Janiszewski, C.; Cristiani, I.; Collin, F.; Courraud, J.; et al. A phase III study evaluating oral glutamine and transforming growth factor-beta 2 on chemotherapy-induced toxicity in patients with digestive neoplasm. *Dig. Liver Dis.* **2016**, *48*, 327–332. [CrossRef] [PubMed]
26. Werner, K.; de Gaudry, D.K.; Taylor, L.A.; Keck, T.; Unger, C.; Hopt, U.T.; Massing, U. Dietary supplementation with n-3-fatty acids in patients with pancreatic cancer and cachexia: Marine phospholipids versus fish oil—A randomized controlled double-blind trial. *Lipids Health Dis.* **2017**, *16*, 104. [CrossRef] [PubMed]
27. Akita, H.; Takahashi, H.; Asukai, K.; Tomokuni, A.; Wada, H.; Marukawa, S.; Yamasaki, T.; Yanagimoto, Y.; Takahashi, Y.; Sugimura, K.; et al. The utility of nutritional supportive care with an eicosapentaenoic acid (EPA)-enriched nutrition agent during pre-operative chemoradiotherapy for pancreatic cancer: Prospective randomized control study. *Clin. Nutr. ESPEN* **2019**, *33*, 148–153. [CrossRef]
28. Kim, S.H.; Lee, S.M.; Jeung, H.C.; Lee, I.J.; Park, J.S.; Song, M.; Lee, D.K.; Lee, S.M. The Effect of Nutrition Intervention with Oral Nutritional Supplements on Pancreatic and Bile Duct Cancer Patients Undergoing Chemotherapy. *Nutrients* **2019**, *11*, 1145. [CrossRef]
29. Lozanovski, V.J.; Polychronidis, G.; Gross, W.; Gharabaghi, N.; Mehrabi, A.; Hackert, T.; Schemmer, P.; Herr, I. Broccoli sprout supplementation in patients with advanced pancreatic cancer is difficult despite positive effects-results from the POUDER pilot study. *Investig. New Drugs* **2020**, *38*, 776–784. [CrossRef]
30. Cañamares-Orbís, P.; García-Rayado, G.; Alfaro-Almajano, E. Nutritional Support in Pancreatic Diseases. *Nutrients* **2022**, *14*, 4570. [CrossRef]
31. Arends, J.; Baracos, V.; Bertz, H.; Bozzetti, F.; Calder, P.C.; Deutz, N.E.P.; Erickson, N.; Laviano, A.; Lisanti, M.P.; Lobo, D.N.; et al. ESPEN expert group recommendations for action against cancer-related malnutrition. *Clin. Nutr.* **2017**, *36*, 1187–1196. [CrossRef]
32. Arends, J.; Bachmann, P.; Baracos, V.; Barthelemy, N.; Bertz, H.; Bozzetti, F.; Fearon, K.; Hütterer, E.; Isenring, E.; Kaasa, S.; et al. ESPEN guidelines on nutrition in cancer patients. *Clin. Nutr.* **2017**, *36*, 11–48. [CrossRef]
33. Vitaloni, M.; Caccialanza, R.; Ravasco, P.; Carrato, A.; Kapala, A.; de van der Schueren, M.; Constantinides, D.; Backman, E.; Chuter, D.; Santangelo, C.; et al. The impact of nutrition on the lives of patients with digestive cancers: A position paper. *Support Care Cancer* **2022**, *30*, 7991–7996. [CrossRef]
34. Nayak, M.G.; George, A.; Vidyasagar, M.S.; Mathew, S.; Nayak, S.; Nayak, B.S.; Shashidhara, Y.N.; Kamath, A. Quality of Life among Cancer Patients. *Indian J. Palliat. Care* **2017**, *23*, 445–450. [CrossRef] [PubMed]
35. Lluch Taltavull, J.I.; Mercadal Orfila, G.; Afonzo Gobbi, Y.S. Improvement of the nutritional status and quality of life of cancer patients through a protocol of evaluation and nutritional intervention. *Nutr. Hosp.* **2018**, *35*, 606–611. [CrossRef]
36. Souza, A.P.S.; Silva, L.C.D.; Fayh, A.P.T. Nutritional Intervention Contributes to the Improvement of Symptoms Related to Quality of Life in Breast Cancer Patients Undergoing Neoadjuvant Chemotherapy: A Randomized Clinical Trial. *Nutrients* **2021**, *13*, 589. [CrossRef]
37. Karnofsky, D.A.; Abelmann, W.H.; Craver, L.F.; Burchenal, J.H. The Use of the Nitrogen Mustards in the Palliative Treatment of Carcinoma—With Particular Reference to Bronchogenic Carcinoma. *Cancer* **1948**, *1*, 634–656. [CrossRef]
38. Plyta, M.; Patel, P.S.; Fragkos, K.C.; Kumagai, T.; Mehta, S.; Rahman, F.; Di Caro, S. Nutritional Status and Quality of Life in Hospitalised Cancer Patients Who Develop Intestinal Failure and Require Parenteral Nutrition: An Observational Study. *Nutrients* **2020**, *12*, 2357. [CrossRef] [PubMed]
39. Dai, W.; Wang, S.A.; Wang, K.; Chen, C.; Wang, J.; Chen, X.; Yan, J. Impact of Nutrition Counseling in Head and Neck Cancer Sufferers Undergoing Antineoplastic Therapy: A Randomized Controlled Pilot Study. *Curr. Oncol.* **2022**, *29*, 6947–6955. [CrossRef] [PubMed]
40. Di Giorgio, A.; Rotolo, S.; Cintoni, M.; Rinninella, E.; Pulcini, G.; Schena, C.A.; Ferracci, F.; Grassi, F.; Raoul, P.; Moroni, R.; et al. The prognostic value of skeletal muscle index on clinical and survival outcomes after cytoreduction and HIPEC for peritoneal metastases from colorectal cancer: A systematic review and meta-analysis. *Eur. J. Surg. Oncol.* **2022**, *48*, 649–656. [CrossRef]
41. Rinninella, E.; Cintoni, M.; Raoul, P.; Ponziani, F.R.; Pompili, M.; Pozzo, C.; Strippoli, A.; Bria, E.; Tortora, G.; Gasbarrini, A.; et al. Prognostic value of skeletal muscle mass during tyrosine kinase inhibitor (TKI) therapy in cancer patients: A systematic review and meta-analysis. *Intern. Emerg. Med.* **2021**, *16*, 1341–1356. [CrossRef]
42. Rinninella, E.; Fagotti, A.; Cintoni, M.; Raoul, P.; Scaletta, G.; Scambia, G.; Gasbarrini, A.; Mele, M.C. Skeletal muscle mass as a prognostic indicator of outcomes in ovarian cancer: A systematic review and meta-analysis. *Int. J. Gynecol. Cancer* **2020**, *30*, 654–663. [CrossRef]
43. Rinninella, E.; Cintoni, M.; Raoul, P.; Pozzo, C.; Strippoli, A.; Bria, E.; Tortora, G.; Gasbarrini, A.; Mele, M.C. Muscle mass, assessed at diagnosis by L3-CT scan as a prognostic marker of clinical outcomes in patients with gastric cancer: A systematic review and meta-analysis. *Clin. Nutr.* **2020**, *39*, 2045–2054. [CrossRef]

44. Rizzo, S.; Scala, I.; Robayo, A.R.; Cefalì, M.; De Dosso, S.; Cappio, S.; Xhepa, G.; Del Grande, F. Body composition as a predictor of chemotherapy-related toxicity in pancreatic cancer patients: A systematic review. *Front. Oncol.* **2022**, *12*, 974116. [CrossRef]

45. Emori, T.; Itonaga, M.; Ashida, R.; Tamura, T.; Kawaji, Y.; Hatamaru, K.; Yamashita, Y.; Shimokawa, T.; Koike, M.; Sonomura, T.; et al. Impact of sarcopenia on prediction of progression-free survival and overall survival of patients with pancreatic ductal adenocarcinoma receiving first-line gemcitabine and nab-paclitaxel chemotherapy. *Pancreatology* **2022**, *22*, 277–285. [CrossRef]

46. Bicakli, D.H.; Uslu, R.; Güney, S.C.; Coker, A. The Relationship Between Nutritional Status, Performance Status, and Survival Among Pancreatic Cancer Patients. *Nutr. Cancer* **2020**, *72*, 202–208. [CrossRef] [PubMed]

47. Dang, C.; Wang, M.; Zhu, F.; Qin, T.; Qin, R. Controlling nutritional status (CONUT) score-based nomogram to predict overall survival of patients with pancreatic cancer undergoing radical surgery. *Asian J. Surg.* **2022**, *45*, 1237–1245. [CrossRef] [PubMed]

48. de van der Schueren, M.A.E.; Laviano, A.; Blanchard, H.; Jourdan, M.; Arends, J.; Baracos, V.E. Systematic review and meta-analysis of the evidence for oral nutritional intervention on nutritional and clinical outcomes during chemo(radio)therapy: Current evidence and guidance for design of future trials. *Ann. Oncol.* **2018**, *29*, 1141–1153. [CrossRef] [PubMed]

49. Prado, C.M.; Orsso, C.E.; Pereira, S.L.; Atherton, P.J.; Deutz, N.E.P. Effects of β-hydroxy β-methylbutyrate (HMB) supplementation on muscle mass, function, and other outcomes in patients with cancer: A systematic review. *J. Cachexia Sarcopenia Muscle* **2022**, *13*, 1623–1641. [CrossRef] [PubMed]

50. Rinninella, E.; Cintoni, M.; Raoul, P.; Pozzo, C.; Strippoli, A.; Bria, E.; Tortora, G.; Gasbarrini, A.; Mele, M.C. Effects of nutritional interventions on nutritional status in patients with gastric cancer: A systematic review and meta-analysis of randomized controlled trials. *Clin. Nutr. ESPEN* **2020**, *38*, 28–42. [CrossRef] [PubMed]

51. Reece, L.; Hogan, S.; Allman-Farinelli, M.; Carey, S. Oral nutrition interventions in patients undergoing gastrointestinal surgery for cancer: A systematic literature review. *Support Care Cancer* **2020**, *28*, 5673–5691. [CrossRef] [PubMed]

52. Tanaka, Y.; Shimokawa, T.; Harada, K.; Yoshida, K. Effectiveness of elemental diets to prevent oral mucositis associated with cancer therapy: A meta-analysis. *Clin. Nutr. ESPEN* **2022**, *49*, 172–180. [CrossRef]

53. Hasegawa, Y.; Ijichi, H.; Saito, K.; Ishigaki, K.; Takami, M.; Sekine, R.; Usami, S.; Nakai, Y.; Koike, K.; Kubota, N. Protein intake after the initiation of chemotherapy is an independent prognostic factor for overall survival in patients with unresectable pancreatic cancer: A prospective cohort study. *Clin. Nutr.* **2021**, *40*, 4792–4798. [CrossRef]

54. Dou, S.; Ding, H.; Jiang, W.; Li, R.; Qian, Y.; Wu, S.; Ling, Y.; Zhu, G. Effect of oral supplements on the nutritional status of nasopharyngeal carcinoma patients undergoing concurrent chemotherapy: A randomized controlled Phase II trial. *J. Cancer Res. Ther.* **2020**, *16*, 1678–1685.

55. Grupińska, J.; Budzyń, M.; Maćkowiak, K.; Brzeziński, J.J.; Kycler, W.; Leporowska, E.; Gryszczyńska, B.; Kasprzak, M.P.; Iskra, M.; Formanowicz, D. Beneficial Effects of Oral Nutritional Supplements on Body Composition and Biochemical Parameters in Women with Breast Cancer Undergoing Postoperative Chemotherapy: A Propensity Score Matching Analysis. *Nutrients* **2021**, *13*, 3549. [CrossRef]

56. Vujasinovic, M.; Valente, R.; Del Chiaro, M.; Permert, J.; Löhr, J.M. Pancreatic exocrine insufficiency in pancreatic cancer. *Nutrients* **2017**, *9*, 183. [CrossRef] [PubMed]

57. Iglesia, D.; Avci, B.; Kiriukova, M.; Panic, N.; Bozhychko, M.; Sandru, V.; de-Madaria, E.; Capurso, G. Pancreatic exocrine insufficiency and pancreatic enzyme replacement therapy in patients with advanced pancreatic cancer: A systematic review and meta-analysis. *United Eur. Gastroenterol. J.* **2020**, *8*, 1115–1125. [CrossRef] [PubMed]

58. Pezzilli, R.; Caccialanza, R.; Capurso, G.; Brunetti, O.; Milella, M.; Falconi, M. Pancreatic Enzyme Replacement Therapy in Pancreatic Cancer. *Cancers* **2020**, *12*, 275. [CrossRef]

59. Domínguez-Muñoz, J.E.; Nieto-Garcia, L.; López-Díaz, J.; Lariño-Noia, J.; Abdulkader, I.; Iglesias-Garcia, J. Impact of the treatment of pancreatic exocrine insufficiency on survival of patients with unresectable pancreatic cancer: A retrospective analysis. *BMC Cancer* **2018**, *18*, 534. [CrossRef]

60. Layer, P.; Kashirskaya, N.; Gubergrits, N. Contribution of pancreatic enzyme replacement therapy to survival and quality of life in patients with pancreatic exocrine insufficiency. *World J. Gastroenterol.* **2019**, *25*, 2430–2441. [CrossRef] [PubMed]

Article

Effects on Serum Hormone Concentrations after a Dietary Phytoestrogen Intervention in Patients with Prostate Cancer: A Randomized Controlled Trial

Rebecca Ahlin [1], Natalja P. Nørskov [2], Sanna Nybacka [3], Rikard Landberg [4], Viktor Skokic [1,5,6], Johan Stranne [7,8], Andreas Josefsson [9,10,11], Gunnar Steineck [1] and Maria Hedelin [1,12,*]

1   Department of Oncology, Division of Clinical Cancer Epidemiology, Institute of Clinical Sciences, Sahlgrenska Academy, University of Gothenburg, 40530 Gothenburg, Sweden; rebecca.ahlin@gu.se (R.A.); viktor.skokic@ki.se (V.S.); gunnar.steineck@oncology.gu.se (G.S.)
2   Department of Animal and Veterinary Sciences, Aarhus University, AU-Foulum, 8830 Tjele, Denmark; natalja.norskov@anis.au.dk
3   Department of Molecular and Clinical Medicine, Institute of Medicine, Sahlgrenska Academy, University of Gothenburg, 41345 Gothenburg, Sweden; sanna.nybacka@gu.se
4   Department of Life Sciences, Division of Food and Nutrition Science, Chalmers University of Technology, 41296 Gothenburg, Sweden; rikard.landberg@chalmers.se
5   Department of Molecular Medicine and Surgery, Karolinska Institute, 17176 Stockholm, Sweden
6   Department of Pelvic Cancer, Karolinska University Hospital, 17176 Stockholm, Sweden
7   Department of Urology, Institute of Clinical Sciences, Sahlgrenska Academy, University of Gothenburg, 40530 Gothenburg, Sweden; johan.stranne@vgregion.se
8   Region Västra Götaland, Department of Urology, Sahlgrenska University Hospital, 41345 Gothenburg, Sweden
9   Sahlgrenska Cancer Center, Department of Urology, Institute of Clinical Sciences, Sahlgrenska Academy, University of Gothenburg, 41345 Gothenburg, Sweden; andreas.josefsson@urology.gu.se
10 Wallenberg Center for Molecular Medicine, Umeå University, 90187 Umeå, Sweden
11 Department of Urology and Andrology, Institute of Surgery and Perioperative Sciences, Umeå University, 90187 Umeå, Sweden
12 Regional Cancer Center West, Sahlgrenska University Hospital, Region Västra Götaland, 41345 Gothenburg, Sweden
*   Correspondence: maria.hedelin@oncology.gu.se or maria.hedelin@gu.se; Tel.: +46-70-7401858

Citation: Ahlin, R.; Nørskov, N.P.; Nybacka, S.; Landberg, R.; Skokic, V.; Stranne, J.; Josefsson, A.; Steineck, G.; Hedelin, M. Effects on Serum Hormone Concentrations after a Dietary Phytoestrogen Intervention in Patients with Prostate Cancer: A Randomized Controlled Trial. *Nutrients* **2023**, *15*, 1792. https://doi.org/10.3390/nu15071792

Academic Editors: Fernando Mendes, Diana Martins and Nuno Borges

Received: 10 March 2023
Revised: 29 March 2023
Accepted: 4 April 2023
Published: 6 April 2023

Abstract: Phytoestrogens have been suggested to have an anti-proliferative role in prostate cancer, potentially by acting through estrogen receptor beta (ERβ) and modulating several hormones. We primarily aimed to investigate the effect of a phytoestrogen intervention on hormone concentrations in blood depending on the ERβ genotype. Patients with low and intermediate-risk prostate cancer, scheduled for radical prostatectomy, were randomized to an intervention group provided with soybeans and flaxseeds (~200 mg phytoestrogens/d) added to their diet until their surgery, or a control group that was not provided with any food items. Both groups received official dietary recommendations. Blood samples were collected at baseline and endpoint and blood concentrations of different hormones and phytoestrogens were analyzed. The phytoestrogen-rich diet did not affect serum concentrations of testosterone, insulin-like growth factor 1, or sex hormone-binding globulin (SHBG). However, we found a trend of decreased risk of increased serum concentration of estradiol in the intervention group compared to the control group but only in a specific genotype of ERβ ($p = 0.058$). In conclusion, a high daily intake of phytoestrogen-rich foods has no major effect on hormone concentrations but may lower the concentration of estradiol in patients with prostate cancer with a specific genetic upset of ERβ.

Keywords: prostate cancer; phytoestrogens; isoflavones; lignans; testosterone; estradiol; sex hormone-binding globulin; insulin-like growth factor 1

---

1. Introduction

The natural cause of prostate cancer is varying and, in some ways, poorly understood, and several hormones are believed to play a role in the prostate and the development of the disease. Firstly, androgens, especially testosterone, have been shown to play a vital function in the prostate [1]. Testosterone is converted to dihydrotestosterone (DHT) in the prostate, and transcriptional activity can be exerted by DHT binding to the androgen receptor, which is important in the progression of prostate cancer [2]. Secondly, sex hormone-binding globulin (SHBG) is a transport glycoprotein for steroid hormones with the highest affinity for androgens in the prostate [3]. Thirdly, insulin-like growth factor 1 (IGF-1) regulates the growth and development of several tissues in the body, including the prostate [4]. Nevertheless, studies have found conflicting results regarding the association between serum concentrations of testosterone [5], SHBG [5,6], and IGF-1 [4] and the development of prostate cancer. Lastly, estrogen receptor alpha and estrogen receptor beta (ERβ) have been associated with proliferative and anti-proliferative effects in prostate cancer, respectively [7].

Phytoestrogens are plant compounds with structural similarities to estrogens, especially estradiol, that can both induce or inhibit estrogenic effects due to their high binding affinity to ERβ [8,9]. By binding to the ERβ, phytoestrogens may increase prostate cancer differentiation [10,11], not only directly, but also by downregulating the androgen receptor and thus androgen-driven proliferation. Phytoestrogens are divided into three main classes: isoflavones (e.g., daidzein, genistein, glycitein), which can be found in soybeans; lignans (e.g., secoisolariciresinol, lariciresinol), which can be found in flaxseeds; and coumestans (e.g., coumestrol), which can be found in bean sprouts [8,12]. Isoflavones and lignans are metabolized by the gut microbiota [12]. Equol is formed from daidzein and secoisolariciresinol, and plant lignans are converted to the mammalian lignan enterodiol, which is subsequently transformed into enterolactone by the gut microbiota. The metabolism of phytoestrogens may depend on factors impacting the gut microbiota, e.g., the intake of antibiotics [13].

An increased intake of phytoestrogens has been associated with a decreased incidence of prostate cancer in some studies [14]. In patients with prostate cancer, several studies suggest an association between an increased intake of phytoestrogens and potentially positive effects in terms of, e.g., reduced proliferation markers [15–18]. Some studies have also found effects of phytoestrogens on hormone blood concentrations, such as testosterone, estradiol, and IGF-1 [18–20]. However, the results are heterogeneous, and the scientific evidence is insufficient to advise patients with prostate cancer to increase their intake of phytoestrogens [21]. In our previous case-control study, we observed that a high intake of phytoestrogens reduced the risk of prostate cancer in men with a specific polymorphic variation (TC/CC carriers) in the promoter region of ERβ [22]. These findings prompted us to investigate the hypothesis that this genotype of ERβ had a favorable effect when patients with prostate cancer increased their intake of phytoestrogens and the potential mechanism of hormones in this. Here, we primarily investigated the effect of a diet rich in phytoestrogens on hormone concentrations in blood depending on the genotype of ERβ. Secondarily, we investigated concentrations of phytoestrogens in the blood and the relationships between phytoestrogen and hormone concentrations in blood.

2. Materials and Methods

2.1. Study Population and Study Design

The design of the PRODICA (the impact of DIet and individual genetic factors on tumor proliferation rate in men with PROstate CAncer) study and the randomization process have been described in detail elsewhere [23]. Men diagnosed with prostate cancer cT1–cT2 (prostate-specific antigen (PSA) < 20, International Society of Urological Pathology (ISUP) grade < 4) and scheduled for radical prostatectomy were invited to participate in the study at the Department of Urology at Sahlgrenska University Hospital in Gothenburg, Sweden. Patients with ongoing hormone therapy, physical or psychiatric disorders, cognitive dysfunction, and allergy to the intervention foods were not included in the study.

During the inclusion meeting, the participants met a dietitian from the study administration, filled out a questionnaire, height and weight were measured, and blood samples were collected before they were randomized to an intervention or a control group (Figure 1). The study dietitian used an envelope containing folded notes, half for the intervention group and the other half for the control group, to randomize the participants. Within seven days before the time of surgery, blood samples were collected, and the participants were instructed to fill out a similar questionnaire in proximity to the time of surgery (preferably 1 to 2 days before). The study period aimed to be at least 6 weeks, but for some patients the surgery was scheduled earlier. Nevertheless, patients with at least two weeks to scheduled surgery were included in the study. However, if already included patients had surgery within two weeks after the inclusion, they were not excluded from the study. The patients were recruited between 1 February 2016, and 12 October 2022, and the last blood sample was collected in November 2022. In the PRODICA study, tumor proliferation is the primary outcome and hormone concentrations are predeclared secondary outcomes [23]. The study was registered at ClinicalTrials.gov (NCT02759380, https://clinicaltrials.gov/ct2/show/NCT02759380?cond=NCT02759380&draw=2&rank=1; accessed on 26 March 2023) on 3 May 2016 after our pilot study (n = 10) was finished. Except for some administrative changes, we made no major changes to the study protocol after the pilot study [23]. The study was approved by the Ethical Review Board in Gothenburg (registration number 410–14, amendment numbers T124-15; 2020-02471; 2021-03320, 2021-05878-02).

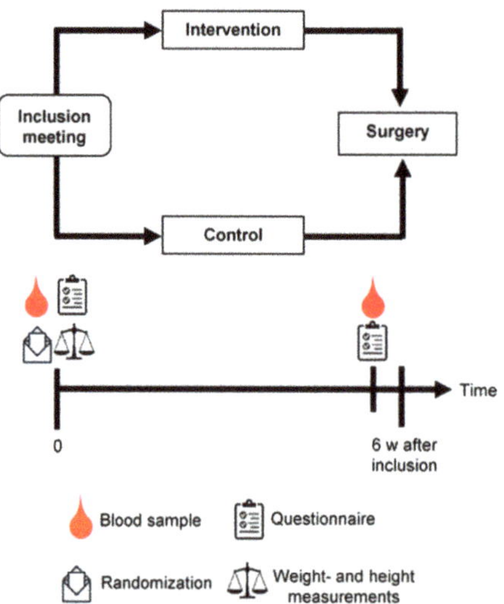

Figure 1. Design of the PRODICA (the impact of DIet and individual genetic factors on tumor proliferation rate in men with PROstate CAncer) study. During the inclusion meeting, participants were randomized to an intervention or a control group, filled out a questionnaire, and blood samples were collected. A similar questionnaire was filled out and blood samples were collected again within seven days before the time of surgery. The intervention was intended to last approximately 6 weeks.

2.2. Intervention and Control Diets

Both groups received written dietary recommendations based on the national dietary guidelines issued by the Swedish National Food Agency [24]. The dietitian went through the guidelines orally with the participants at the inclusion meeting. The participants were instructed to avoid dietary supplements, but no other dietary restrictions were given. During the inclusion meeting, the participants in the intervention group were provided with the amounts of soybeans and flaxseeds that were planned to suffice until the scheduled surgery. The participants received a schedule on the amounts of the intervention foods to eat [23], serving suggestions, and recipes. Intake of the food items was gradually increased during the first nine days and thereafter included a daily intake of 28 g flaxseeds, 47 g green soybeans, and 28 g roasted yellow soybeans (corresponding to an estimated amount of 100 mg isoflavones and 100 mg of lignans and thus 200 mg of phytoestrogens [25]). Participants randomized to the intervention groups among the first 18 participants received crushed flaxseeds, but thereafter participants received whole flaxseeds instead due to the content of cyanogenic glycosides in flaxseeds [26], as explained in detail elsewhere [23]. Both groups were aware of which group they were allocated to, but the control group did not know what the intervention diet consisted of.

2.3. Blood Samples

Blood samples were collected and handled according to standard procedures [23] and were thereafter stored at $-80\ ^{\circ}$C before being sent for genotyping analysis (whole blood), analysis of hormones (serum), and analysis of phytoestrogens (plasma). Analysis and selection of single nucleotide polymorphisms of the ERβ gene were performed in whole blood to assign each participant to the genotype of either TT, TC, or CC, as described elsewhere [22,23]. Plasma concentrations of phytoestrogens were analyzed at Aarhus University (Aarhus, Denmark) using LC-MS/MS measurements performed on a microLC 200 series (Eksigent/AB Sciex, Redwood City, CA, USA) and QTrap 5500 mass spectrometer (AB Sciex, Framingham, MA, USA) [27,28] with a coefficient of variation (CV) varying between 4.6% and 8.6% depending on the analyte. Quality control samples were used to calculate intra- and inter-batch CV. The chemical structures of the analyzed phytoestrogens are shown in Figure 2. The concentrations of estradiol, testosterone, SHBG, and IGF-1 were analyzed using the serum samples at the Department of Clinical Chemistry (Halland Hospital in Halmstad and Varberg, Sweden) according to their standard clinical protocol, described elsewhere [23]. However, the standard protocol for IGF-1 changed during the study period, and the first 104 participants were analyzed using sandwich enzyme-linked immunosorbent assay (ELISA), and the rest of the participants by sandwich assay on a Cobas 8000 (Hitachi High-Tech Corporation, Tokyo, Japan) analyzer series (reagent: Roche Diagnostics GmbH, Mannheim, Germany). All executors of the analyses received coded samples and were blinded to whether the samples belonged to the intervention or the control groups.

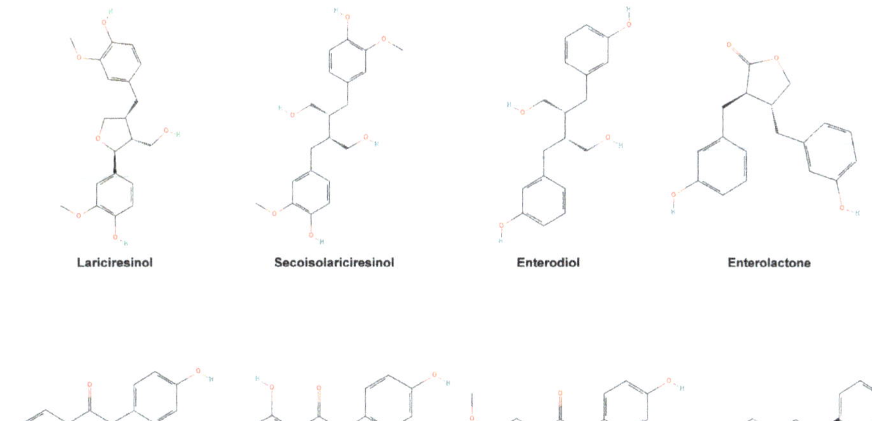

Figure 2. Chemical structures of the analyzed phytoestrogens in the study. Collected with permission from PubChem, URL: pubchem.ncbi.nlm.nih.gov [29].

2.4. Statistical Analysis

Stata/SE version 17.0 (StataCorp LLC, College Station, TX, USA) was used for statistical analysis. Analyses of demographics, serum concentrations of hormones, and plasma concentrations of phytoestrogens were stratified according to the genotype of ERβ, and differences were tested between the genotypes within the intervention and control groups and between the intervention and the control groups with the same genotype. An independent t-test was used to test differences in normally distributed data and the Mann–Whitney U test or the Kruskal–Wallis test for non-normally distributed data. The Shapiro–Wilk test was used for guidance to test if the data were normally distributed. All analyses included only participants with data available from both baseline and endpoint blood samples.

To investigate changes in hormone concentrations between baseline and endpoint, we dichotomized hormone changes into increased concentrations (1) and unchanged or decreased concentrations (0) between baseline and endpoint. Then, the dichotomized variables were used as outcomes in a generalized linear model providing estimates of the risk difference (RDs) and corresponding 95% confidence intervals (CIs) of the difference between the intervention and control groups. These analyses were stratified by ERβ genotype and adjusted for body mass index (BMI), age, and smoking. Additive interactions between the ERβ genotype and intake of phytoestrogens on increased hormone concentrations were tested.

A linear regression was used to investigate the relationships between plasma concentrations of phytoestrogens (explanatory variables) and serum concentrations of hormones (outcomes). The regression model was stratified according to the intervention and control groups and adjusted for body mass index (BMI), age, and smoking. Due to skewed data, the hormone and phytoestrogen concentrations were logarithmized in the linear regression using the natural logarithm.

In the analysis of plasma concentrations of phytoestrogens, we compared users and non-users (including participants who reported "do not know") of antibiotics over the last five years and different intervention lengths (<28, 28–56, >56 days). In a subgroup analysis, the concentrations of different lignans were stratified in participants receiving crushed and whole flaxseed. For considered confounding variables, BMI was categorized as underweight (<18.5 kg/m^2), normal weight (18.5–24.9 kg/m^2), overweight (25–29.9 kg/m^2), obese class 1 (30–34.9 kg/m^2), and obese class ≥ 2 (≥35 kg/m^2) [30]. Age was categorized in ≥median of the study population and <median of the study population. Intake of antibiotics was categorized as 0 (non-users) if the participants reported in the questionnaire that they did not know or had no intake of antibiotics during the intervention and the recent five years. A reported intake of antibiotics, at least once during the intervention or the recent five years, was categorized as 1 (users). Smoking was categorized as current smoker (1) and nonsmoker (0). If a participant had quit smoking ≤5 years ago he was categorized as a current smoker, and if he quit smoking >5 years ago he was categorized as a nonsmoker.

3. Results

3.1. Population and Baseline Characteristics

Of 195 invited men, 55 patients declined to participate, with the main reasons being occupied or unwillingness to participate in the inclusion meeting (mainly due to long travel times) (Figure 3). In total, 140 participants were randomized to either the intervention ($n = 71$) or the control ($n = 69$) groups. Of these, 135 participants completed the blood sample at endpoint (intervention $n = 68$, control $n = 67$). Five participants (intervention $n = 3$, control $n = 2$) did not complete the intervention; two participants in the intervention group experienced gastrointestinal problems from the intervention foods and the participants in the control group did not state a reason. In total, seven participants in the intervention group reported gastrointestinal symptoms, and of those, five completed the intervention. Other adverse effects of the intervention foods were reported by three participants and were of different kinds.

The participants' median age was 66 years (IQR 11; range 40–76) and the median study period was 47 days (IQR 33, range 7–583; Table 1). At the time of diagnosis, most participants had an ISUP grade of 2 and a tumor stage of T1c. At baseline, participants in the intervention group had a higher BMI, a lower level of physical activity, and a higher tumor stage, and there were less users of antibiotics in the recent years compared to participants in the control group (Table 1).

Table 1. Demographics of the patients included in the PRODICA [1] study.

	Intervention (*n* = 68)				Control (*n* = 67)			
	Genotype TT [2] (*n* = 35)		Genotype TC/CC [2] (*n* = 33)		Genotype TT [2] (*n* = 26)		Genotype TC/CC [2] (*n* = 41)	
	Median (IQR)	Range	Median (IQR)	Range	Median (IQR)	Range	Median (IQR)	Range
Age, years	65 (13)	51–76	67 (8)	43–76	66 (10)	51–74	65 (10)	40–75
Intervention period, d	47 (46)	12–189	48 (28)	7–146	46 (27)	8–213	47 (29)	14–583
BMI, kg/m^2	27.8 (5.2)	21.7–37.4	28.1 (4.7)	21.3–35.1	26.0 (3.9)	20.0–40.0	25.5 (3.8)	20.6–33.3
	n(%)		*n*(%)		*n*(%)		*n*(%)	
Tumor stage at diagnosis								
cT1	20 (57)		20 (61)		14 (54)		31 (76)	
cT2	15 (43)		12 (36)		10 (38)		10 (24)	
cTX	0 (0)		1 (3)		2 (8)		0 (0)	
ISUP grade at diagnosis								
1	11 (31)		14 (42)		10 (38)		19 (46)	
2	19 (54)		15 (45)		11 (42)		19 (46)	
3	5 (14)		4 (12)		5 (19)		3 (7)	
Physical activity [3]								
Low	6 (17)		8 (24)		2 (8)		4 (10)	
Moderate	19 (54)		16 (48)		17 (65)		19 (46)	
High	10 (29)		9 (27)		7 (27)		18 (44)	
Heredity								
Yes	13 (37)		12 (36)		7 (27)		14 (34)	
No	9 (26)		10 (30)		4 (15)		14 (34)	
Do not know	13 (37)		11 (33)		15 (58)		13 (32)	
Antibiotic treatment last year								
Yes	12 (34)		7 (21)		11 (42)		11 (27)	
No	22 (63)		25 (76)		15 (58)		29 (71)	
Do not know	1 (3)		1 (3)		0 (0)		1 (2)	
Antibiotic treatment last 2–5 years								
Yes	14 (40)		10 (30)		13 (50)		22 (54)	
No	19 (54)		17 (52)		12 (46)		14 (34)	
Do not know	2 (6)		6 (18)		1 (4)		5 (12)	
Antibiotic treatment during the intervention, *n* (%)								
Yes	1 (3)		3 (9)		3 (12)		4 (10)	
No	34 (97)		30 (91)		23 (88)		36 (88)	
Missing, *n* (%)	0 (0)		0 (0)		0 (0)		1 (2)	

Table 1. *Cont.*

	Intervention (*n* = 68)				Control (*n* = 67)			
	Genotype TT [2] (*n* = 35)		Genotype TC/CC [2] (*n* = 33)		Genotype TT [2] (*n* = 26)		Genotype TC/CC [2] (*n* = 41)	
	Median (IQR)	Range	Median (IQR)	Range	Median (IQR)	Range	Median (IQR)	Range
Smoking								
Currently	2 (6)		1 (3)		1 (4)		2 (5)	
Previously	16 (46)		17 (52)		12 (46)		21 (51)	
Never	17 (49)		15 (45)		13 (50)		18 (44)	

[1] The impact of DIet and individual genetic factors on tumor proliferation rate in men with PROstate CAncer. [2] Participants were assigned to the genotype TT, TC, or CC of the estrogen receptor beta. [3] Activity in the daytime: sedentary (100 p); partly sedentary, sitting, and walking (200 p); mostly standing and walking (300 p), physical labor (400 p). Physical activity in the evening time: sedentary (1 p), slight activity—equal to a 30-min walk (2 p); moderately strenuous activity—equal to a bike ride of $\geq$30 min (3 p); sports activity (4 p). Low physical activity: 101–103, 201 p; moderate physical activity: 104, 202–203, 301–302 p; high physical activity: 204, 303–304, 401–404 p. Abbreviations: ISUP, International Society of Urological Pathology.

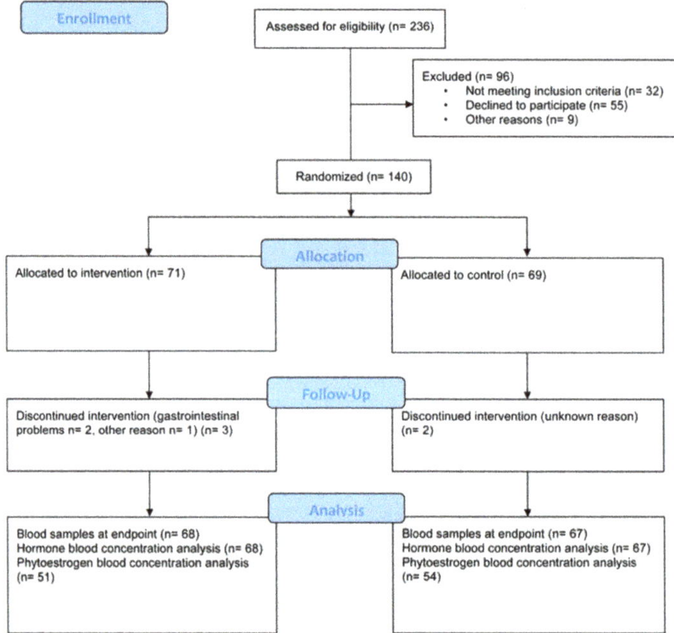

Figure 3. Flowchart of the PRODICA study.

3.2. Effects of the Intervention Diet on Hormone Concentrations

Besides higher concentrations of estradiol at baseline in the intervention group compared to the control group in participants with the TT genotype of ERβ, there were no statistically significant differences in hormone serum concentrations between the intervention and control groups at any time points (Table 2). Within the intervention group, we found a decreased concentration of SHBG in participants with the TC/CC genotype in comparison with participants with the TT genotype who increased their concentrations. We found no effect of the intervention diet on the risk of increasing different hormone serum concentrations between baseline and endpoint, except for estradiol. There was a trend of decreased risk of increased serum concentration of estradiol in the intervention group compared to the control group but only in participants with the TC/CC genotype (RD −22%, p = 0.058, Table 3).

Table 2. Serum concentrations of hormones in the intervention and control groups in patients with prostate cancer, stratified by the genotype of estrogen receptor beta.

| | Intervention (n = 68) | | | | | Control (n = 67) | | | | | p [1] | |
	Genotype TT [2] (n = 35)		Genotype TC/CC [2] (n = 33)			Genotype TT [2] (n = 26)		Genotype TC/CC [2] (n = 41)			TT	TC/CC
	Median (IQR)	Range	Median (IQR)	Range	p [3]	Median (IQR)	Range	Median (IQR)	Range	p [3]		
Testosterone (nmol/L)												
Baseline	15.3 (6.6)	9.6–30.1	15.0 (6.4)	5.1–26.2	0.710 [5]	15.6 (6.3)	8.0–22.7	14.9 (7.5)	8.4–32.6	0.889	0.521 [5]	0.699 [5]
Endpoint	14.8 (4.9)	6.9–22.6	14.2 (6.4)	5.1–30.0	0.566 [5]	14.4 (6.7)	7.3–26.0	15.8 (8.8)	5.4–30.5	0.331 [5]	0.483 [5]	0.362 [5]
Change [4]	−0.5 (3.7)	−10.1–13.0	0.2 (4.8)	−9.0–9.0	0.854 [5]	−0.8 (2.0)	−8.1–9.6	−0.6 (3.8)	−7.7–18.3	0.648	0.814	0.865
Estradiol (nmol/L)												
Baseline	0.099 (0.034)	0.069–0.18	0.10 (0.029)	0.040–0.18	0.273	0.090 (0.040)	0.049–0.15	0.092 (0.040)	0.030–0.32	0.500	0.0137 [5]	0.546
Endpoint	0.10 (0.048)	0.045–0.18	0.10 (0.034)	0.048–0.16	0.973 [5]	0.088 (0.035)	0.023–0.15	0.10 (0.061)	0.033–0.35	0.149	0.0625 [5]	0.727
Change [4]	−0.0050 (0.027)	−0.050–0.055	0.0 (0.026)	−0.038–0.048	0.0725	−0.0015 (0.027)	−0.026–0.034	0.0050 (0.032)	−0.046–0.089	0.159	0.679 [5]	0.657 [5]
SHBG (nmol/L)												
Baseline	49.0 (25.0)	28.0–114.0	45.0 (25.0)	21.0–110.0	0.640	50.0 (22.0)	19.0–94.0	48.0 (32.0)	22.0–126.0	0.947	0.848	0.744
Endpoint	51.0 (22.0)	28.0–113.0	41.0 (22.0)	20.0–96.0	0.0982	50.5 (29.0)	24.0–96.0	51.0 (33.0)	20.0–112.0	0.797	0.980	0.273
Change [4]	1.0 (8.0)	−25.0–16.0	−2.0 (10.0)	−29.0–13.0	0.00390	0.0 (6.0)	−12.0–21.0	−2.0 (8.0)	−20.0–25.0	0.189	0.502	0.418
IGF−1 (μg/L)												
Baseline	126.0 (62.0)	75.0–280.0	138.0 (56.0)	60.0–276.0	0.915	166.5 (44.0)	90.0–274.0	146.0 (64.0)	68.0–381.0	0.173	0.0973	0.536
Endpoint	135.0 (57.0)	82.0–304.0	151.0 (64.0)	59.0–266.0	0.819	161.0 (81.0)	90.0–253.0	153.0 (85.0)	63.0–340.0	0.864	0.301	0.499 [5]
Change [4]	−1.0 (18.0)	−25.0–124.0	7.0 (23.0)	−66.0–41.0	0.281	−0.5 (26.0)	−65.0–54.0	7.0 (31.0)	−45.0–55.0	0.232	0.656	0.955 [5]
Testosterone/SHBG ratio												
Baseline	0.33 (0.10)	0.14–0.55	0.32 (0.080)	0.16–0.49	0.826 [5]	0.30 (0.099)	0.18–0.67	0.32 (0.12)	0.17–0.56	0.426	0.319	0.828 [5]
Endpoint	0.30 (0.12)	0.17–0.58	0.33 (0.12)	0.19–0.59	0.215 [5]	0.28 (0.11)	0.17–0.47	0.32 (0.12)	0.17–0.60	0.0705	0.380	0.773
Change [4]	0.011 (0.090)	−0.19–0.17	−0.0023 (0.088)	−0.12–0.094	0.0734 [5]	0.012 (0.068)	−0.092–0.26	0.00043 (0.067)	−0.29–0.12	0.145	0.789	0.462
Testosterone/estradiol ratio												
Baseline	134.9 (37.4)	0.09–311.1	140.5 (77.8)	0.09–348.5	0.637	166.1 (75.6)	0.1–346.2	155.0 (12.6)	0.09–380.0	0.280	0.0697	0.854
Endpoint	149.3 (69.9)	0.1–322.2	140.4 (84.0)	0.083–321.7	0.678	164.1 (102.1)	0.1–487.0	138.8 (127.1)	0.09–356.4	0.242	0.107	0.681
Change [4]	0.07 (46.4)	−70.1–68.9	−6.4 (40.7)	−95.0–54.6	0.127 [5]	−5.2 (48.4)	−106.3–225.7	−2.8 (26.1)	−116.4–108.3	0.934	0.438	0.417

[1] Difference between the intervention and control groups within the same genotype of the estrogen receptor beta. [2] Participants were assigned to the genotype of either TT, TC, or CC of the estrogen receptor beta. [3] Difference between genotypes within the intervention and control groups. [4] The median difference between endpoint and baseline. [5] An independent T-test test was used to compare groups depending on the normal distribution. The Mann–Whitney U test was used to compare groups, except when noted otherwise. Abbreviations: IGF-1, insulin-like growth factor 1; SHBG, sex hormone binding globulin.

Table 3. Risk differences (RDs) with 95% confidence intervals (CIs) for the risk of increasing different hormone concentrations between baseline and endpoint, in relation to intake of phytoestrogens, stratified by estrogen receptor beta genotype (TT or TC/CC).

Hormone Concentrations (nmol/L)		RD	95% CI	Adjusted [1] RD	Adjusted [1] 95% CI	*p* Additive Interaction
Testosterone	All cases (*n* = 135)	0.083	−0.84, 0.25	0.067	−0.10, 0.23	0.792
	TT (*n* = 61)	0.12	−0.12, 0.36	0.099	−0.14, 0.34	
	TC/CC (*n* = 74)	0.076	−0.15, 0.30	0.073	−0.16, 0.31	
Estradiol	All cases (*n* = 135)	−0.11	−0.28, 0.057	−0.13	−0.30, 0.045	0.424
	TT (*n* = 61)	−0.013	−0.26, 0.23	−0.027	−0.27, 0.22	
	TC/CC (*n* = 74)	−0.15	−0.37, 0.076	−0.22	−0.45, 0.071	
SHBG	All cases (*n* = 135)	0.038	−0.13, 0.20	0.030	−0.14, 0.20	0.149
	TT (*n* = 61)	0.15	−0.10, 0.40	0.13	−0.12, 0.37	
	TC/CC (*n* = 74)	−0.093	−0.30, 0.12	−0.11	−0.33, 0.011	
IGF-1	All cases (*n* = 135)	−0.038	−0.21, 0.13	−0.028	−0.20, 0.14	0.386
	TT (*n* = 61)	−0.090	−0.34, 0.16	−0.076	−0.31, 0.15	
	TC/CC (*n* = 74)	0.057	−0.16, 0.28	0.068	−0.16, 0.30	
Testosterone/SHBG ratio	All cases (*n* = 135)	0.0061	−0.16, 0.17	0.027	−0.14, 0.20	0.632
	TT (*n* = 61)	0.022	−0.21, 0.25	0.028	−0.20, 0.25	
	TC/CC (*n* = 74)	−0.058	−0.29, 0.17	−0.083	−0.31, 0.15	
Testosterone/estradiol ratio	All cases (*n* = 135)	0.097	−0.068, 0.26	0.081	−0.090, 0.25	0.458
	TT (*n* = 61)	0.15	−0.10, 0.40	0.13	−0.12, 0.39	
	TC/CC (*n* = 74)	0.022	−0.20, 0.24	−0.0050 [2] 0.0011 [3] 0.024 [4]	−0.21, 0.22 [2] −0.21, 0.21 [3] −0.21, 0.25 [4]	

[1] Analyses were adjusted for BMI (kg/m2) (≤18.5; 18.5 to <25; 25 to <30; 30 to <35; ≥35), age (≥median, <median), and smoking (1 = current smoker or quit smoking ≤5 years ago; 0 = nonsmoker or quit smoking >5 years ago). [2] The analysis did not converge and was therefore only adjusted for BMI and smoking. [3] The analysis did not converge and was therefore only adjusted for age and smoking. [4] The analysis did not converge and was therefore only adjusted for BMI and age. Abbreviations: IGF-1, insulin-like growth factor 1; SHBG, sex hormone binding globulin.

3.3. Plasma Concentrations of Phytoestrogens

There were no differences in the plasma concentrations of phytoestrogens at baseline between the intervention and control groups (Figure 4). The plasma concentrations of enterolactone, enterodiol, secoisolariciresinol, daidzein, genistein, glycitein, and equol were statistically significantly higher in the intervention group compared to the control group at endpoint. Participants in the intervention group increased their concentrations of these phytoestrogens during the study period compared to participants in the control group whose concentrations were maintained or reduced (Table 4). None of the participants had detectable concentrations of equol at baseline, and only ten participants (20%) in the intervention group and one participant in the control group (2%) had detectable concentrations at endpoint (Table 4).

Non-users of antibiotics had higher median values of genistein and daidzein at baseline. We found no differences in median values between users and non-users of antibiotics at endpoint or for the change between endpoint and baseline. When the change in different concentrations of phytoestrogens was compared depending on different intervention durations, no difference was found between the three different durations. Stratified analyses of participants receiving crushed and whole flaxseed showed no difference in plasma concentrations for enterodiol and enterolactone at any time point (Table S1). However, participants who received crushed flaxseeds had a higher change between baseline and endpoint in plasma concentration of secoisolariciresinol compared to those receiving whole flaxseeds.

Table 4. Plasma concentrations of phytoestrogens in the intervention and control groups in patients with prostate cancer, stratified by the genotype of estrogen receptor beta.

| Plasma Concentrations (nmol/L) | Intervention (n = 51) | | | | | Control (n = 54) | | | | | p^1 | |
| | Genotype TT [2] (n = 26) | | Genotype TC/CC [2] (n = 25) | | | Genotype TT [2] (n = 24) | | Genotype TC/CC [2] (n = 30) | | | TT | TC/CC |
	Median (IQR)	Range	Median (IQR)	Range	p^3	Median (IQR)	Range	Median (IQR)	Range	p^3		
Lariciresinol												
Baseline	4.9 (4.1)	1.8–17.0	5.2 (4.2)	1.7–20.6	0.797	5.9 (2.5)	3.1–12.2	5.7 (6.5)	2.0–17.9	0.833	0.718	0.654
Endpoint	6.5 (4.9)	2.6–12.0	6.0 (6.0)	2.4–16.4	0.374	6.1 (5.5)	1.2–18.4	4.6 (4.9)	0.0–15.6	0.419	0.473	0.463
Change [4]	1.9 (5.8)	−11.0–8.5	−0.02 (4.3)	−10.4–7.1	0.449	0.02 (5.8)	−6.7–11.8	−1.3 (5.1)	−12.4–5.3	0.138	0.481	0.164
Secoisolariciresinol												
Baseline	6.5 (7.1)	0.8–26.7	5.7 (5.5)	0.6–103.1	0.664	6.0 (8.6)	1.5–61.1	6.2 (6.9)	1.8–31.0	0.883	0.859	0.507
Endpoint	13.9 (17.9)	2.2–80.8	10.4 (9.9)	1.2–108.3	0.275	5.5 (11.0)	1.8–29.6	4.8 (4.5)	0.9–41.6	0.766	0.0094	0.0314
Change [4]	5.0 (21.4)	−17.2–78.5	4.2 (11.1)	−90.3–98.5	0.647	0.08 (6.5)	−50.1–26.8	−0.8 (5.2)	−25.4–18.9	0.506	0.0068	0.0045
Enterodiol												
Baseline	7.0 (10.2)	0.8–122.0	5.7 (15.1)	0.6–584.5	0.507	7.7 (23.0)	1.2–175.0	4.5 (6.9)	0.4–81.9	0.0932	0.683	0.694
Endpoint	25.5 (37.5)	4.3–880.3	32.0 (122.4)	1.2–1526.1	0.604	8.8 (11.6)	0.5–691.3	4.5 (9.4)	0.2–82.6	0.244	<0.001	<0.001
Change [3]	11.7 (32.6)	−87.9–876.8	17.1 (92.3)	−330.8–1501.2	0.395	−1.6 (14.6)	−114.1–683.5	−0.2 (5.7)	−52.0–77.4	0.264	0.0014	0.0014
Enterolactone												
Baseline	113.6 (123.0)	10.7–348.7	71.2 (107.5)	5.2–321.8	0.207	107.4 (144.3)	8.6–517.7	66.0 (69.0)	8.4–393.7	0.115	0.981	0.923
Endpoint	213.0 (404.8)	0.8–1348.2	175.4 (371.7)	5.3–1172.1	0.344	78.8 (123.6)	4.5–578.8	78.1 (91.0)	15.8–231.9	0.572	<0.001	0.0048
Change [4]	109.1 (364.4)	−218.2–1264.5	109.1 (270.7)	−215.0–1052.0	0.993	−29.2 (92.9)	−168.3–446.0	−22.2 (61.9)	−213.9–203.8	0.714	0.0013	<0.001
Daidzein												
Baseline	40.3 (112.1)	0.0–477.8	32.4 (53.0)	0.0–1257.5	0.426	16.4 (46.7)	0.0–349.0	33.3 (94.7)	0.0–453.0	0.776	0.372	0.792
Endpoint	579.2 (796.6)	0.0–2096.7	250.0 (819.4)	0.0–3688.4	0.384	10.2 (23.4)	0.0–153.0	32.3 (38.3)	0.0–157.7	0.124	<0.001	<0.001
Change [4]	499.1 (790.5)	−205.0–2093.2	132.3 (680.6)	−89.1–3655.4	0.472	−7.6 (33.5)	−308.9–135.7	−3.2 (66.9)	−365.9–142.5	0.554	<0.001	<0.001
Genistein												
Baseline	147.5 (325.1)	0.0–8798.1	68.1 (163.4)	0.0–2378.8	0.486	34.3 (150.6)	0.0–2335.0	43.8 (157.1)	0.0–2091.8	0.675	0.290	0.782
Endpoint	1055.5 (3141.7)	4.4–8075.1	474.9 (1378.7)	10.3–13,070.3	0.129	18.5 (127.4)	0.0–667.7	29.3 (79.9)	0.0–621.7	0.382	<0.001	<0.001
Change [4]	668.6 (3003.5)	−5060.9–8041.4	421.7 (692.8)	−322.1–12,911.8	0.438	−13.6 (144.4)	−1667.3–201.1	−14.2 (86.5)	−1470.1–438.2	0.982	<0.001	<0.001
Glycitein												
Baseline	0.0 (4.4)	0.0–29.2	0.0 (3.4)	0.0–174.3	0.853	0.0 (3.3)	0.0–29.6	0.0 (4.4)	0.0–40.0	0.905	0.834	0.678
Endpoint	20.9 (43.3)	0.0–87.9	14.4 (42.7)	0.0–287.4	0.406	0.0 (0.0)	0.0–6.1	0.0 (2.2)	0.0–13.2	0.130	<0.001	<0.001
Change [4]	17.3 (47.5)	−10.9–83.7	14.4 (40.0)	−5.0–287.4	0.604	0.0 (3.3)	−25.0–3.5	0.0 (4.1)	−40.0–13.2	0.816	<0.001	<0.001
Equol												
Baseline	0.0 (0.0)	0.0–0.0	0.0 (0.0)	0.0–0.0	1.000	0.0 (0.0)	0.0–0.0	0.0 (0.0)	0.0–0.0	1.000	1.000	1.000
Endpoint	0.0 (0.0)	0.0–1204.9	0.0 (0.0)	0.0–616.1	0.404	0.0 (0.0)	0.0–0.0	0.0 (0.0)	0.0–289.0	1.000	0.0290	0.0742
Change [4]	0.0 (0.0)	0.0–1204.9	0.0 (0.0)	0.0–616.1	0.404	0.0 (0.0)	0.0–0.0	0.0 (0.0)	0.0–289.0	1.000	0.0290	0.0742

[1] Difference between the intervention and control groups within the same genotype of the estrogen receptor beta. [2] Participants were assigned to the genotype of either TT, TC, or CC of the estrogen receptor beta. [3] Difference between genotypes of the estrogen receptor beta within the intervention- and control groups. [4] The median difference between endpoint and baseline.

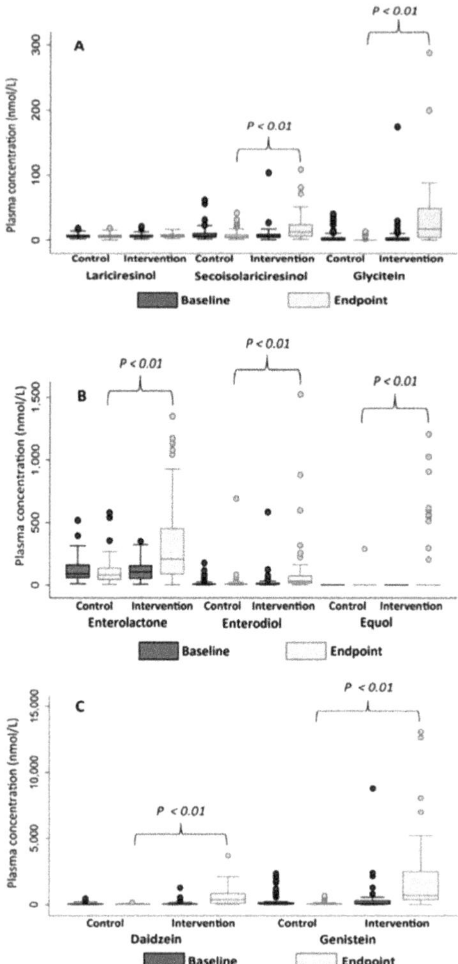

Figure 4. Boxplots showing plasma concentrations of different phytoestrogens (nmol/L) in the intervention- (n = 51) and control groups (n = 54) in patients with prostate cancer at baseline and endpoint. (**A**) Plasma concentrations of lariciresinol, secoisolariciresinol, and glycitein. (**B**) Plasma concentrations of enterolactone, enterodiol, and equol. (**C**) Plasma concentrations of daidzein and genistein. Concentrations of secoisolariciresinol, glycitein, enterolactone, enterodiol, equol, daidzein, and genistein were statistically significantly higher in the intervention group compared to the control group at the endpoint. The Mann–Whitney U test was used to test differences between groups.

3.4. The Relationship between Blood Concentrations of Phytoestrogens and Hormones

We found a relationship between higher plasma concentrations of lignans and higher serum concentrations of SHBG, but it did not remain statistically significant after adjusting for confounders. A 10% increase in plasma concentrations of lignans was associated with a 55% increase in serum concentrations of SHBG (p = 0.11; Table 5).

Table 5. Linear regression analyses between plasma concentrations of phytoestrogens (explanatory variables) and serum concentrations of hormones (outcomes) in patients with prostate cancer (n = 105).

Hormone Concentrations (nmol/L)		Plasma Concentrations of Lignans [1] (nmol/L) β (95% CI)	Plasma Concentrations of Isoflavones [2] (nmol/L) β (95% CI)	Plasma Concentrations of Phytoestrogens [3] (nmol/L) β (95% CI)
Testosterone	Unadjusted	0.035 (−0.028, 0.099)	0.0071 (−0.024, 0.039) [4]	0.013 (−0.034, 0.059)
	Adjusted [5]	0.029 (−0.034, 0.092)	0.0099 (−0.021, 0.041) [4]	0.014 (−0.031, 0.059)
Estradiol	Unadjusted	0.019 (−0.051, 0.089)	−0.012 (−0.046, 0.022) [4]	−0.013 (−0.064, 0.037)
	Adjusted [5]	0.026 (−0.044, 0.097)	−0.015 (−0.050, 0.019) [4]	−0.014 (−0.65, 0.036)
SHBG	Unadjusted	0.071 (0.0013, 0.14)	−0.0090 (−0.044, 0.026) [4]	0.011 (−0.040, 0.062)
	Adjusted [5]	0.055 (−0.013, 0.12)	−0.0048 (−0.038, 0.029) [4]	0.010 (−0.039, 0.059)
IGF-1	Unadjusted	0.0065 (−0.049, 0.062)	0.012 (−0.015, 0.038) [4]	0.0085 (−0.031, 0.048)
	Adjusted [5]	0.015 (−0.041, 0.071)	0.0095 (−0.017, 0.036) [4]	0.0093 (−0.031, 0.049)
Testosterone/SHBG ratio	Unadjusted	−0.036 (−0.85, 0.014)	0.016 (−0.0082, 0.040) [4]	0.0017 (−0.034, 0.038)
	Adjusted [5]	−0.026 (−0.075, 0.023)	0.015 (−0.0093, 0.039) [4]	0.0038 (−0.031, 0.039)
Testosterone/estradiol ratio	Unadjusted	0.016 (−0.057, 0.089)	0.019 (−0.016, 0.055) [4]	0.026 (−0.026, 0.078)
	Adjusted [5]	0.0031 (−0.064, 0.070)	0.025 (−0.0071, 0.058) [4]	0.028 (−0.019, 0.076)

[1] Include lariciresinol, secoisolariciresinol enterolactone, and enterodiol. [2] Include daidzein, genistein, glycitein, and equol. [3] Include isoflavones and lignans. [4] One participant is missing. [5] Analyses were adjusted for BMI (kg/m2) (≤18.5; 18.5 to <25; 25 to <30; 30 to <35; ≥35), age (≥median, <median), and smoking (1 = current smoker or quit smoking ≤5 years ago; 0 = nonsmoker or quit smoking >5 years ago). Samples of plasma and serum were collected at endpoint. Hormone concentrations and phytoestrogen concentrations were logarithmized using the natural logarithm. Abbreviations: β, beta-coefficient; IGF-1, insulin-like growth factor 1; SHBG, sex hormone binding globulin.

4. Discussion

In this randomized controlled dietary intervention study of patients with prostate cancer, an increased intake of phytoestrogens did not affect the serum of concentrations of testosterone, SHBG, and IGF-1. However, a trend of decreased risk of increased concentration of estradiol was found in participants with the TC/CC genotype of ERβ. In the intervention group, participants with the TC/CC genotype decreased serum concentrations of SHBG during the intervention compared to participants with the TT genotype, who increased their concentrations.

In one of the genotype groups of ERβ, we found a trend of a decreased risk of increasing estradiol concentration, comparing the intervention and control groups. This is in contrast to another study that demonstrated an increase in plasma concentrations of phytoestrogens and increased concentrations of serum estradiol, although the blood concentrations of phytoestrogens were lower than in our study [19]. Hamilton-Reeves et al. found higher urinary excretion of estradiol and 2-hydroxy estrogens to 16α-hydroxyestrone (2:16 OH-E1) ratio with an isoflavone supplement compared to a control group [31]. A higher 2:16 OH-E1 ratio has been associated with a reduced risk of prostate cancer [32]. However, 2:16 OH-E1 ratio was not analyzed in our study. A potential mechanism of our result is that the increased intake of phytoestrogens resulted in negative feedback on estradiol. To our knowledge, there is no study confirming this mechanism.

We did not find any effect of the phytoestrogen intervention on serum concentrations of testosterone, IGF-1, and SHBG. This is in line with several other studies finding no effect on blood concentrations of testosterone, IGF-1, and SHBG [15–17]. Other investigations challenge these results with favorable effects on testosterone [18,19] and IGF-1 in African American men [20]. These conflicting results can depend on the different doses and sources of phytoestrogens used in the studies, varying effect-modifying factors, short duration of interventions, and small sample sizes [21].

In the intervention group, the two genotype groups of ERβ had opposites effects on SHBG concentration, participants with the genotype of TC/CC decreased their concentrations during the intervention, and participants with TT genotype increased their concentrations. This suggests that the genotype of ERβ may affect the serum concentrations of hormones. To our knowledge, this is the first study to investigate the effect of phytoestrogens on serum hormones depending on the genotype of ERβ. ERβ has an important role in, e.g., hormonal and protein regulation and transcription, and thus the genotype of ERβ could play a role in these different responses [11]. Lee et al. found that postoperative biochemical recurrence-free survival was worse for patients with higher SHBG concentrations [33], suggesting a beneficial effect of the TC/CC genotype in prostate cancer. Nevertheless, the impact of the genotype of ERβ needs to be further examined in future studies.

We found that the intervention diet increased the plasma concentrations of several phytoestrogens. Elevated plasma phytoestrogen concentrations have been confirmed in similar intervention studies [19,34]. Daidzein and genistein are the main isoflavones in soybeans [12], which explains why these compounds had the highest concentrations in participants in the intervention group at endpoint. For isoflavones, few participants in our study had detectable measures of equol. Previous studies have observed a higher proportion (35%) of equol producers than ours in a general Caucasian population [35,36]. Our population's reduced ability to metabolize equol may be caused by the fact that equol or the ability to produce equol could be related to the development of prostate cancer [37]. We noticed that the intake of antibiotics affected some phytoestrogen concentrations at baseline. This was probably because the intake of antibiotics impacted the intestinal microbiota and negatively affected the phytoestrogen metabolism [13].

The strengths of this study include the randomized design and the fact that blood concentrations of phytoestrogens were measured. The results are also based on a large clinical study with a low dropout rate. Even if the number of dropouts would be large enough to affect the results, we do not expect the results of the participants who dropped out from the study to differ from the rest of the study population. A limitation of the present study is the change from crushed to whole flaxseeds, which probably decreased the absorption of lignans [38]. This is confirmed by the higher change in concentrations of secoisolariciresinol in participants who received crushed flaxseeds compared to those who received whole flaxseeds. Another limitation is the wide range (1 to 83 weeks) of the duration of the intervention depending on when the surgery was scheduled, as both very short and very long intervention durations could have influenced the blood concentrations of phytoestrogens. However, we did not find any difference in plasma concentrations when we stratified the analysis after the intervention duration. Previous research observed half-lives of 2–11 h in plant lignans and isoflavones and longer half-lives in their mammalian conversion products, likely because of continuous transformation by the gut microbiota [39,40]. Moreover, the gut microbiota is important for the formation of enterolactone, enterodiol, and equol, but we did not collect any stool samples or information on additional factors influencing the metabolism of phytoestrogens (e.g., diseases or drugs influencing gut microbiota and intake of prebiotics and probiotics) [12,41]. For the generalizability of the result, we lack data on patients with high-grade prostate cancer, men without prostate cancer, and women.

In conclusion, our findings suggest that a high intake of phytoestrogens may lower the concentration of estradiol in patients with prostate cancer with a specific genetic upset of ERβ but does not affect serum concentrations of testosterone, IGF-1, and SHBG. However, the effect on SHBG concentration differed across the ERβ genotype groups. The effect of the genotype of ERβ on hormone concentrations in patients with prostate cancer should be confirmed in future studies. Further research is needed to investigate whether elevated plasma concentrations of phytoestrogens have a beneficial effect in terms of reduced tumor proliferation and prolonged survival for patients with prostate cancer.

Supplementary Materials: The following supporting information can be downloaded at: https://www.mdpi.com/article/10.3390/nu15071792/s1, Table S1: Plasma concentrations of phytoestrogens in the intervention group receiving crushed or whole flaxseeds in the PRODICA study.

Author Contributions: Conceptualization, G.S. and M.H.; methodology, G.S. and M.H.; project administration, S.N.; validation, S.N.; formal analysis, R.A. and V.S.; investigation, R.A.; resources, G.S. and N.P.N.; data curation, R.A.; writing—original draft preparation, R.A.; writing—review and editing, A.J., G.S., J.S., N.P.N., M.H., R.A., R.L., S.N. and V.S.; visualization, A.J., G.S., J.S., N.P.N., M.H., R.A., R.L., S.N. and V.S.; supervision, G.S., M.H. and R.L.; project administration, M.H. and R.A.; funding acquisition, A.J., G.S. and M.H. All authors have read and agreed to the published version of the manuscript.

Funding: This research was funded by the LUA/ALF agreement in the West of Sweden health care region (grant number ALFGBG-727821, ALFGBG-966360), King Gustav V Jubilee Clinic Research Foundation at Sahlgrenska University Hospital (grant number 2021:353, 2020:320, 2019:262, 2018:206, 2017:145, 2016:76), Dr. P. Håkansson's Foundation, Eslöv, Sweden (grant number Maria Hedelin 2014), Patientföreningen för prostatacancer ProLiv Väst, Lions Cancerfond Väst (grant number 2013;10), Knut and Alice Wallenberg Foundation (grant number KAW 2015.0114).

Institutional Review Board Statement: The study was conducted in accordance with the Declaration of Helsinki and approved by the Ethical Review Board in Gothenburg (protocol code 410-14, amendment codes T124-15, 2020-02471, 2021-03320, and 2021-05878-02; date of approval, 30 June 2014).

Informed Consent Statement: Informed consent was obtained from all subjects involved in the study.

Data Availability Statement: The data presented in this study are available on request from the corresponding author. The data are not publicly available due to analyses that are ongoing for future publications.

Acknowledgments: We want to thank all participants that contributed to this study. The staff at the Department of Urology at Sahlgrenska Hospital in Gothenburg recruited patients for the study. The staff at the Department of Urology and the Center for Blood Sample Collection at Sahlgrenska Hospital helped us with blood sampling, and the staff at the Department of Clinical Chemistry centrifuged the blood samples. The staff at the Department of Clinical Chemistry at Halland Hospital in Halmstad and Varberg and Robert Buckland at the Department of Surgical and Perioperative Sciences at Umeå University analyzed the blood samples. Ida Sigvardsson and Daniel Munkberg were involved in the data collection. Finally, we also want to thank Anna Turesson Wadell and Toshima Parris for their valuable input on the manuscript.

Conflicts of Interest: The authors declare no conflict of interest. The funders had no role in the design of the study; in the collection, analyses, or interpretation of data; in the writing of the manuscript; or in the decision to publish the results.

References

1. Huggins, C.; Hodges, C.V. Studies on prostatic cancer. I. The effect of castration, of estrogen and androgen injection on serum phosphatases in metastatic carcinoma of the prostate. *CA Cancer J. Clin.* **1972**, *22*, 232–240. [CrossRef] [PubMed]
2. Heinlein, C.A.; Chang, C. Androgen receptor in prostate cancer. *Endocr. Rev.* **2004**, *25*, 276–308. [CrossRef] [PubMed]
3. De Nunzio, C.; Lombardo, R.; Albisinni, S.; Gacci, M.; Tubaro, A. Serum levels of Sex Hormone Binding Globulin (SHBG) are not predictive of prostate cancer diagnosis and aggressiveness: Results from an italian biopsy cohort. *Int. Braz. J. Urol.* **2013**, *39*, 793–799. [CrossRef] [PubMed]
4. Kojima, S.; Inahara, M.; Suzuki, H.; Ichikawa, T.; Furuya, Y. Implications of insulin-like growth factor-I for prostate cancer therapies. *Int. J. Urol.* **2009**, *16*, 161–167. [CrossRef]
5. Roddam, A.W.; Allen, N.E.; Appleby, P.; Key, T.J. Endogenous sex hormones and prostate cancer: A collaborative analysis of 18 prospective studies. *J. Natl. Cancer Inst.* **2008**, *100*, 170–183. [CrossRef] [PubMed]
6. García-Cruz, E.; Carrion Puig, A.; García-Larrosa, A.; Sallent, A.; Castañeda-Argáiz, R.; Piqueras, M.; Ribal, M.J.; Leibar-Tamayo, A.; Romero-Otero, J.; Alcaraz, A. Higher sex hormone-binding globulin and lower bioavailable testosterone are related to prostate cancer detection on prostate biopsy. *Scand. J. Urol.* **2013**, *47*, 282–289. [CrossRef]
7. Dobbs, R.W.; Malhotra, N.R.; Greenwald, D.T.; Wang, A.Y.; Prins, G.S.; Abern, M.R. Estrogens and prostate cancer. *Prostate Cancer Prostatic Dis.* **2019**, *22*, 185–194. [CrossRef]
8. Kurzer, M.S.; Xu, X. Dietary phytoestrogens. *Annu. Rev. Nutr.* **1997**, *17*, 353–381. [CrossRef]
9. Mueller, S.O.; Simon, S.; Chae, K.; Metzler, M.; Korach, K.S.J.T.S. Phytoestrogens and their human metabolites show distinct agonistic and antagonistic properties on estrogen receptor α (ERα) and ERβ in human cells. *Toxicol. Sci.* **2004**, *80*, 14–25. [CrossRef]
10. Hartman, J.; Strom, A.; Gustafsson, J.A. Current concepts and significance of estrogen receptor beta in prostate cancer. *Steroids* **2012**, *77*, 1262–1266. [CrossRef]

11. Li, J.; Liu, Q.; Jiang, C. Signal Crosstalk and the Role of Estrogen Receptor beta (ERβ) in Prostate Cancer. *Med. Sci. Monit.* **2022**, *28*, e935599. [CrossRef]

12. Setchell, K.; Adlercreutz, H. Mammalian Lignans and Phytooestrogens Recent Studies on their Formation, Metabolism and Biological Role in Health and Disease. *Role Gut Flora Toxic. Cancer* **1988**, *14*, 315–345.

13. Kilkkinen, A.; Pietinen, P.; Klaukka, T.; Virtamo, J.; Korhonen, P.; Adlercreutz, H. Use of oral antimicrobials decreases serum enterolactone concentration. *Am. J. Epidemiol.* **2002**, *155*, 472–477. [CrossRef]

14. Zhang, Q.; Feng, H.; Qluwakemi, B.; Wang, J.; Yao, S.; Cheng, G.; Xu, H.; Qiu, H.; Zhu, L.; Yuan, M. Phytoestrogens and risk of prostate cancer: An updated meta-analysis of epidemiologic studies. *Int. J. Food Sci. Nutr.* **2017**, *68*, 28–42. [CrossRef]

15. Bylund, A.; Lundin, E.; Zhang, J.-X.; Nordin, A.; Kaaks, R.; Stenman, U.; Åman, P.; Adlercreutz, H.; Nilsson, T.; Hallmans, G. Randomised controlled short-term intervention pilot study on rye bran bread in prostate cancer. *Eur. J. Cancer Prev.* **2003**, *12*, 407–415. [CrossRef]

16. Demark-Wahnefried, W.; Polascik, T.J.; George, S.L.; Switzer, B.R.; Madden, J.F.; Ruffin, M.T.; Snyder, D.C.; Owzar, K.; Hars, V.; Albala, D.M. Flaxseed supplementation (not dietary fat restriction) reduces prostate cancer proliferation rates in men presurgery. *Cancer Epidemiol. Prev. Biomark.* **2008**, *17*, 3577–3587. [CrossRef]

17. Dalais, F.S.; Meliala, A.; Wattanapenpaiboon, N.; Frydenberg, M.; Suter, D.A.; Thomson, W.K.; Wahlqvist, M.L. Effects of a diet rich in phytoestrogens on prostate-specific antigen and sex hormones in men diagnosed with prostate cancer. *Urology* **2004**, *64*, 510–515. [CrossRef]

18. Kumar, N.B.; Cantor, A.; Allen, K.; Riccardi, D.; Besterman-Dahan, K.; Seigne, J.; Helal, M.; Salup, R.; Pow-Sang, J. The specific role of isoflavones in reducing prostate cancer risk. *Prostate* **2004**, *59*, 141–147. [CrossRef]

19. Kumar, N.B.; Kang, L.; Pow-Sang, J.; Xu, P.; Allen, K.; Riccardi, D.; Besterman-Dahan, K.; Krischer, J.P. Results of a randomized phase I dose-finding trial of several doses of isoflavones in men with localized prostate cancer: Administration prior to radical prostatectomy. *J. Soc. Integr. Oncol.* **2010**, *8*, 3–13.

20. Kumar, N.B.; Pow-Sang, J.; Spiess, P.; Dickinson, S.; Schell, M.J. A phase II randomized clinical trial using aglycone isoflavones to treat patients with localized prostate cancer in the pre-surgical period prior to radical prostatectomy. *Oncotarget* **2020**, *11*, 1218–1234. [CrossRef]

21. Grammatikopoulou, M.G.; Gkiouras, K.; Papageorgiou, S.; Myrogiannis, I.; Mykoniatis, I.; Papamitsou, T.; Bogdanos, D.P.; Goulis, D.G. Dietary Factors and Supplements Influencing Prostate Specific-Antigen (PSA) Concentrations in Men with Prostate Cancer and Increased Cancer Risk: An Evidence Analysis Review Based on Randomized Controlled Trials. *Nutrients* **2020**, *12*, 2985. [CrossRef] [PubMed]

22. Hedelin, M.; Bälter, K.A.; Chang, E.T.; Bellocco, R.; Klint, Å.; Johansson, J.E.; Wiklund, F.; Thellenberg-Karlsson, C.; Adami, H.O.; Grönberg, H. Dietary intake of phytoestrogens, estrogen receptor-beta polymorphisms and the risk of prostate cancer. *Prostate* **2006**, *66*, 1512–1520. [CrossRef] [PubMed]

23. Ahlin, R.; Nybacka, S.; Josefsson, A.; Stranne, J.; Steineck, G.; Hedelin, M. The effect of a phytoestrogen intervention and impact of genetic factors on tumor proliferation markers among Swedish patients with prostate cancer: Study protocol for the randomized controlled PRODICA trial. *Trials* **2022**, *23*, 1041. [CrossRef] [PubMed]

24. Swedish Food Agency. *Find Your Way to Eat Greener Not Too Much and Be Active*; Livsmedelsverket: Uppsala Lan, Sweden. Available online: https://www.livsmedelsverket.se/globalassets/publikationsdatabas/andra-sprak/kostraden/kostrad-eng.pdf (accessed on 26 March 2023).

25. Hedelin, M.; Löf, M.; Andersson, T.M.; Adlercreutz, H.; Weiderpass, E. Dietary phytoestrogens and the risk of ovarian cancer in the women's lifestyle and health cohort study. *Cancer Epidemiol. Biomarkers Prev.* **2011**, *20*, 308–317. [CrossRef]

26. Chain, E.P.o.C.i.t.F.; Schrenk, D.; Bignami, M.; Bodin, L.; Chipman, J.K.; del Mazo, J.; Grasl-Kraupp, B.; Hogstrand, C.; Hoogenboom, L.; Leblanc, J.C. Evaluation of the health risks related to the presence of cyanogenic glycosides in foods other than raw apricot kernels. *EFSA J.* **2019**, *17*, e05662. [CrossRef]

27. Nørskov, N.P.; Olsen, A.; Tjønneland, A.; Bolvig, A.K.; Lærke, H.N.; Knudsen, K.E. Targeted LC-MS/MS Method for the Quantitation of Plant Lignans and Enterolignans in Biofluids from Humans and Pigs. *J. Agric. Food Chem.* **2015**, *63*, 6283–6292. [CrossRef]

28. Nørskov, N.P.; Givens, I.; Purup, S.; Stergiadis, S. Concentrations of phytoestrogens in conventional, organic and free-range retail milk in England. *Food Chem.* **2019**, *295*, 1–9. [CrossRef]

29. National Library of Medicine. PubChem—Explore Chemistry. Available online: https://pubchem.ncbi.nlm.nih.gov/ (accessed on 23 March 2023).

30. World Health Organization. Body Mass Index—BMI. Available online: https://www.euro.who.int/en/health-topics/disease-prevention/nutrition/a-healthy-lifestyle/body-mass-index-bmi (accessed on 5 April 2023).

31. Hamilton-Reeves, J.M.; Rebello, S.A.; Thomas, W.; Slaton, J.W.; Kurzer, M.S. Soy protein isolate increases urinary estrogens and the ratio of 2: 16 α-hydroxyestrone in men at high risk of prostate cancer. *J. Nutr.* **2007**, *137*, 2258–2263. [CrossRef]

32. Muti, P.; Westerlind, K.; Wu, T.; Grimaldi, T.; De Berry, J., 3rd; Schünemann, H.; Freudenheim, J.L.; Hill, H.; Carruba, G.; Bradlow, L. Urinary estrogen metabolites and prostate cancer: A case-control study in the United States. *Cancer Causes Control.* **2002**, *13*, 947–955. [CrossRef]

33. Lee, J.K.; Byun, S.S.; Lee, S.E.; Hong, S.K. Preoperative Serum Sex Hormone-Binding Globulin Level Is an Independent Predictor of Biochemical Outcome After Radical Prostatectomy. *Medicine* **2015**, *94*, e1185. [CrossRef]

34. Kumar, N.B.; Krischer, J.P.; Allen, K.; Riccardi, D.; Besterman-Dahan, K.; Salup, R.; Kang, L.; Xu, P.; Pow-Sang, J. A Phase II randomized, placebo-controlled clinical trial of purified isoflavones in modulating steroid hormones in men diagnosed with localized prostate cancer. *Nutr. Cancer* **2007**, *59*, 163–168. [CrossRef]
35. Lampe, J.W.; Karr, S.C.; Hutchins, A.M.; Slavin, J.L. Urinary equol excretion with a soy challenge: Influence of habitual diet. *Proc. Soc. Exp. Biol. Med.* **1998**, *217*, 335–339. [CrossRef]
36. Rowland, I.R.; Wiseman, H.; Sanders, T.A.; Adlercreutz, H.; Bowey, E.A. Interindividual variation in metabolism of soy isoflavones and lignans: Influence of habitual diet on equol production by the gut microflora. *Nutr. Cancer* **2000**, *36*, 27–32. [CrossRef]
37. Akaza, H.; Miyanaga, N.; Takashima, N.; Naito, S.; Hirao, Y.; Tsukamoto, T.; Fujioka, T.; Mori, M.; Kim, W.J.; Song, J.M.; et al. Comparisons of percent equol producers between prostate cancer patients and controls: Case-controlled studies of isoflavones in Japanese, Korean and American residents. *Jpn. J. Clin. Oncol.* **2004**, *34*, 86–89. [CrossRef]
38. Kuijsten, A.; Arts, I.C.; van't Veer, P.; Hollman, P.C. The relative bioavailability of enterolignans in humans is enhanced by milling and crushing of flaxseed. *J. Nutr.* **2005**, *135*, 2812–2816. [CrossRef]
39. Bolvig, A.K.; Nørskov, N.P.; van Vliet, S.; Foldager, L.; Curtasu, M.V.; Hedemann, M.S.; Sørensen, J.F.; Lærke, H.N.; Bach Knudsen, K.E. Rye Bran Modified with Cell Wall-Degrading Enzymes Influences the Kinetics of Plant Lignans but Not of Enterolignans in Multicatheterized Pigs. *J. Nutr.* **2017**, *147*, 2220–2227. [CrossRef]
40. Watanabe, S.; Yamaguchi, M.; Sobue, T.; Takahashi, T.; Miura, T.; Arai, Y.; Mazur, W.; Wähälä, K.; Adlercreutz, H. Pharmacokinetics of soybean isoflavones in plasma, urine and feces of men after ingestion of 60 g baked soybean powder (kinako). *J. Nutr.* **1998**, *128*, 1710–1715. [CrossRef]
41. Cani, P.D. Human gut microbiome: Hopes, threats and promises. *Gut* **2018**, *67*, 1716–1725. [CrossRef]

Review

Nutritional Management of Patients with Head and Neck Cancer—A Comprehensive Review

Dinko Martinovic [1], Daria Tokic [2], Ema Puizina Mladinic [1], Mislav Usljebrka [1], Sanja Kadic [1], Antonella Lesin [1], Marino Vilovic [3], Slaven Lupi-Ferandin [1], Sasa Ercegovic [1], Marko Kumric [3], Josipa Bukic [4] and Josko Bozic [3,*]

1   Department of Maxillofacial Surgery, University Hospital of Split, 21000 Split, Croatia
2   Department of Anesthesiology and Intensive Care, University Hospital of Split, 21000 Split, Croatia
3   Department of Pathophysiology, University of Split School of Medicine, 21000 Split, Croatia
4   Department of Pharmacy, University of Split School of Medicine, 21000 Split, Croatia
*   Correspondence: josko.bozic@mefst.hr

Abstract: While surgical therapy for head and neck cancer (HNC) is showing improvement with the advancement of reconstruction techniques, the focus in these patients should also be shifting to supportive pre and aftercare. Due to the highly sensitive and anatomically complex region, these patients tend to exhibit malnutrition, which has a substantial impact on their recovery and quality of life. The complications and symptoms of both the disease and the therapy usually make these patients unable to orally intake food, hence, a strategy should be prepared for their nutritional management. Even though there are several possible nutritional modalities that can be administrated, these patients commonly have a functional gastrointestinal tract, and enteral nutrition is indicated over the parenteral option. However, after extensive research of the available literature, it seems that there is a limited number of studies that focus on this important issue. Furthermore, there are no recommendations or guidelines regarding the nutritional management of HNC patients, pre- or post-operatively. Henceforth, this narrative review summarizes the nutritional challenges and management modalities in this particular group of patients. Nonetheless, this issue should be addressed in future studies and an algorithm should be established for better nutritional care of these patients.

Keywords: head and neck surgery; oromaxillofacial surgery; head and neck cancer; clinical nutrition

Citation: Martinovic, D.; Tokic, D.; Puizina Mladinic, E.; Usljebrka, M.; Kadic, S.; Lesin, A.; Vilovic, M.; Lupi-Ferandin, S.; Ercegovic, S.; Kumric, M.; et al. Nutritional Management of Patients with Head and Neck Cancer—A Comprehensive Review. *Nutrients* **2023**, *15*, 1864. https://doi.org/10.3390/nu15081864

Academic Editors: Fernando Mendes, Diana Martins and Nuno Borges

Received: 22 February 2023
Revised: 7 April 2023
Accepted: 11 April 2023
Published: 13 April 2023

1. Introduction

Head and neck cancers (HNC) represent epithelial malignancies in the paranasal sinuses, nasal cavity, oral cavity, pharynx, and larynx, all of which significantly impact the morbidity and mortality of the affected population [1]. Hence, the National Cancer Institute's (NCI's) Surveillance, Epidemiology, and End Results (SEERs) refer to more than 65,000 new cases annually, with more than 14,500 death cases associated with the disease [2].

Most cases of HNC are head and neck squamous cell carcinoma (HNSCC), which is the sixth most widespread malignancy worldwide with an estimated 3% of all cancer cases [3]. The most significant risk factors that contribute to its pathogenesis are tobacco usage, alcohol consumption, and infection with a high-risk human papilloma virus [4]. Furthermore, the most lethal among the HNSCC is the oral squamous cell carcinoma (OSCC), which accounts for 90% of all oral malignancies and has an estimated 2–3% death rate of all cancer-related deaths [5].

Even though there are several different treatment modalities for HNC, due to the complex anatomy along with the highly important functions in the head and neck region, management of HNC should be multidisciplinary, with the focus not only on therapy but also on supportive pre and aftercare [6]. One of the significant reasons for the high HNC disease burden is malnutrition and nutritional deficits in these patients, which have a substantial impact on health outcomes as well as the overall quality of life [7]. Malnutrition can be referred to by following several different criteria, including more than 5% weight loss in three months or more than 10% in a six-month period, or a BMI less than 20 kg/m^2, while albumin levels less than 35 g/L can be suggestive of malnutrition as well [8 10].

Problems with nutrition already start with disease onset, with various studies showing 25–65% of HNC patients presenting themselves with malnutrition, with more than 10% weight loss from the normal body mass [11–13]. High levels of variability and difficulties to obtain exact data are probably driven by different malnutrition definitions and different methods of malnutrition assessment [14]. However, there could be several reasons for the still significant percentages of malnutrition in these patients even before treatment, including chronic malnutrition, which is associated with alcohol and tobacco usage, trismus, obstruction of the respiratory and digestive system with aspiration, odynophagia, and dysphagia [6,13]. Hence, Kubrak et al. showed in a large cohort of HNC patients that independent predictors of weight loss in a naive population were tumor stage, dietary intake categories, and performance status [15].

Different treatment modalities can only further worsen the nutritional status of HNC patients. These modalities include chemotherapy, radiation therapy, surgical procedures, specialized targeted therapies, and different combination therapies [16]. While surgical procedures are a potential cause of disrupted food intake, mostly based on tumor location and resection type and size, radiotherapy, and chemotherapy are connected to various symptoms that further impair oral food intake and often cause treatment withdrawal [17,18]. Furthermore, according to several studies, malnutrition in HNC patients during therapy can be present in substantially high percentages, up to 80% [19–21]. Mucositis is commonly associated with chemotherapy and radiotherapy and is characterized by inflamed lesions in the area of the mouth and throat. These complications are connected to infections, dysphagia, and pain, among others [22]. Xerostomia and impaired parathyroid gland function are two of the most common adverse effects of radiotherapy that significantly reduce the quality of life of patients and are often associated with thick saliva. This causes further negative effects on chewing, swallowing, and speaking, with an accompanying increase in infections [23,24]. Other adverse effects of therapeutic modalities that significantly impair weight maintenance and the proper healing process include nausea, vomiting, constipation, depression, dysgeusia, and odynophagia [6].

Adequate nutritional support in HNC patients is of vital importance, as malnutrition is associated with a decreased therapeutic and immunological response with a higher incidence of infections and post-surgical complications. Furthermore, it causes treatment breaks with a higher economic burden and decreased functional performances that cumulatively lead to decreased quality of life and higher mortality rates. Interestingly, overweight and obese people with higher BMIs are at the same risk of all of the mentioned complications [6,13,25–27]. Hence, studies have shown that weight loss before treatment was one of the main independent survival predictors, while nutritional support before surgery can lead to significant beneficial changes in quality of life with fewer post-operative infectious complications [28,29].

These considerations were leading to the development of various screening tools in order to diagnose vulnerable patients and to prevent treatment complications and negative health outcomes. Moreover, the Nutritional Risk Screening 2002 (NRS 2002) and Malnutrition Universal Screening Tool (MUST) were accepted by the European Society for Clinical Nutrition and Metabolism (ESPEN) for hospital usage [19]. The newest edition of the ESPEN clinical guidelines for clinical nutrition in cancer, which is dedicated to all healthcare professionals that participate in cancer patient care, has a total of 43 recommendations and short commentaries regarding cancer patient management. Even though general recommendations are included that can be of use in HNC patient care, with some specialized parts that address this population specifically, there is still a need for comprehensive, detailed guidelines that involve only the HNC patient population [30].

Another important practice that should be implemented regularly in HNC patient management is involving enhanced recovery after surgery (ERAS) protocols. These specialized, evidence-based protocols were initially presented in the perioperative care of colorectal surgery, and up to today investigations have shown numerous beneficial health outcomes with managed patients in different surgical branches [31,32]. The ERAS protocol with published recommendations was introduced for HNC surgical procedures with free flap reconstruction in 2017, with a special focus on perioperative nutritional care [33]. According to these recommendations, Moore et al. performed a clinical investigation on 25 HNC surgery patients that received perioperative nutritional supplementation. Results have shown that processes still need some modifications to ensure specialized approaches for patient care improvement [34].

These data suggest that even though some nutritional guidelines for HNC surgery exist, there is still a potential need for new investigations, modifications, and better clinical utilizations.

2. Nutritional Modalities

Patients with head and neck cancer, due to the location of the tumor in the upper airway and digestive tract, can have difficulties masticating and swallowing food. Tumor-related dysphagia can significantly compromise the nutrition status of those patients and put them at risk of malnutrition at the time of diagnosis. Due to inadequate nutrition intake, these patients are prone to weight loss, decrease in muscle mass, fatigue, and anemia, which ultimately leads to cancer cachexia syndrome. Studies indicate that impaired nutrition status is noticed in 25–27% of these patients at the time of diagnosis and before the start of treatment [35–38]. It was also shown that pretreatment weight loss is a strong survival predictor and pretreatment cachexia is connected to poor survival [29,38,39]. Additionally, sarcopenia that occurs as a result of cancer cachexia has been connected with unfavorable treatment outcomes in patients with head and neck cancer [40]. Cancer treatment can further aggravate treatment-related dysphagia, resulting in reduced food intake and deteriorating nutrition status [35–37]. Severe weight loss (more than 10% of body weight) during treatment has been observed in the absence of intensive nutrition support in up to 58% of patients [38,41,42].

Nutrition support is a crucial part of head and neck cancer control, supporting better disease outcomes. Oral intake of food is the preferred method of nutrition, but in cases when adequate nutritional intake cannot be maintained by mouth, enteral or parenteral nutrition is necessary. Enteral nutrition is preferred over parenteral nutrition, being physiologically natural. It also maintains gastrointestinal integrity, protecting it from atrophy, and also supports gut immune function (Table 1). Enteral nutrition implies the administration of food into the gastrointestinal tract through a tube or stoma.

Table 1. Main characteristics and differences between enteral and parenteral nutrition.

Enteral Nutrition	Parenteral Nutrition
Cheaper	Expensive
Lower infection rate	Higher infection rate
Need of monitoring for optimal nutrition	Delivery of optimal nutrition
No gut atrophy	Gut atrophy
Shorter hospitalization rate Fewer complications	Longer hospitalization rate More complications

Furthermore, it is important to plan the nutritional management of these patients during pre-treatment care. Although head and neck cancer patients have restricted oral intake, their gastrointestinal tract is usually functional. Therefore, enteral nutrition is indicated in patients that are unable to feed orally but have a functioning and accessible gastrointestinal tract. It can be short-termed or long-termed and can be administrated into the stomach—gastric or into the intestine—post-pyloric [43]. It is a safe and effective nutrition manner to provide nutrition that can be easily administered by patients and their families at home. The main contraindication for applying enteral nutrition is the non-functional gastrointestinal (GI) tract. Numerous studies have shown that administrating feeding tubes before the start of treatment in these patients ensures better overall outcomes. It acts beneficially in the prevention of weight loss and sarcopenia, dehydration, reduced hospital admission, and improved quality of life [44–49]. Improvements in nutrition parameters, including anthropometrics and laboratory data, have also been observed [50].

3. Enteral Nutrition

Several factors must be considered in determining which type and modality of enteral nutrition are administrated. The site of feeding and expected duration is primary, but it is also important to consider timing and rate of initiation, feeding modality, risk of complications, and possible contraindications. An individual approach and careful assessment are key in ensuring clinically appropriate and nutritionally complete enteral nutrition. The following options are available: nasogastric tube, nasojejunal tube, gastrostomy, and jejunostomy (Table 2) [51].

Table 2. Enteral nutrition modalities and their characteristics.

Enteral Nutrition Method	Term	Advantages	Disadvantages
Nasogastric tube	Short term	Simple placement; cheap	Discomfort; potential displacement; risk of aspiration
Nasojejunal tube	Short term	Can be used in patients with impaired gastric motility; reduced aspiration risk; cheap	Discomfort; potential displacement; require skilled specialist; no bolus feeding
Gastrostomy			
PEG	Long term	Can be used for a prolonged time; minimally invasive technique	Needs sedation; invasive technique; potential displacement; stoma complications; tumor seeding
RIG	Long term	Can be used for a prolonged time; minimally invasive technique; no sedation needed	Invasive technique; stoma complications; potential displacement due to smaller tubes
surgical	Long term	Can be used for a prolonged time; no tube displacement	Requires surgery under general anesthesia; stoma complications
Jejunostomy	Long term	Can be used for a prolonged time; no risk of aspiration; less discomfort	By-passes the stomach; continuous slow rate of feeding; potential bowel obstruction

Abbreviations: PEG—percutaneous endoscopic gastrostomy; RIG—radiologically inserted gastrostomy.

3.1. Nasogastric Tube

A nasogastric tube (NGT) is a plastic catheter inserted through the nose, passing the oropharynx and esophagus to the stomach [52]. It is the most commonly used type of enteral nutrition (EN) in HNC patients with a functional gastrointestinal tract if the tumor is obstructive, thereby impacting swallow function [53]. Generally, NGTs are used for a shorter period (<4 weeks) and can be administered perioperatively (prophylactic) or postoperatively (reactive) [54].

Prophylactic enteral feeding is used when nutrition support is anticipated for an extended time after more invasive surgical procedures and in severely malnourished patients. The National Comprehensive Cancer Network (NCCN) published indications for prophylactic tube placement, summarized in Table 1 [55].

Nutritional status seems to be maintained or enhanced with tube implantation, according to Langius et al.'s systematic study, which found that prophylactic tube feeding increases nutritional intake and nutritional status compared to oral consumption alone [56].

If oral intake is not possible after 4–6 weeks, percutaneous endoscopic gastrostomy (PEG) is indicated [54]. The literature showed similar nutritional and clinical outcomes in patients with NGTs and PEGs. However, complications such as increased tube dislodgement and chest infection were noticed after a long period in patients with NGTs. Additionally, compared to patients with gastrostomies, patients with nasogastric tubes reported greater body image problems, difficulty with feeding, and considerable social activity interruption [57,58].

Still, NGTs remain a routine modality of enteral nutrition because there are no notable differences in overall complication rates. They are easily placed, significantly less expensive, and the transition from enteral nutrition back to oral nutrition is shorter than PEG tubes [59].

3.2. Nasojejunal Tube

Oroenteric or nasoenteric feeding tubes are intended for enteral nutrition into small intestine regions, meaning the duodenum and jejunum. Application of nutrients in the small intestine is indicated for short-term feeding in patients that have impaired gastric motility since they are at risk for gastric emptying delay, reflux, and aspiration. It is also instructed to avoid gastric feeding in patients with severe acute pancreatitis and promote enteral nutrition [60].

Enteric feeding tubes can be placed blindly at the bedside without the use of any technology. When placing the tube, the distal end must surpass the duodenojejunal flexure, otherwise, it can inevitably retract back to the stomach. Nasojejunal feeding tubes can be of various sizes. Usually, they are small-bore with flexible tips that are intended to ensure spontaneous passage into the small intestine and protect from injury of mucosa or perforation. The tubes are also provided with stylets or guide wires providing structure for easier placement [61,62]. Studies suggest that blinded positioning of a small-bore feeding tube by a well-trained and experienced clinician has a success rate of 80% or more [63].

Possible complications of the application of nasojejunal tube and nasojejunal tube feeding are epistaxis, sinusitis, oesophageal ulceration or strictures, blockage, malposition or injury, and perforation of the gastrointestinal (GI) tract. Malposition implies placing the tube outside the GI tract. Placing it in the bronchopulmonary tree can cause infection, effusion, or empyema [51,63–66]. As a rare occurrence, but a very severe complication, it can malposition intracranially. To reduce the risk of malposition or injury, tubes can be placed using endoscopy or fluoroscopy, or magnetic guidance. Several studies show that using technology for placement increases success rates by 90% [66–69]. In addition, it is important to verify the proper position of the tube after placement using an abdominal X-ray.

There are indications that jejunal feeding may assist in the increased delivery of nutrients since there are fewer interruptions compared to gastric feeding, which is often interrupted by the application of medication and lavage [70,71]. It is propounded that post-pyloric tubes are more comfortable for patients than gastric tubes, so they are a good alternative for patients that show discomfort with NGT, decreasing removal of the tube by the patients [72]. Preferred modality of post-pyloric feeding should be continuous to avoid discomfort and dumping syndrome that occurs with bolus feeding. However, for HNC patients, a NGT is still the preferred option over the nasojejunal tubes, and the latter are usually only reserved for patients who have impaired gastric motility.

3.3. Percutaneous Endoscopic Gastrostomy

Percutaneous endoscopic gastrostomy (PEG) is nowadays considered the "golden standard" for feeding and nutrition support in oral/maxillofacial surgery patients undergoing extensive surgery treatments. PEG tube placement is a well-accepted and frequently conducted procedure worldwide that has replaced surgical and radiological gastrostomy techniques [73]. Morbidity linked with PEG can vary from 5 to 10.3%, of which 3% are major adverse events [74].

The main indication for PEG in oral and maxillofacial patients is providing oral intake and meeting metabolic requirements due to the closeness of the cancer and organs that are in charge of normal food intake [6,75]. It is important to highlight that patients with maxillofacial region tumors are usually excessive smokers and/or alcohol consumers, with inadequate eating habits prior to diagnosis or treatment [76]. This group of patients often faces postoperative swallowing disorders, changed anatomy of the oral cavity due to surgery, impaired function of the tongue, dysphagia, odynophagia, dysgeusia, xerostomia, and nausea [6]. Malnutrition and weight loss are connected with poorer treatment outcomes, poor quality of life, and consequently elevated rates of morbidity and mortality [77].

The PEG can be permanent or temporary depending on the patient's requirements. Since the nasoenteric tubes are usually used in patients with preserved airway reflexes who require enteral feeding for less than 30 days, PEG is currently the method of choice for medium to long-term enteral feeding [36]. It is important to emphasize the fact that preventively placed PEG tubes resulted in lower complication rates in comparison to therapeutically inserted PEGs [78].

Regarding the insertion techniques, there are three most commonly used methods: the "pull", "push" (guide wire), and introducer (Russel) methods. All of them have the same concept of insertion through the abdominal wall at the spot where the abdominal wall and stomach are the closest. According to several studies, the success rate of PEG is 84–96% [79,80]. Since it could be unable to perform oral percutaneous gastrostomy for 4–7% of head and neck cancer patients, transnasal endoscopy is a method of choice. Oropharyngeal obstruction, severe trismus, or airway endangerment are the main causes of the transnasal approach.

Nasoenteric tubes contribute to a higher number of complications such as irritation, nasal decubitus, patient discomfort, ulceration, bleeding, esophageal reflux, and aspiration pneumonia [81]. They are also connected to poorer acceptance and psychological or social problems. Interestingly, compared to PEG, nasoenteric tubes have a lower feeding efficacy [82]. There is some strong evidence in the literature that the initiation of PEG feeding, as soon as the medical indication has been set up, can prevent further weight loss and contribute to patients' quality of life.

The most common contraindications for the PEG placement are systemic, such as coagulation disorders, hemodynamic instability, or sepsis; disturbances at the site of placement in the abdomen, such as abdominal wall infection, ascites, peritonitis; and peritoneal carcinomatosis; and interposed organs such as the colon or liver [75,83].

The PEG tube insertion process is commonly considered harmless; however, some complications can take place. The mortality rate is expectedly higher in patients with underlying comorbidities [84]. Minor complications consist of wound infection, which is, according to the literature, somewhere between 5% and 25%. Since this probably occurs due to contamination by oral flora, antibiotic prophylaxis is recommended [85,86]. Granuloma formation is the next most common complication and is due to friction and excess moisture around the tube [87]. Moreover, some of the other possible minor complications are tube leakage into the abdominal cavity and consequent development of peritonitis; stoma leakage; tube obstruction; pneumoperitoneum; gastric outlet obstruction, and peritonitis [84,88].

Major complications are aspiration pneumonia; however, that is more often in neurologic patients due to feeding a large amount of content and being in the prone position than in OMFS patients; hemorrhage (retroperitoneal bleeding due to injuries of the gastric artery, splenic and mesenteric veins, and rectus sheath hematoma); necrotizing fasciitis, which is a potentially lethal complication; buried bumper syndrome (characterized by excessive tension amid internal and external bumpers that cause ischemia and necrosis of the gastric wall and consequently, the tube moves toward the abdominal wall), and perforation of the bowel [75,88–90].

Furthermore, it is important to highlight an interesting and unusual complication of PEG placement in HNC patients: the seeding of the tumor. It was observed that the seeding occurs during the "pull" or "push" method of PEG insertion when the tube collides with the tumor, directly transferring tumor cells [91]. Unfortunately, diagnosis is set when the tumor metastasis is large and visible. However, questions regarding this complication are still emerging and some studies imply that the hematogenous or lymphatic spread is actually responsible for the metastasis [92].

3.4. Radiologically Inserted Gastrostomy

Endoscopically placed percutaneous gastrostomy was very soon followed by the development of radiological techniques for fluoroscopic percutaneous placement [93]. Radiologically inserted gastrostomy implies a procedure where gastrostomy is inserted directly into the stomach under X-ray guidance. Indications and complications are similar to PEG, with the only difference being in the performed technique.

A preliminary CT (computed tomography) scan or ultrasound examination can be performed to rule out an overlapping colon or left hepatic lobe. Sometimes, 100–200 mL of barium sulfate can be orally administered to patients the night before the procedure to sketch the colon. On the day of the procedure, air or CO_2 gas is insufflated in the stomach via a nasogastric tube to extend the stomach. Afterward, gastric fixation is required and a gastropexy is conducted, which is followed by a gastric puncture. Gastropexy can reduce the spilling of the gastric contents around gastrostomy into the peritoneum and is performed in 93% of cases according to a prospective multicenter survey of the United Kingdom [94,95]. The position in the stomach is confirmed by aspirating air or injecting a contrast medium. After the tube placement, contrast is injected to prove the proper position [96,97].

Minor complications are similar to PEG, except for tube displacement, which happens in 1% of cases. According to Cherian et al., there are no differences in major complications between PEG and RIG at 6.8% and 8.5%, respectively [98]. Moreover, a meta-analysis from Wollman et al. found that the success rate of the tube placement is higher in RIG compared to PEG, while also major complications occurred less often in RIG [99]. Another meta-analysis proved that major complications were more frequent in PEG patients (2.19%) than RIG patients (0.07%), but with no significant statistical difference [100]. Additionally, RIG can be placed in cases of esophageal/oropharyngeal obstruction and since there is no gastroscopy involved, there is no need for antibiotic prophylaxis.

3.5. Surgical Gastrostomy

Surgical gastrostomy is nowadays considered a rudimentary technique due to safer and less invasive characteristics of PEG. However, in certain situations, especially in HNC patients, surgical gastrostomy is still a viable option. It is conducted for indications such as when HNC prevents access for gastroscopy, severe strictures of the esophagus or impossibility to set stomach to adjacent to the abdominal wall due to obesity, previous surgery, or hepatosplenomegaly [101].

Surgical gastrostomy can be conducted in two ways: with open laparotomy or by laparoscopy. Laparoscopy is the less invasive and surgically better method due to the better exposure of the stomach since in open laparotomy the incision is commonly very small [102]. Nevertheless, the literature has shown that even laparoscopy gastrostomy placement has significantly more complications compared to PEG [103,104].

3.6. Jejunostomy

The jejunostomy feeding tube is a method of nutrition through access to the jejunum. It is used when a gastrostomy tube is not technically possible or contraindicated. It can be administrated endoscopically, radiologically, or by surgery. When placing these feeding tubes, minimally invasive techniques are always preferred [105].

Jejunostomy, as gastrostomy, is preferred for long-term enteral nutrition greater than six weeks. Benefits to this technique compared to the NGT and PEG are lessening the risk of tube malposition, lower risk of aspiration since the gastroesophageal sphincter is not held open by a tube, no nasal discomfort, and pressure injury of the nares [72].

Despite having some advantages, there are also complications including infections, nausea, vomiting, diarrhea, abdominal distension, bowel obstruction, and metabolic abnormalities [106,107]. This method of enteral nutrition is not suitable for patients with bowel obstruction distal to the placement site. Regarding HNC patients, jejunostomy is usually indicated in those patients that have other conditions that prevent using oral or gastric nutrition. When feeding jejunely, a continuous and slow rate is advised due to the loss of the stomach reservoir [51].

4. Parenteral Nutrition

Parenteral nutrition represents an intravenously delivered nutritionally balanced synthetic mixture of sterile nutrients [108]. It has to be slowly introduced, starting with 15–20 calories per kg of body weight a day with a maximum of 1000 calories a day. As it is initiated and the caloric levels rise toward the target level, the patient needs to be monitored because of possible drops in potassium, phosphorous, and magnesium levels [109]. Multiple meta-analyses have shown more infectious complications in patients using parenteral nutrition, yet estimated caloric requirements can be more easily accomplished [110]. Moreover, several studies imply that better outcomes occur when at least one part of a patient's nutrition is via the enteral route [6,111,112]. However, when the intake of food and nutrient absorption becomes almost impossible due to a defective, unavailable, or ruptured gastrointestinal tract, parenteral nutrition is the best and only choice [111]. Hence, the main indication of PN is when there is no other way to meet the required calorie intake. Several other indications are severe catabolic state due to the severity of the patient's condition (such as sepsis and polytrauma), gastrointestinal malformations (congenital in children or after surgery or cancer), severe vomiting, and diarrhea [113–116]. Moreover, parenteral nutrition can be used in cases of prolonged chyle leaks that can take place after extensive left neck surgery (the thoracic duct is located nearby) [117]. This type of nutrition is also recommended for terminal care patients [118].

Parenteral nutrition is contraindicated in patients with a functional gastrointestinal tract, patients that do not have access to an intravenous line, and in patients that need therapy for fewer than 5 days if they are not severely malnourished [119]. As soon as the adequate gastrointestinal function is reached, patients are gradually transitioned from parenteral to enteral or oral nutrition to avoid biliary sludge and GI mucosal atrophy. When enteral or oral intake reaches 500 calories a day, parenteral nutrition ought to be decreased, while when more than 60% of caloric needs are achieved via oral or enteral intake, parenteral nutrition should be canceled [119]. Parenteral nutrition is rarely applied to patients with HNC because, in most cases, the lower GI tract is functional. A study performed by Ryu et al. revealed that total parenteral nutrition was USD 11.81 more expensive than certain types of enteral nutrition (feeding via NGT) on a daily basis [120]. Research conducted by Scolapio et al. revealed that many patients would prefer the parenteral type of feeding more than enteral (NGT feeding) if they were unable to eat [118].

When it comes to HNC, patients often deal with morphological malformations, caused both by malignancy and/or treatment [112]. While severe illness such as cancer causes metabolic changes and malnutrition, demanding and extensive surgeries impact and exhaust organism as well [111]. Therefore, intensive follow-up treatment is required in order to enable the patient to recover as soon and as best as possible. When it comes to the nutritional deficits of patients after head and neck surgery, the problem that often occurs is in the upper gastrointestinal area. Some of the surgeries require manipulation and extensive resection in the intraoral space and that can lead to several complications. As previously mentioned, the most common are odynophagia, dysphagia, mucositis, xerostomia, and edema [6]. These complications can induce malnutrition, in which case it is of vital importance to adequately approach this problem. When malnutrition occurs for more than two weeks, there is a risk of developing refeeding syndrome; metabolic changes followed by hypokalemia, hypomagnesemia, and hypophosphatemia; and life-threatening conditions such as cardiac and respiratory distress [121]. To avoid this dangerous state, it is necessary to recognize the nutritional needs of the patient at the time and adjust if enteral feeding is in any way compromised. For parenteral nutrition, an interdisciplinary approach is needed (physicians, nutritionists, and educated staff in the surgical department) in order to reduce potential complications and give the patient the necessary care [119]. In the current literature, there are not many studies regarding this subject. Ackerman et al. evaluated and presented nutrition management for HNC patients, where they discussed and pointed out that prevention of undernutrition is the key to avoiding poor treatment outcomes [6]. On the other hand, there are no specific guidelines when it comes to the nutrition of head and neck cancer patients. While ESPEN brings relevant guidelines on nutrition covering most of the conditions and diseases, on the other hand, when it comes to parenteral nutrition in surgery, ESPEN recommends the formula of 25 kcal/kg ideal body weight [122].

5. Conclusions

HNC patients are, due to the anatomical region involved, a highly sensitive group of patients regarding their nutritional management. While oral intake is the most superior way of feeding, that is rarely possible in this particular group. Since they most commonly have a functional gastrointestinal tract, enteral nutrition should be indicated over the parenteral possibility. However, according to the relevant literature, it is still debatable which enteral modality should be administrated. While PEG is starting to be deemed the "golden standard" for these patients, on the other hand, NGT is still considered a highly viable choice. Moreover, even though the literature is very scarce regarding the importance of the pre-surgery preparation and planning of the nutritional route, we believe that it is an integral part of a faster recovery after surgery. Furthermore, none of the relevant sources give any recommendations or guidelines for the nutritional management of these patients. Henceforth, this issue should be given more focus in future studies and an algorithm should be established for an improvement of nutritional support in this particular group of patients.

Author Contributions: Conceptualization, D.M., D.T., E.P.M., S.L.-F. and J.B. (Josko Bozic); software, D.T., M.U., S.K., A.L. and M.K.; investigation, D.M., D.T., M.K., S.E. and J.B. (Josipa Bukic); writing—original draft preparation, D.M., D.T., E.P.M., M.U., S.K., A.L. and M.V.; writing—review and editing, D.M., D.T., E.P.M., M.U., S.K., A.L. and J.B. (Josipa Bukic); visualization, M.K., S.E., M.V. and J.B. (Josipa Bukic); supervision, J.B. (Josko Bozic), S.L.-F. and S.E.; project administration, J.B. (Josko Bozic), S.L.-F. and M.V. All authors have read and agreed to the published version of the manuscript.

Funding: This research received no external funding.

Institutional Review Board Statement: Not applicable.

Informed Consent Statement: Not applicable.

Conflicts of Interest: The authors declare no conflict of interest.

References

1. Argiris, A.; Eng, C. Epidemiology, staging, and screening of head and neck cancer. *Cancer Treat Res.* **2003**, *114*, 15–60. [PubMed]
2. Howlader, N.N.A.; Krapcho, M.; Miller, D.; Brest, A.; Yu, M.; Ruhl, J.; Tatalovich, Z.; Mariotto, A.; Lewis, D.R.; Chen, H.S.; et al. EER Cancer Statistics Review, 1975–2016. Available online: https://seer.cancer.gov/csr/1975_2016/ (accessed on 1 February 2023).
3. Siegel, R.L.; Miller, K.D.; Wagle, N.S.; Jemal, A. Cancer statistics, 2023. *CA Cancer J. Clin.* **2023**, *73*, 17–48. [CrossRef]
4. Johnson, D.E.; Burtness, B.; Leemans, C.R.; Lui, V.W.Y.; Bauman, J.E.; Grandis, J.R. Head and neck squamous cell carcinoma. *Nat. Rev. Dis. Primers* **2020**, *6*, 92. [CrossRef]
5. Bugshan, A.; Farooq, I. Oral squamous cell carcinoma: Metastasis, potentially associated malignant disorders, etiology and recent advancements in diagnosis. *F1000Research* **2020**, *9*, 229. [CrossRef] [PubMed]
6. Ivkovic, N.; Martinovic, D.; Kozina, S.; Lupi-Ferandin, S.; Tokic, D.; Usljebrka, M.; Kumric, M.; Bozic, J. Quality of Life and Aesthetic Satisfaction in Patients Who Underwent the "Commando Operation" with Pectoralis Major Myocutaneus Flap Reconstruction-A Case Series Study. *Healthcare* **2022**, *10*, 1737. [CrossRef] [PubMed]
7. Ackerman, D.; Laszlo, M.; Provisor, A.; Yu, A. Nutrition Management for the Head and Neck Cancer Patient. *Cancer Treat Res.* **2018**, *174*, 187–208.

8. Bonilha, S.F.; Tedeschi, L.O.; Packer, I.U.; Razook, A.G.; Nardon, R.F.; Figueiredo, L.A.; Alleoni, G.F. Chemical composition of whole body and carcass of Bos indicus and tropically adapted Bos taurus breeds. *J. Anim. Sci.* **2011**, *89*, 2859–2866. [CrossRef]

9. Platek, M.E.; Popp, J.V.; Possinger, C.S.; Denysschen, C.A.; Horvath, P.; Brown, J.K. Comparison of the prevalence of malnutrition diagnosis in head and neck, gastrointestinal, and lung cancer patients by 3 classification methods. *Cancer Nurs.* **2011**, *34*, 410–416. [CrossRef]

10. Hickson, M. Malnutrition and ageing. *Postgrad. Med. J.* **2006**, *82*, 2–8. [CrossRef]

11. van Wayenburg, C.A.; Rasmussen-Conrad, E.L.; van den Berg, M.G.; Merkx, M.A.; van Staveren, W.A.; van Weel, C.; van Binsbergen, J.J. Weight loss in head and neck cancer patients little noticed in general practice. *J. Prim. Health Care* **2010**, *2*, 16–21. [CrossRef]

12. Eley, K.A.; Shah, R.; Bond, S.E.; Watt-Smith, S.R. A review of post-operative feeding in patients undergoing resection and reconstruction for oral malignancy and presentation of a pre-operative scoring system. *Br. J. Oral Maxillofac. Surg.* **2012**, *50*, 601–605. [CrossRef]

13. Alshadwi, A.; Nadershah, M.; Carlson, E.R.; Young, L.S.; Burke, P.A.; Daley, B.J. Nutritional considerations for head and neck cancer patients: A review of the literature. *J. Oral Maxillofac. Surg.* **2013**, *71*, 1853–1860. [CrossRef]

14. Saroul, N.; Pastourel, R.; Mulliez, A.; Farigon, N.; Dupuch, V.; Mom, T.; Boirie, Y.; Gilain, L. Which Assessment Method of Malnutrition in Head and Neck Cancer? *Otolaryngol. Head Neck Surg.* **2018**, *158*, 1065–1071. [CrossRef]

15. Kubrak, C.; Martin, L.; Gramlich, L.; Scrimger, R.; Jha, N.; Debenham, B.; Chua, N.; Walker, J.; Baracos, V.E. Prevalence and prognostic significance of malnutrition in patients with cancers of the head and neck. *Clin. Nutr.* **2020**, *39*, 901–909. [CrossRef]

16. Anderson, G.; Ebadi, M.; Vo, K.; Novak, J.; Govindarajan, A.; Amini, A. An Updated Review on Head and Neck Cancer Treatment with Radiation Therapy. *Cancers* **2021**, *13*, 4912. [CrossRef]

17. Larsson, M.; Hedelin, B.; Johansson, I.; Athlin, E. Eating problems and weight loss for patients with head and neck cancer: A chart review from diagnosis until one year after treatment. *Cancer Nurs.* **2005**, *28*, 425–435. [CrossRef] [PubMed]

18. Trotti, A.; Bellm, L.A.; Epstein, J.B.; Frame, D.; Fuchs, H.J.; Gwede, C.K.; Komaroff, E.; Nalysnyk, L.; Zilberberg, M.D. Mucositis incidence, severity and associated outcomes in patients with head and neck cancer receiving radiotherapy with or without chemotherapy: A systematic literature review. *Radiother. Oncol.* **2003**, *66*, 253–262. [CrossRef] [PubMed]

19. Gorenc, M.; Kozjek, N.R.; Strojan, P. Malnutrition and cachexia in patients with head and neck cancer treated with (chemo)radiotherapy. *Rep. Pract. Oncol. Radiother.* **2015**, *20*, 249–258. [CrossRef] [PubMed]

20. Unsal, D.; Mentes, B.; Akmansu, M.; Uner, A.; Oguz, M.; Pak, Y. Evaluation of Nutritional Status in Cancer Patients Receiving Radiotherapy: A Prospective Study. *Am. J. Clin. Oncol.* **2006**, *29*, 183–188. [CrossRef] [PubMed]

21. Langius, J.A.E.; Doornaert, P.; Spreeuwenberg, M.D.; Langendijk, J.A.; Leemans, C.R.; van Bokhorst-de van der Schueren, M.A.E. Radiotherapy on the neck nodes predicts severe weight loss in patients with early stage laryngeal cancer. *Radiother. Oncol.* **2010**, *97*, 80–85. [CrossRef]

22. Sroussi, H.Y.; Epstein, J.B.; Bensadoun, R.J.; Saunders, D.P.; Lalla, R.V.; Migliorati, C.A.; Heaivilin, N.; Zumsteg, Z.S. Common oral complications of head and neck cancer radiation therapy: Mucositis, infections, saliva change, fibrosis, sensory dysfunctions, dental caries, periodontal disease, and osteoradionecrosis. *Cancer Med.* **2017**, *6*, 2918–2931. [CrossRef]

23. Deasy, J.O.; Moiseenko, V.; Marks, L.; Chao, K.S.; Nam, J.; Eisbruch, A. Radiotherapy dose-volume effects on salivary gland function. *Int. J. Radiat. Oncol. Biol. Phys.* **2010**, *76*, 090. [CrossRef]

24. Paterson, C.; Caldwell, B.; Porteous, S.; McLean, A.; Messow, C.M.; Thomson, M. Radiotherapy-induced xerostomia, pre-clinical promise of LMS-611. *Support Care Cancer* **2016**, *24*, 629–636. [CrossRef]

25. Bozzetti, F.; Cotogni, P. Nutritional Issues in Head and Neck Cancer Patients. *Healthcare* **2020**, *8*, 102. [CrossRef] [PubMed]

26. Langius, J.A.; van Dijk, A.M.; Doornaert, P.; Kruizenga, H.M.; Langendijk, J.A.; Leemans, C.R.; Weijs, P.J.; Verdonck-de Leeuw, I.M. More than 10% weight loss in head and neck cancer patients during radiotherapy is independently associated with deterioration in quality of life. *Nutr. Cancer* **2013**, *65*, 76–83. [CrossRef]

27. Farhangfar, A.; Makarewicz, M.; Ghosh, S.; Jha, N.; Scrimger, R.; Gramlich, L.; Baracos, V. Nutrition impact symptoms in a population cohort of head and neck cancer patients: Multivariate regression analysis of symptoms on oral intake, weight loss and survival. *Oral Oncol.* **2014**, *50*, 877–883. [CrossRef] [PubMed]

28. Bertrand, P.C.; Piquet, M.A.; Bordier, I.; Monnier, P.; Roulet, M. Preoperative nutritional support at home in head and neck cancer patients: From nutritional benefits to the prevention of the alcohol withdrawal syndrome. *Curr. Opin. Clin. Nutr. Metab. Care* **2002**, *5*, 435–440. [CrossRef]

29. Mick, R.; Vokes, E.E.; Weichselbaum, R.R.; Panje, W.R. Prognostic factors in advanced head and neck cancer patients undergoing multimodality therapy. *Otolaryngol. Head Neck Surg.* **1991**, *105*, 62–73. [CrossRef] [PubMed]

30. Muscaritoli, M.; Arends, J.; Bachmann, P.; Baracos, V.; Barthelemy, N.; Bertz, H.; Bozzetti, F.; Hütterer, E.; Isenring, E.; Kaasa, S.; et al. ESPEN practical guideline: Clinical Nutrition in cancer. *Clin. Nutr.* **2021**, *40*, 2898–2913. [CrossRef] [PubMed]

31. Halawa, A.; Rowe, S.; Roberts, F.; Nathan, C.; Hassan, A.; Kumar, A.; Suvakov, B.; Edwards, B.; Gray, C. A Better Journey for Patients, a Better Deal for the NHS: The Successful Implementation of an Enhanced Recovery Program after Renal Transplant Surgery. *Exp. Clin. Transpl.* **2018**, *16*, 127–132.

32. Muller, S.; Zalunardo, M.P.; Hubner, M.; Clavien, P.A.; Demartines, N. A fast-track program reduces complications and length of hospital stay after open colonic surgery. *Gastroenterology* **2009**, *136*, 842–847. [CrossRef]

33. Dort, J.C.; Farwell, D.G.; Findlay, M.; Huber, G.F.; Kerr, P.; Shea-Budgell, M.A.; Simon, C.; Uppington, J.; Zygun, D.; Ljungqvist, O.; et al. Optimal Perioperative Care in Major Head and Neck Cancer Surgery with Free Flap Reconstruction: A Consensus Review and Recommendations from the Enhanced Recovery after Surgery Society. *JAMA Otolaryngol. Head Neck Surg.* **2017**, *143*, 292–303. [CrossRef]

34. Moore, C.; Pegues, J.; Narisetty, V.; Spankovich, C.; Jackson, L.; Jefferson, G.D. Enhanced Recovery after Surgery Nutrition Protocol for Major Head and Neck Cancer Surgery. *OTO Open* **2021**, *5*, 2473974X211021100. [CrossRef] [PubMed]

35. Schattner, M. Enteral nutritional support of the patient with cancer: Route and role. *J. Clin. Gastroenterol.* **2003**, *36*, 297–302. [CrossRef]

36. Mekhail, T.M.; Adelstein, D.J.; Rybicki, L.A.; Larto, M.A.; Saxton, J.P.; Lavertu, P. Enteral nutrition during the treatment of head and neck carcinoma: Is a percutaneous endoscopic gastrostomy tube preferable to a nasogastric tube? *Cancer* **2001**, *91*, 1785–1790. [CrossRef]

37. Lees, J. Incidence of weight loss in head and neck cancer patients on commencing radiotherapy treatment at a regional oncology centre. *Eur. J. Cancer Care* **1999**, *8*, 133–136. [CrossRef] [PubMed]

38. Orell-Kotikangas, H.; Österlund, P.; Mäkitie, O.; Saarilahti, K.; Ravasco, P.; Schwab, U.; Mäkitie, A.A. Cachexia at diagnosis is associated with poor survival in head and neck cancer patients. *Acta Otolaryngol.* **2017**, *137*, 778–785. [CrossRef] [PubMed]

39. Economopoulou, P.; de Bree, R.; Kotsantis, I.; Psyrri, A. Diagnostic Tumor Markers in Head and Neck Squamous Cell Carcinoma (HNSCC) in the Clinical Setting. *Front. Oncol.* **2019**, *9*, 827. [CrossRef] [PubMed]

40. Brown, T.E.; Banks, M.D.; Hughes, B.G.M.; Lin, C.Y.; Kenny, L.M.; Bauer, J.D. Comparison of Nutritional and Clinical Outcomes in Patients with Head and Neck Cancer Undergoing Chemoradiotherapy Utilizing Prophylactic versus Reactive Nutrition Support Approaches. *J. Acad. Nutr. Diet.* **2018**, *118*, 627–636. [CrossRef]

41. Newman, L.A.; Vieira, F.; Schwiezer, V.; Samant, S.; Murry, T.; Woodson, G.; Kumar, P.; Robbins, K.T. Eating and weight changes following chemoradiation therapy for advanced head and neck cancer. *Arch. Otolaryngol. Head Neck Surg.* **1998**, *124*, 589–592. [CrossRef]

42. Beaver, M.E.; Matheny, K.E.; Roberts, D.B.; Myers, J.N. Predictors of weight loss during radiation therapy. *Otolaryngol. Head Neck Surg.* **2001**, *125*, 645–648. [CrossRef]

43. Doley, J. Enteral Nutrition Overview. *Nutrients* **2022**, *14*, 2180. [CrossRef] [PubMed]

44. Yanni, A.; Dequanter, D.; Lechien, J.R.; Loeb, I.; Rodriguez, A.; Javadian, R.; Van Gossum, M. Malnutrition in head and neck cancer patients: Impacts and indications of a prophylactic percutaneous endoscopic gastrostomy. *Eur. Ann. Otorhinolaryngol. Head Neck Dis.* **2019**, *136*, S27–S33. [CrossRef] [PubMed]

45. Silander, E.; Nyman, J.; Bove, M.; Johansson, L.; Larsson, S.; Hammerlid, E. Impact of prophylactic percutaneous endoscopic gastrostomy on malnutrition and quality of life in patients with head and neck cancer: A randomized study. *Head Neck* **2012**, *34*, 1–9. [CrossRef] [PubMed]

46. Rutter, C.E.; Yovino, S.; Taylor, R.; Wolf, J.; Cullen, K.J.; Ord, R.; Athas, M.; Zimrin, A.; Strome, S.; Suntharalingam, M. Impact of early percutaneous endoscopic gastrostomy tube placement on nutritional status and hospitalization in patients with head and neck cancer receiving definitive chemoradiation therapy. *Head Neck* **2011**, *33*, 1441–1447. [CrossRef]

47. Salas, S.; Baumstarck-Barrau, K.; Alfonsi, M.; Digue, L.; Bagarry, D.; Feham, N.; Bensadoun, R.J.; Pignon, T.; Loundon, A.; Deville, J.L.; et al. Impact of the prophylactic gastrostomy for unresectable squamous cell head and neck carcinomas treated with radio-chemotherapy on quality of life: Prospective randomized trial. *Radiother. Oncol.* **2009**, *93*, 503–509. [CrossRef]

48. Raykher, A.; Schattner, M.; Friedman, A. Safety and efficacy of pretreatment placement of PEG tubes in head and neck cancer patients undergoing chemoradiation treatment. *Clin. Nutr.* **2004**, *23*, 757.

49. Nguyen, N.P.; North, D.; Smith, H.J.; Dutta, S.; Alfieri, A.; Karlsson, U.; Lee, H.; Martinez, T.; Lemanski, C.; Nguyen, L.M.; et al. Safety and effectiveness of prophylactic gastrostomy tubes for head and neck cancer patients undergoing chemoradiation. *Surg. Oncol.* **2006**, *15*, 199–203. [CrossRef]

50. Anwander, T.; Ber`e, S.; Appel, T.; von Lindern, J.J.; Martini, M.; Mommsen, J.; Kipnowski, J.; Niederhagen, B. Percutaneous endoscopic gastrostomy for long-term feeding of patients with oropharyngeal tumors. *Nutr. Cancer* **2004**, *50*, 40–45. [CrossRef]

51. Scott, R.; Bowling, T.E. Enteral tube feeding in adults. *J. R. Coll. Physicians Edinb.* **2015**, *45*, 49–54. [CrossRef]

52. Sánchez-Sánchez, E.; Ruano-Álvarez, M.A.; Díaz-Jiménez, J.; Díaz, A.J.; Ordonez, F.J. Enteral Nutrition by Nasogastric Tube in Adult Patients under Palliative Care: A Systematic Review. *Nutrients* **2021**, *13*, 1562. [CrossRef]

53. Bossola, M. Nutritional interventions in head and neck cancer patients undergoing chemoradiotherapy: A narrative review. *Nutrients* **2015**, *7*, 265–276. [CrossRef] [PubMed]

54. Arends, J.; Bodoky, G.; Bozzetti, F.; Fearon, K.; Muscaritoli, M.; Selga, G.; van Bokhorst-de van der Schueren, M.A.; von Meyenfeldt, M.; Zürcher, G.; Fietkau, R.; et al. ESPEN Guidelines on Enteral Nutrition: Non-surgical oncology. *Clin. Nutr.* **2006**, *25*, 245–259. [CrossRef] [PubMed]

55. Pfister, D.G.; Ang, K.K.; Brizel, D.M.; Burtness, B.A.; Cmelak, A.J.; Colevas, A.D.; Dunphy, F.; Eisele, D.W.; Gilbert, J.; Gillison, M.L.; et al. Head and neck cancers. *J. Natl. Compr. Cancer Netw.* **2011**, *9*, 596–650. [CrossRef]

56. Langius, J.A.; Zandbergen, M.C.; Eerenstein, S.E.; van Tulder, M.W.; Leemans, C.R.; Kramer, M.H.; Weijs, P.J. Effect of nutritional interventions on nutritional status, quality of life and mortality in patients with head and neck cancer receiving (chemo)radiotherapy: A systematic review. *Clin. Nutr.* **2013**, *32*, 671–678. [CrossRef]

57. Bossola, M.; Antocicco, M.; Pepe, G. Tube feeding in patients with head and neck cancer undergoing chemoradiotherapy: A systematic review. *JPEN J. Parenter. Enter. Nutr.* **2022**, *46*, 1258–1269. [CrossRef]
58. Soria, A.; Santacruz, E.; Vega-Piñeiro, B.; Gión, M.; Molina, J.; Villamayor, M.; Mateo, R.; Riveiro, J.; Nattero, L.; Botella-Carretero, J.I. Gastrostomy vs nasogastric tube feeding in patients with head and neck cancer during radiotherapy alone or combined chemoradiotherapy. *Nutr. Hosp.* **2017**, *34*, 512–516. [CrossRef]
59. Corry, J.; Poon, W.; McPhee, N.; Milner, A.D.; Cruickshank, D.; Porceddu, S.V.; Rischin, D.; Peters, L.J. Randomized study of percutaneous endoscopic gastrostomy versus nasogastric tubes for enteral feeding in head and neck cancer patients treated with (chemo)radiation. *J. Med. Imaging Radiat. Oncol.* **2008**, *52*, 503–510. [CrossRef]
60. Compher, C.; Bingham, A.L.; McCall, M.; Patel, J.; Rice, T.W.; Braunschweig, C.; McKeever, L. Guidelines for the provision of nutrition support therapy in the adult critically ill patient: The American Society for Parenteral and Enteral Nutrition. *JPEN J. Parenter. Enter. Nutr.* **2022**, *46*, 12–41. [CrossRef] [PubMed]
61. Vanek, V.W. Ins and outs of enteral access. Part 1: Short-term enteral access. *Nutr. Clin. Pract.* **2002**, *17*, 275–283. [CrossRef]
62. Bankhead, R.; Boullata, J.; Brantley, S.; Corkins, M.; Guenter, P.; Krenitsky, J.; Lyman, B.; Metheny, N.A.; Mueller, C.; Robbins, S.; et al. Enteral nutrition practice recommendations. *JPEN J. Parenter. Enter. Nutr.* **2009**, *33*, 122–167. [CrossRef] [PubMed]
63. Miller, K.R.; McClave, S.A.; Kiraly, L.N.; Martindale, R.G.; Benns, M.V. A tutorial on enteral access in adult patients in the hospitalized setting. *JPEN J. Parenter. Enter. Nutr.* **2014**, *38*, 282–295. [CrossRef] [PubMed]
64. Baskin, W.N. Acute complications associated with bedside placement of feeding tubes. *Nutr. Clin. Pract.* **2006**, *21*, 40–55. [CrossRef]
65. Sparks, D.A.; Chase, D.M.; Coughlin, L.M.; Perry, E. Pulmonary complications of 9931 narrow-bore nasoenteric tubes during blind placement: A critical review. *JPEN J. Parenter. Enter. Nutr.* **2011**, *35*, 625–629. [CrossRef] [PubMed]
66. Kozeniecki, M.; Fritzshall, R. Enteral Nutrition for Adults in the Hospital Setting. *Nutr. Clin. Pract.* **2015**, *30*, 634–651. [CrossRef]
67. Bear, D.E.; Champion, A.; Lei, K.; Smith, J.; Beale, R.; Camporota, L.; Barrett, N.A. Use of an Electromagnetic Device Compared with Chest X-ray to Confirm Nasogastric Feeding Tube Position in Critical Care. *JPEN J. Parenter. Enter. Nutr.* **2016**, *40*, 581–586. [CrossRef]
68. Welpe, P.; Frutiger, A.; Vanek, P.; Kleger, G.R. Jejunal feeding tubes can be efficiently and independently placed by intensive care unit teams. *JPEN J. Parenter. Enter. Nutr.* **2010**, *34*, 121–124. [CrossRef]
69. Stayner, J.L.; Bhatnagar, A.; McGinn, A.N.; Fang, J.C. Feeding tube placement: Errors and complications. *Nutr. Clin. Pract.* **2012**, *27*, 738–748. [CrossRef]
70. Li, Z.; Qi, J.; Zhao, X.; Lin, Y.; Zhao, S.; Zhang, Z.; Li, X.; Kissoon, N. Risk-Benefit Profile of Gastric vs Transpyloric Feeding in Mechanically Ventilated Patients: A Meta-Analysis. *Nutr. Clin. Pract.* **2016**, *31*, 91–98. [CrossRef]
71. Mahadeva, S.; Malik, A.; Hilmi, I.; Qua, C.S.; Wong, C.H.; Goh, K.L. Transnasal endoscopic placement of nasoenteric feeding tubes: Outcomes and limitations in non-critically ill patients. *Nutr. Clin. Pract.* **2008**, *23*, 176–181. [CrossRef]
72. Boullata, J.I.; Carrera, A.L.; Harvey, L.; Escuro, A.A.; Hudson, L.; Mays, A.; McGinnis, C.; Wessel, J.J.; Bajpai, S.; Beebe, M.L.; et al. ASPEN Safe Practices for Enteral Nutrition Therapy [Formula: See text]. *JPEN J. Parenter. Enter. Nutr.* **2017**, *41*, 15–103. [CrossRef] [PubMed]
73. Lucendo, A.J.; Friginal-Ruiz, A.B. Percutaneous endoscopic gastrostomy: An update on its indications, management, complications, and care. *Rev. Esp. Enferm. Dig.* **2014**, *106*, 529–539. [PubMed]
74. McClave, S.A.; Chang, W.K. Complications of enteral access. *Gastrointest. Endosc.* **2003**, *58*, 739–751. [CrossRef] [PubMed]
75. Rahnemai-Azar, A.A.; Rahnemaiazar, A.A.; Naghshizadian, R.; Kurtz, A.; Farkas, D.T. Percutaneous endoscopic gastrostomy: Indications, technique, complications and management. *World J. Gastroenterol.* **2014**, *20*, 7739–7751. [CrossRef]
76. Koehler, J.; Buhl, K. Percutaneous endoscopic gastrostomy for postoperative rehabilitation after maxillofacial tumor surgery. *Int. J. Oral Maxillofac. Surg.* **1991**, *20*, 38–39. [CrossRef]
77. Platek, M.E. The role of dietary counseling and nutrition support in head and neck cancer patients. *Curr. Opin. Support Palliat. Care* **2012**, *6*, 438–445. [CrossRef]
78. Baschnagel, A.M.; Yadav, S.; Marina, O.; Parzuchowski, A.; Lanni, T.B., Jr.; Warner, J.N.; Parzuchowski, J.S.; Ignatius, R.T.; Akervall, J.; Chen, P.Y.; et al. Toxicities and costs of placing prophylactic and reactive percutaneous gastrostomy tubes in patients with locally advanced head and neck cancers treated with chemoradiotherapy. *Head Neck* **2014**, *36*, 1155–1161. [CrossRef]
79. Barkmeier, J.M.; Trerotola, S.O.; Wiebke, E.A.; Sherman, S.; Harris, V.J.; Snidow, J.J.; Johnson, M.S.; Rogers, W.J.; Zhou, X.H. Percutaneous radiologic, surgical endoscopic, and percutaneous endoscopic gastrostomy/gastrojejunostomy: Comparative study and cost analysis. *Cardiovasc. Interv. Radiol.* **1998**, *21*, 324–328. [CrossRef]
80. Laasch, H.U.; Wilbraham, L.; Bullen, K.; Marriott, A.; Lawrance, J.A.; Johnson, R.J.; Lee, S.H.; England, R.E.; Gamble, G.E.; Martin, D.F. Gastrostomy insertion: Comparing the options—PEG, RIG or PIG? *Clin. Radiol.* **2003**, *58*, 398–405. [CrossRef]
81. Sigmon, D.F.; An, J. Nasogastric Tube. In *StatPearls*; StatPearls Publishing LLC.: Treasure Island, FL, USA, 2022.
82. Baeten, C.; Hoefnagels, J. Feeding via nasogastric tube or percutaneous endoscopic gastrostomy. A comparison. *Scand. J. Gastroenterol. Suppl.* **1992**, *194*, 95–98. [CrossRef]
83. Foutch, P.G.; Talbert, G.A.; Waring, J.P.; Sanowski, R.A. Percutaneous endoscopic gastrostomy in patients with prior abdominal surgery: Virtues of the safe tract. *Am. J. Gastroenterol.* **1988**, *83*, 147–150. [PubMed]
84. Zopf, Y.; Maiss, J.; Konturek, P.; Rabe, C.; Hahn, E.G.; Schwab, D. Predictive factors of mortality after PEG insertion: Guidance for clinical practice. *JPEN J. Parenter. Enter. Nutr.* **2011**, *35*, 50–55. [CrossRef]

85. Ahmad, I.; Mouncher, A.; Abdoolah, A.; Stenson, R.; Wright, J.; Daniels, A.; Tillett, J.; Hawthorne, A.B.; Thomas, G. Antibiotic prophylaxis for percutaneous endoscopic gastrostomy–a prospective, randomised, double-blind trial. *Aliment. Pharm.* **2003**, *18*, 209–215. [CrossRef] [PubMed]

86. Allison, M.C.; Sandoe, J.A.; Tighe, R.; Simpson, I.A.; Hall, R.J.; Elliott, T.S. Antibiotic prophylaxis in gastrointestinal endoscopy. *Gut* **2009**, *58*, 869–880. [CrossRef] [PubMed]

87. Warriner, L.; Spruce, P. Managing overgranulation tissue around gastrostomy sites. *Br. J. Nurs.* **2012**, *21*, S14–S16, S18, S20 passim. [CrossRef]

88. Schurink, C.A.; Tuynman, H.; Scholten, P.; Arjaans, W.; Klinkenberg-Knol, E.C.; Meuwissen, S.G.; Kuipers, E.J. Percutaneous endoscopic gastrostomy: Complications and suggestions to avoid them. *Eur. J. Gastroenterol. Hepatol.* **2001**, *13*, 819–823. [CrossRef]

89. Lau, G.; Lai, S.H. Fatal retroperitoneal haemorrhage: An unusual complication of percutaneous endoscopic gastrostomy. *Forensic. Sci. Int.* **2001**, *116*, 69–75. [CrossRef]

90. Klein, S.; Heare, B.R.; Soloway, R.D. The "buried bumper syndrome": A complication of percutaneous endoscopic gastrostomy. *Am. J. Gastroenterol.* **1990**, *85*, 448–451.

91. Sinclair, J.J.; Scolapio, J.S.; Stark, M.E.; Hinder, R.A. Metastasis of head and neck carcinoma to the site of percutaneous endoscopic gastrostomy: Case report and literature review. *JPEN J. Parenter. Enter. Nutr.* **2001**, *25*, 282–285. [CrossRef]

92. Brown, M.C. Cancer metastasis at percutaneous endoscopic gastrostomy stomata is related to the hematogenous or lymphatic spread of circulating tumor cells. *Am. J. Gastroenterol.* **2000**, *95*, 3288–3291. [CrossRef]

93. Hoffer, E.K.; Cosgrove, J.M.; Levin, D.Q.; Herskowitz, M.M.; Sclafani, S.J. Radiologic gastrojejunostomy and percutaneous endoscopic gastrostomy: A prospective, randomized comparison. *J. Vasc. Interv. Radiol.* **1999**, *10*, 413–420. [CrossRef]

94. Ryan, J.M.; Hahn, P.F.; Boland, G.W.; McDowell, R.K.; Saini, S.; Mueller, P.R. Percutaneous gastrostomy with T-fastener gastropexy: Results of 316 consecutive procedures. *Radiology* **1997**, *203*, 496–500. [CrossRef]

95. Lowe, A.S.; Laasch, H.U.; Stephenson, S.; Butterfield, C.; Goodwin, M.; Kay, C.L.; Glancy, S.; Jackson, S.; Brown, D.; McLean, P.; et al. Multicentre survey of radiologically inserted gastrostomy feeding tube (RIG) in the UK. *Clin. Radiol.* **2012**, *67*, 843–854. [CrossRef] [PubMed]

96. Shin, J.H.; Park, A.W. Updates on percutaneous radiologic gastrostomy/gastrojejunostomy and jejunostomy. *Gut Liver* **2010**, *4* (Suppl. S1), S25–S31. [CrossRef] [PubMed]

97. Ho, S.G.; Marchinkow, L.O.; Legiehn, G.M.; Munk, P.L.; Lee, M.J. Radiological percutaneous gastrostomy. *Clin. Radiol.* **2001**, *56*, 902–910. [CrossRef] [PubMed]

98. Cherian, P.; Blake, C.; Appleyard, M.; Clouston, J.; Mott, N. Outcomes of radiologically inserted gastrostomy versus percutaneous endoscopic gastrostomy. *J. Med. Imaging Radiat. Oncol.* **2019**, *63*, 610–616. [CrossRef] [PubMed]

99. Wollman, B.; D'Agostino, H.B.; Walus-Wigle, J.R.; Easter, D.W.; Beale, A. Radiologic, endoscopic, and surgical gastrostomy: An institutional evaluation and meta-analysis of the literature. *Radiology* **1995**, *197*, 699–704. [CrossRef]

100. Yuan, T.W.; He, Y.; Wang, S.B.; Kong, P.; Cao, J. Technical success rate and safety of radiologically inserted gastrostomy versus percutaneous endoscopic gastrostomy in motor neuron disease patients undergoing: A systematic review and meta-analysis. *J. Neurol. Sci.* **2020**, *410*, 116622. [CrossRef]

101. Anselmo, C.B.; Tercioti Junior, V.; Lopes, L.R.; Coelho Neto Jde, S.; Andreollo, N.A. Surgical gastrostomy: Current indications and complications in a university hospital. *Rev. Col. Bras. Cir.* **2013**, *40*, 458–462. [CrossRef]

102. Murayama, K.M.; Schneider, P.D.; Thompson, J.S. Laparoscopic gastrostomy: A safe method for obtaining enteral access. *J. Surg. Res.* **1995**, *58*, 1–5. [CrossRef]

103. Bravo, J.G.; Ide, E.; Kondo, A.; de Moura, D.T.; de Moura, E.T.; Sakai, P.; Bernardo, W.M.; de Moura, E.G. Percutaneous endoscopic versus surgical gastrostomy in patients with benign and malignant diseases: A systematic review and meta-analysis. *Clinics* **2016**, *71*, 169–178. [CrossRef]

104. Kim, J.; Lee, M.; Kim, S.C.; Joo, C.U.; Kim, S.J. Comparison of Percutaneous Endoscopic Gastrostomy and Surgical Gastrostomy in Severely Handicapped Children. *Pediatr. Gastroenterol. Hepatol. Nutr.* **2017**, *20*, 27–33. [CrossRef] [PubMed]

105. Bakhos, C.; Patel, S.; Petrov, R.; Abbas, A. Jejunostomy-technique and controversies. *J. Vis. Surg.* **2019**, *5*, 3.

106. Holmes, J.H.T.; Brundage, S.I.; Yuen, P.; Hall, R.A.; Maier, R.V.; Jurkovich, G.J. Complications of surgical feeding jejunostomy in trauma patients. *J. Trauma* **1999**, *47*, 1009–1012. [CrossRef] [PubMed]

107. Myers, J.G.; Page, C.P.; Stewart, R.M.; Schwesinger, W.H.; Sirinek, K.R.; Aust, J.B. Complications of needle catheter jejunostomy in 2,022 consecutive applications. *Am. J. Surg.* **1995**, *170*, 547–550. [CrossRef]

108. Inayet, N.; Neild, P. Parenteral nutrition. *J. R. Coll. Physicians Edinb.* **2015**, *45*, 45–48. [CrossRef]

109. Kraft, M.D.; Btaiche, I.F.; Sacks, G.S. Review of the refeeding syndrome. *Nutr. Clin. Pract.* **2005**, *20*, 625–633. [CrossRef]

110. Ridley, E.J.; Davies, A.R.; Parke, R.; Bailey, M.; McArthur, C.; Gillanders, L.; Cooper, D.J.; McGuinness, S.; for the Supplemental Parenteral Nutrition Clinical Investigators. Supplemental parenteral nutrition versus usual care in critically ill adults: A pilot randomized controlled study. *Crit. Care* **2018**, *22*, 12. [CrossRef] [PubMed]

111. Muthanandam, S.; Muthu, J. Understanding Cachexia in Head and Neck Cancer. *Asia Pac. J. Oncol. Nurs.* **2021**, *8*, 527–538. [CrossRef]

112. Mäkitie, A.A.; Alabi, R.O.; Orell, H.; Youssef, O.; Almangush, A.; Homma, A.; Takes, R.P.; López, F.; de Bree, R.; Rodrigo, J.P.; et al. Managing Cachexia in Head and Neck Cancer: A Systematic Scoping Review. *Adv. Ther.* **2022**, *39*, 1502–1523. [CrossRef]

113. Gotthardt, D.N.; Gauss, A.; Zech, U.; Mehrabi, A.; Weiss, K.H.; Sauer, P.; Stremmel, W.; Büchler, M.W.; Schemmer, P. Indications for intestinal transplantation: Recognizing the scope and limits of total parenteral nutrition. *Clin. Transpl.* **2013**, *25*, 49–55. [CrossRef] [PubMed]

114. Maudar, K.K. Total Parenteral Nutrition. *Med. J. Armed Forces India* **1995**, *51*, 122–126. [CrossRef] [PubMed]

115. Chowdary, K.V.; Reddy, P.N. Parenteral nutrition: Revisited. *Indian J. Anaesth.* **2010**, *54*, 95–103. [CrossRef]

116. Weimann, A.; Ebener, C.; Holland-Cunz, S.; Jauch, K.W.; Hausser, L.; Kemen, M.; Kraehenbuehl, L.; Kuse, E.R.; Laengle, F. Surgery and transplantation—Guidelines on Parenteral Nutrition, Chapter 18. *Ger. Med. Sci.* **2009**, *18*, 000069.

117. Shah, R.D.; Luketich, J.D.; Schuchert, M.J.; Christie, N.A.; Pennathur, A.; Landreneau, R.J.; Nason, K.S. Postesophagectomy chylothorax: Incidence, risk factors, and outcomes. *Ann. Thorac. Surg.* **2012**, *93*, 897–903. [CrossRef]

118. Scolapio, J.S.; Picco, M.F.; Tarrosa, V.B. Enteral versus parenteral nutrition: The patient's preference. *J. Parenter. Enter. Nutr.* **2002**, *26*, 248–250. [CrossRef]

119. Druyan, M.E.; Compher, C.; Boullata, J.I.; Braunschweig, C.L.; George, D.E.; Simpser, E.; Worthington, P.A.; American Society for Parenteral and Enteral Nutrition Board of Directors. Clinical Guidelines for the Use of Parenteral and Enteral Nutrition in Adult and Pediatric Patients: Applying the GRADE system to development of A.S.P.E.N. clinical guidelines. *JPEN J. Parenter. Enter. Nutr.* **2012**, *36*, 77–80. [CrossRef] [PubMed]

120. Ryu, J.; Nam, B.H.; Jung, Y.S. Clinical outcomes comparing parenteral and nasogastric tube nutrition after laryngeal and pharyngeal cancer surgery. *Dysphagia* **2009**, *24*, 378–386. [CrossRef] [PubMed]

121. Marinella, M.A. Refeeding syndrome: An important aspect of supportive oncology. *J. Support Oncol.* **2009**, *7*, 11–16.

122. Braga, M.; Ljungqvist, O.; Soeters, P.; Fearon, K.; Weimann, A.; Bozzetti, F. ESPEN Guidelines on Parenteral Nutrition: Surgery. *Clin. Nutr.* **2009**, *28*, 378–386. [CrossRef] [PubMed]

Article

Impact of Nutritional Status on Postoperative Outcomes in Cancer Patients following Elective Pancreatic Surgery

Renata Menozzi [1,*], Filippo Valoriani [1], Roberto Ballarin [2], Luca Alemanno [3,4], Martina Vinciguerra [1], Riccardo Barbieri [1], Riccardo Cuoghi Costantini [5], Roberto D'Amico [5], Pietro Torricelli [3,4] and Annarita Pecchi [3,4]

[1]   Division of Metabolic Diseases and Clinical Nutrition, Department of Specialistic Medicines, University Hospital of Modena, 41100 Modena, Italy
[2]   Hepato-Pancreato-Biliary Surgery and Liver Transplantation Unit, Policlinico Modena Hospital, 41125 Modena, Italy
[3]   Department of Radiology, University Hospital of Modena, 41224 Modena, Italy
[4]   Department of Medical and Surgical Sciences of Children and Adults, University Hospital of Modena, 41224 Modena, Italy
[5]   Unit of Clinical Statistics, University Hospital of Modena, 41224 Modena, Italy
*   Correspondence: rmenozzi@unimore.it; Tel.: +39-59-4225545; Fax: +39-59-4222760

Abstract: Background: Pancreatic surgery has been associated with important postoperative morbidity, mortality and prolonged length of hospital stay. In pancreatic surgery, the effect of poor preoperative nutritional status and muscle wasting on postsurgery clinical outcomes still remains unclear and controversial. Materials and Methods: A total of 103 consecutive patients with histologically proven carcinoma undergoing elective pancreatic surgery from June 2015 through to July 2020 were included and retrospectively studied. A multidimensional nutritional assessment was performed before elective surgery as required by the local clinical pathway. Clinical and nutritional data were collected in a medical database at diagnosis and after surgery. Results: In the multivariable analysis, body mass index (OR 1.25, 95% CI 1.04–1.59, p = 0.039) and weight loss (OR 1.16, 95% CI 1.06–1.29, p = 0.004) were associated with Clavien score I–II; weight loss (OR 1.13, 95% CI 1.02–1.27, p = 0.027) affected postsurgery morbidity/mortality, and reduced muscle mass was identified as an independent, prognostic factor for postsurgery digestive hemorrhages (OR 0.10, 95% CI 0.01 0.72, p = 0.03) and Clavien score I–II (OR 7.43, 95% CI 1.53–44.88, p = 0.018). No association was identified between nutritional status parameters before surgery and length of hospital stay, 30 days reintervention, 30 days readmission, pancreatic fistula, biliary fistula, Clavien score III–IV, Clavien score V and delayed gastric emptying. Conclusions: An impaired nutritional status before pancreatic surgery affects many postoperative outcomes. Assessment of nutritional status should be part of routine preoperative procedures in order to achieve early and appropriate nutritional support in pancreatic cancer patients. Further studies are needed to better understand the effect of preoperative nutritional therapy on short-term clinical outcomes in patients undergoing pancreatic elective surgery.

Keywords: pancreatic surgery; cancer; nutritional status; reduced muscle mass; sarcopenia; BMI; weight loss; malnutrition

Citation: Menozzi, R.; Valoriani, F.; Ballarin, R.; Alemanno, L.; Vinciguerra, M.; Barbieri, R.; Cuoghi Costantini, R.; D'Amico, R.; Torricelli, P.; Pecchi, A. Impact of Nutritional Status on Postoperative Outcomes in Cancer Patients following Elective Pancreatic Surgery. *Nutrients* **2023**, *15*, 1958. https://doi.org/10.3390/nu15081958

Academic Editors: Fernando Mendes, Diana Martins and Nuno Borges

Received: 7 March 2023
Revised: 31 March 2023
Accepted: 17 April 2023
Published: 19 April 2023

1. Introduction

Despite the large number of efforts to enhance the efficacy of varied therapeutic opportunities over the past decade, pancreatic tumors are still one of the most deadly cancers, and five-year survival rates are currently within the range of 6% to 10% [1,2]. Worldwide, exocrine pancreatic cancer is the seventh leading cause of cancer death in both sexes [3].

Among pancreatic cancer patients, survival rates are much better in those who have undergone surgery than those who are unresectable [4]. Regrettably, less than 20% of pancreatic cancer patients will be eligible for resectable surgery [5]. This low resection rate is strongly linked to an advanced cancer stage, the location of the tumor, patients' comorbidities and an impaired performance status [4]. Indeed, in pancreatic cancer patients, poor oral nutritional intake, catabolism due to malignancy and reduced intestinal absorption because of obstruction or exocrine insufficiency can synergically affect nutritional status and lead to malnutrition and loss of muscle mass [6]. These in turn worsen the patients' nutritional and performance status and their suitability for surgery.

Pancreatic surgery has been associated with important postoperative morbidity, mortality and prolonged length of hospital stay [7–10], primarily linked to pancreatic anastomotic leak [11]. Even though technological advances in surgical techniques and perioperative management have greatly improved the mortality rate after pancreatic resection, postoperative morbidity continues to be a significant critical issue [12–14].

Pancreatic resection has been identified as one of the most complex surgical procedures as a result of the extended resection, the resulting metabolic stress and the comparatively high rate of complications. This specific kind of surgery strongly modifies metabolic activities and nutritional conditions by triggering inflammation, stress hormones and cytokines. In this specific clinical setting, nutritional status before surgery can also affect postsurgery clinical outcomes. Indeed, in cancer patients, impaired muscle mass before pancreatic surgery is associated with worse long-term survival [15–18]. However, the effect of preoperative poor nutritional status and muscle wasting on postoperative complications, in-hospital mortality and length of stay still remains unclear and controversial; furthermore, studies on heterogeneity and risk of bias limit the strength of this conclusion [16–19]. Further research in this area is needed to obtain a definitive answer.

Our retrospective observational study aims to investigate the association between nutritional status before pancreatic elective surgery and short-term clinical outcomes in cancer patients.

2. Materials and Methods

This single-center, retrospective study was approved by the local Ethics Committee (no: 67/2022/OSS/AOUMO), and all living patients provided written informed consent.

Patients with histologically proven carcinoma undergoing elective pancreatic surgery in University Hospital of Modena from June 2015 through July 2020 were consecutively included and retrospectively studied. No neoadjuvant chemotherapy was administered to enrolled patients. Nutritional assessment was performed before elective surgery as required by local clinical pathway. Oral food intake was assessed by 24 h recall in order to define energy and protein intake. A 24 h dietary recall (24HR) is a structured interview that aims to collect detailed informations and knowledges about all foods, beverages and oral nutritional supplements consumed by patients in the last 24 h. Food models, images and other visual aids were used to support patients in judging and describing volume of portions. Energy requirement was defined in line with European Society for Clinical Nutrition and Metabolism (ESPEN) guidelines on nutrition in cancer patients [20]. Before surgery, computed tomography (CT) scan was performed in order to stage cancer disease and define muscle mass. Diagnosis of cancer-related malnutrition (CRM) was detected in line with Global Leadership Initiative on Malnutrition (GLIM) criteria [21], which include three phenotypic criteria (unintentional weight loss, reduced body mass index and loss of muscle mass) and two etiologic criteria (inflammation and reduced energy intake or absorption). To diagnose malnutrition, at least one phenotypic and one etiologic criterion must be identified [21]. Phenotypic metrics for staging severity of malnutrition as Stage 1 (moderate) and Stage 2 (severe) were available [21].

Clinical and nutritional data were collected in medical records and the hospital electronic medical database at diagnosis and after surgery, including the following variables: age, gender, height, weight, body mass index (BMI), unintentional weight loss %, American Society of Anesthesiology (ASA) score, kind of surgery, operation time and vascular resection.

Postsurgery clinical outcomes (length of hospital stay, morbidity, in-hospital mortality, 30 days reintervention, 30 days readmission, pancreatic fistula, biliary fistula, delayed gastric emptying and digestive hemorrhage) were recorded for all patients. The Clavien score was collected in order to grade adverse events that occur as a result of surgical procedures.

2.1. Muscle Mass Measurement

The CT scans performed by the patients for pancreatic disease staging were also used for the evaluation of the body composition, and in particular for the muscle mass analysis. CT investigations were performed with two pieces of equipment: General Electric VCT 64 slice CT scanner (Milwaukee) and General Electric Optima 64 slice CT scanner (Milwaukee). For the reconstructions and the acquisition of the anthropometric parameters, a GE Healthcare AW Volume Share 7 workstation was used, with software that allows one to selectively visualize certain tissues, such as that of muscle, by setting threshold values of density typical of the tissue, and in this case between -29 and $+150$ Hounsfield units (HU).

Thanks to selective visualization, it was possible to perform a more precise segmentation of the skeletal muscle tissue at the level of the third lumbar vertebra (L3), in which both transverse processes were clearly visible. Areas of interest (ROI) were then drawn using a software tool corresponding to the compartments to be analyzed, within which the area expressed in cm2 and the average density value were calculated automatically.

Total lumbar muscle area (TLA) (cm^2), including paraspinal and abdominal wall muscles at the L3 level, was calculated by the software after manually tracing an ROI including the psoas muscles, paraspinal muscles (erector spine, quadratus lumborum, multifidus) and wall muscles (transversus, internal and external oblique, rectus abdominis). The skeletal muscle index (SMI) is the parameter obtained from the ratio between the total area of the lumbar muscles (TLA) and the square of the height (cm^2/m^2)Figure 1. It is an index of normalization of skeletal muscle mass with respect to the patient's height. Reduced muscle mass was defined using default sex-specific SMI cutoff values: 52.4 cm^2/m^2 for men and 38.5 cm^2/m^2 for women [22].

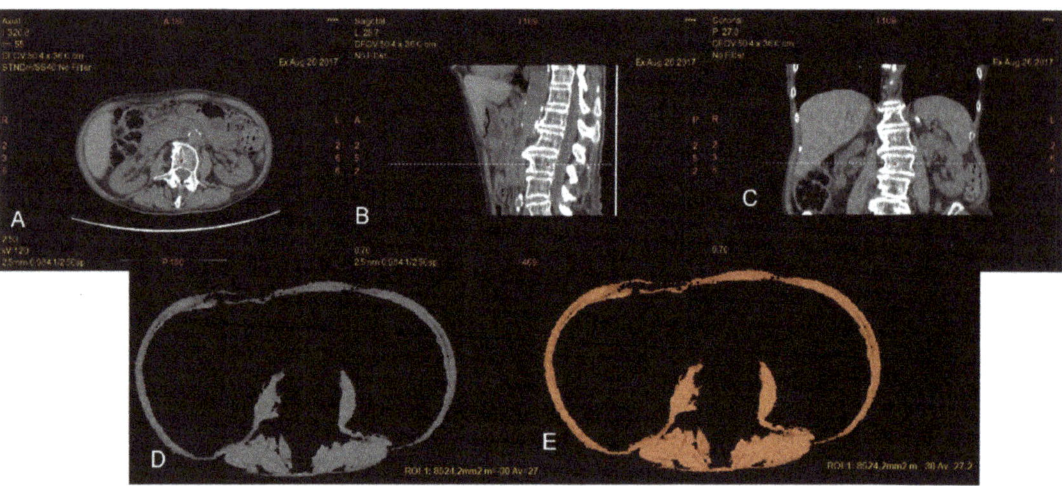

Figure 1. Patient with reduced muscle mass (**A–E**): in (**A**), axial CT image at the level of L3 as confirmed by the corresponding reference line in the sagittal (**B**) and coronal (**C**) planes. By applying the threshold $-29/+150$ HU, it is possible to selectively choose the muscle component and draw a ROI including all the musculature at the level of L3 (TLA). Automatically, it is possible to read the overall value of the traced area in cm^2 and the average value of the density.

2.2. Statistical Analysis

Continuous variables were expressed using mean and standard deviation or median and interquartile range; binary and categorical data were reported as frequencies and percentages.

The associations between length of hospital stay and the patients' characteristics were assessed using linear regression models, whereas logistic regression models were adopted to investigate the associations with respect to the other outcome measures. In the first place, for each outcome, we performed univariable analyses, and then, when appropriate, a multivariable model was estimated, considering all subjects with nonmissing data. The covariates included in the multivariable models were selected based on the results obtained from the univariable analysis and their clinical importance. In particular, for each outcome, all variables that were statistically associated with that outcome (nominal p-value less than 0.05) were selected; furthermore, the main clinical variables of this study, such as the reduced muscle mass indicator, were included. Subsequently, covariates with high association with respect to other covariates were excluded from the models to avoid issues of multicollinearity. Regarding propensity scores, they were used to estimate the probability that a subject has reduced muscle mass, holding other covariates constant. The selection of covariates for the propensity scores was carried out using the same methods described above. A multivariable logistic regression model was then estimated by including in the propensity score model that the variables potentially associated with reduced muscle mass.

Results were reported as the mean difference (MD) or odds ratio (OR) with 95% confidence intervals (CI). Analyses were carried out using R 4.2.1 statistical software.

3. Results

3.1. Patients' Characteristics

A total of 103 consecutive patients with a confirmed diagnosis of carcinoma and treated with elective pancreatic surgery in the University Hospital of Modena from June 2015 to July 2020 were retrospectively selected and included in the study. The main characteristics of the enrolled patients are summarized in Table 1. The mean age was 68.7 ($\pm$11.2)years, and 59.2% were male. The ASA score was 2 in 61.6% of patients. Pancreatic adenocarcinoma was the most prevalent (76.6%) cancer diagnosis and Vater papilla adenocarcinoma was the second most frequent (10.5%) histological diagnosis. Concerning surgery, pancreaticoduodenectomy was the most common (61.2%) procedure. A vascular resection was performed in 21.2% of surgery. Mean operation time was about 430 min.

Regarding nutritional status before surgery, the mean BMI was 24.6 kg/m^2 (DS $\pm$ 4.6), and unintentional weight loss was detected in 70.5% of the population. An unintentional weight loss higher than 5% was detected in 56.8% of our population. The mean SMI was 43.24 cm^2/m^2 (DS $\pm$ 13.3), the mean SMI in females was 39.89 cm^2/m^2 (DS $\pm$ 10.7) and the mean SMI in males was 48.24 cm^2/m^2 (DS $\pm$ 9.96). A condition of reduced muscle mass was observed in 56.3% of patients. A total of 48% of patients showed an energy oral intake of less than 75% of the daily energy nutritional requirement; differently52% of patients showed an energy oral intake of more than 75% of the daily energy nutritional requirement. Cancer-related malnutrition (CRM) was recognized in 75.7% of patients, in line with GLIM criteria (one phenotypic and one etiologic criterion). Nutritional parameters are summarized in Table 2.

Regarding postsurgery outcomes, the mean length of hospital stay was 16.2 ($\pm$11.8) days. We observed a prevalence of 73% (number 73) for postsurgery morbidity and 1.9% (number 2 events) for in-hospital mortality. A Clavien score of I–II was recognized in 53.5% (number 54 events) of patients, and the prevalence of digestive hemorrhage was 10% (number 10 events).

Table 1. General characteristics.

		Number	Percentage	Mean	±SD
Gender	Male	61	59.2		
	Female	42	40.8		
Age	Years			68.7	11.2
ASA Score	1	2	2		
	2	59	59.6		
	3	38	38.4		
Site of Cancer	Pancreatic adenocarcinoma	79	76.6		
	NET	6	5.9		
	Vater papilla carcinoma	10	10.5		
	Biliary carcinoma	8	8		
Type of Surgery	Pancreaticoduodenectomy	62	61.2		
	Distal pancreasectomy	6	5.8		
	Total pancreasectomy	35	34		
Operation Time	Minutes			430	107.1
Vascular Resection Performing		21	21.2		

Missing values were excluded from calculations. Abbreviations: ASA: American Society of Anesthesiologists; NET: neuroendocrine tumor; SD: standard deviation.

Table 2. Nutritional parameters before surgery.

		Number	Percentage	Mean	±SD
BMI	kg/m^2			24.6	4.6
Unintentional Weight Loss %	No weight Loss	28	29.5		
	<5%	13	13.7		
	5–10%	22	23.2		
	≥10%	32	33.6		
Oral Intake	>75%	51	52		
	75–50%	23	23.5		
	<50%	24	24.5		
SMI	Total—cm^2/m^2			43.24	13.34
	Female—cm^2/m^2			39.89	10.76
	Male—cm^2/m^2			48.24	9.96
Reduced Muscle Mass	Yes	58	56.3		
Diagnosis of Malnutrition (GLIM)	Yes	78	75.7		

Missing values were excluded from calculations. Abbreviations: BMI: body mass index; SMI: skeletal muscle index; GLIM: Global Leadership Initiative on Malnutrition.

3.2. Role of Nutritional Status before Surgery

Our analysis was aimed at searching for clinical and nutritional prognostic parameters. Following adjustment for significantly prognostic covariates at the univariate analysis, a multivariable analysis was performed, which confirmed weight loss before surgery (OR 1.13, 95% CI 1.02–1.27, $p = 0.027$) and total pancreasectomy (OR 8.91, 95% CI 1.79–70.71, $p = 0.016$) as independent prognostic factors in terms of postsurgery morbidity/mortality (Table 3).

Table 3. Multivariable analysis for the risk of adverse clinical outcomes.

		Length of Hospital Stay		Morbidity–Mortality		Clavien I–II	
	Category	MD (95% CI)	*p*-Value	OR (95% CI)	*p*-Value	OR (95% CI)	*p*-Value
Age		0.27 (−0.03; 0.56)	0.083	1.01 (0.95; 1.08)	0.694	1.02 (0.96; 1.08)	0.558
Pancreaticoduodenectomy	1	Reference		Reference		Reference	
Distal pancreasectomy	2	−7.88 (−22.16; 6.41)	0.284	0.83 (0.06; 12.54)	0.891	1.33 (0.09; 21.98)	0.830
Total pancreasectomy	3	7.08 (0.83; 13.34)	0.030	8.91 (1.79; 70.71)	0.016	1.07 (0.30; 3.83)	0.913
BMI (kg/m^2)		−0.14 (−0.84; 0.57)	0.702	1.13 (0.99; 1.35)	0.103	1.25 (1.04; 1.59)	0.039
Unintentional weight loss %		−0.08 (−0.48; 0.32)	0.705	1.13 (1.02; 1.27)	0.027	1.16 (1.06; 1.29)	0.004
ASA	1–2	Reference		Reference		Reference	
ASA	3	1.13 (−5.36; 7.62)	0.734	0.35 (0.07; 1.57)	0.177	0.70 (0.20; 2.43)	0.575
Operation time (min)		0.01 (−0.02; 0.04)	0.682	1.00 (0.99; 1.01)	0.709	0.99 (0.99; 1.00)	0.155
Vascular resection	No	Reference		Reference		Reference	
	Yes	2.61 (−4.43; 9.65)	0.470	0.74 (0.14; 4.24)	0.720	1.35 (0.32; 6.02)	0.685
Reduced muscle mass	No	Reference		Reference		Reference	
	Yes	−2.39 (−10.45; 5.68)	0.564	2.28 (0.37; 18.03)	0.395	7.43 (1.53; 44.88)	0.018

Abbreviations: MD: mean difference; CI: confidence interval; OR: odds ratio; BMI: body mass index; ASA: American Society of Anesthesiologists; min: minutes.

We also evaluated the prognostic impact of anthropometric measures before surgery. Following adjustment for significantly prognostic covariates at univariate analysis, a multivariable analysis was performed, which confirmed a significant and independent interaction between BMI (OR 1.25, 95% CI 1.04–1.59, *p* = 0.039), weight loss (OR 1.16, 95% CI 1.06–1.29, *p* = 0.004), reduced muscle mass (OR 7.43, 95% CI 1.53–44.88, *p* = 0.018) and Clavien score I–II. Table 3. Overall, no correlation was highlighted between nutritional status parameters before surgery and length of hospital stay, 30 days reintervention, 30 days readmission and pancreatic fistula (Table 3).

To define the association between reduced muscle mass and each outcome when the outcome's characteristics did not meet the requirements to estimate the desirable multivariable model, propensity scores were performed to adjust possible differences in covariates. Loss of muscle mass before surgery was associated with postsurgery digestive hemorrhage (OR 0.10, 95%CI 0.01 0.72, *p* = 0.03) (Table 4). No association was identified between reduced muscle mass before surgery, pancreatic fistula, biliary fistula, delayed gastric emptying, Clavien Score III–V, 30 days reintervention and 30 days readmission (Table 4).

Table 4. Multivariable analysis with propensity scores for the risk of adverse clinical outcomes.

	OR	95% CI		*p*-Value
Pancreatic Fistula				
Reduced muscle mass	1.79	0.35	10.88	0.50
Propensity score	0.28	0.02	2.92	0.30
Biliary Fistula				
Reduced muscle mass	0.41	0.01	18.69	0.62
Propensity score	4.51	0.01	5207.77	0.62
Delayed Gastric Emptying				
Reduced muscle mass	1.78	0.42	8.49	0.45
Propensity score	1.08	0.10	11.21	0.95
Digestive Hemorrhage				
Reduced muscle mass	0.10	0.01	0.72	0.03
Propensity score	101.86	3.07	10,309.36	0.02
Clavien Score III IV				
Reduced muscle mass	0.27	0.06	1.08	0.07
Propensity score	10.24	1.11	119.37	0.05
Clavien Score V				
Reduced muscle mass	0.55	0.01	27.12	0.74
Propensity score	2.13	0.01	1119.19	0.79
30 days Reintervention				
Reduced muscle mass	2.14	0.43	13.45	0.38
Propensity score	2.14	0.16	31.86	0.57
30 days Readmission				
Reduced muscle mass	0.40	0.06	2.75	0.35
Propensity score	1.03	0.05	19.74	0.98

Abbreviations: OR: odds ratio; CI: confidence interval.

4. Discussion

Pancreatic resection has been identified as one of the most complex surgical procedures as a result of the extended resection, the resulting metabolic stress and the comparatively high rate of complications. This specific kind of surgery strongly modifies metabolism and nutritional status by triggering inflammation, stress hormones and cytokines [23].

In order to support proper tissue healing and recovery or maintenance of organ functions after surgery, an effective anabolic response and adequate qualitative and quantitative nutritional substrates are required. Malnourished patients deplete their nutritional reserves quickly, which thereby affects their recovery and healing [23]. The development and progression of CRM can be associated with reduced oral nutritional intake and/or increased catabolism [24,25]. Recently, malnutrition has been defined through (one phenotypic and one etiologic criterion) weight loss, low body mass index, muscle wasting, poor energy intake and increased catabolism, in line with GLIM criteria [21].

In our study, the prevalence of CRM before surgery was very high (75.7%); as supposed by some preliminary publications [26–28], unintentional weight loss, low BMI, loss of muscle mass and Onodera's prognostic nutrition index (PNI) have been identified as possible independent prognostic factors for several adverse clinical outcomes after pancreatic surgery.

In particular, BMI (OR 1.25, 95% CI 1.04–1.59, $p = 0.039$) and weight loss (OR 1.16, 95% CI 1.06–1.29, $p = 0.004$) were associated with Clavien score I–II, while weight loss before surgery (OR 1.13, 95% CI 1.02–1.27, $p = 0.027$) affected postsurgery morbidity/mortality.

Our results also highlight the effect of reduced muscle mass before pancreatic surgery on postoperative clinical outcomes, since muscle mass before surgery has been identified as an independent, negative prognostic factor for postsurgery digestive hemorrhages (OR 0.10, 95% CI 0.01 0.72, $p = 0.03$) and Clavien score I–II (OR 7.43, 95% CI 1.53–44.88, $p = 0.018$).

Many publications improperly define sarcopenia only as a condition of reduced muscle mass without performing muscle function measurements as required [29–34]. In addition, different tools are available for the assessment of sarcopenia, and the interpretation of results across studies is particularly difficult [16,34]. In pancreatic cancer patients, the impact of preoperative loss of muscle mass on the surgical outcome is still unclear and controversial [16–19], and available publications show several limitations. In particular, studies included patients receiving pancreatic surgery for both benign and malignant diseases, and not all studies used comparable parameters (different methods, tools and/or cutoffs) to define reduced muscle mass; moreover, a comprehensive nutritional assessment was not performed [16–19] as required [29,34]. Otherwise, in order to reduce the bias described above, in our study, all included patients had a histologically proven carcinoma, and a global assessment of nutritional status was achieved for each patient before elective pancreatic surgery to diagnose CRM. Furthermore, a quantitative CT analysis of muscle mass was performed by applying the most widely used cutoff for reduced muscle mass as parameter to define malnutrition as recommended [17,19–22,34]. Indeed, our study was carried out in a high-volume institution for pancreatic surgery, and all the main adverse clinical outcomes after pancreatic surgery were taken into account.

Unfortunately, the retrospective design of this study limits the strength of its conclusions, and for this reason, these findings definitely need to be confirmed in a larger prospective study. Our findings strongly support the relationship between poor nutritional status before pancreatic surgery and short-term adverse clinical outcomes, since not only CT-detected reduced muscle mass but also unintentional weight loss and BMI could negatively affect several short-term clinical outcomes in pancreatic surgery. Further research is needed to better evaluate the effect of severity of malnutrition before surgery on short-term clinical outcomes.

Although they overlap, sarcopenia, reduced muscle mass and CRM are different conditions, the term sarcopenia is unfortunately extensively used to define two different clinical situations: muscle wasting alone and reduced muscle mass associated with an impaired muscle function [29]. This is a significant source of doubts, confusion and mistakes in the research field and in many clinical settings.

Some recent studies have investigated the single effect of the depletion of skeletal muscle mass on short-term clinical outcomes after pancreatic surgery, achieving unclear and controversial results [15–19]. Our findings greatly highlight the need to take not only muscle wasting into account, but all the diagnostic parameters for malnutrition as required by GLIM criteria [21], as part of a nutritional assessment before elective pancreatic surgery.

It should also be remembered that a large number of nutritional assessment tools and scores are available to properly identify cancer patients with malnutrition in surgical settings. Nevertheless, it is still unclear which of these tools are the most appropriate and careful in predicting postoperative adverse outcomes in pancreatic cancer patients [35]. For a long time, hematological biomarkers of status of visceral proteins and liver function have also been used as indicators of impaired nutritional status. Nevertheless, the real predictive efficacy of these biomarkers still remains unclear [36]. Additional research is needed in this area.

Notably, it is strictly recommended that all cancer patients undergoing pancreatic surgery should receive an early, comprehensive and multidimensional evaluation of their nutritional status before elective surgery [23]. Our research supports the advice to assess nutritional status before and after major pancreatic surgery using a validated tool. A multidimensional and comprehensive nutritional assessment is required in order to detect early muscle wasting and/or malnutrition, in line with GLIM criteria, which include three phenotypic criteria (unintentional weight loss, reduced body mass index and loss of muscle mass) and two etiologic criteria (inflammation and reduced energy intake or absorption) [21].

In malnourished patients and in patients at risk of malnutrition, nutritional therapy should be started prior to major cancer surgery, even if operations must be delayed. A period from 7 to 14 days can be suitable [37].

In addition, after elective pancreatic surgery, adjuvant chemotherapy is often indicated to reduce the risk of cancer recurrence. A large number of available publications in this specific clinical setting have reported an impaired response, a reduced tolerance and worse survival rates in pancreatic cancer patients with reduced muscle mass [16]. In light of this, during pancreatic surgery, the early prevention of malnutrition and/or proper perioperative nutritional therapy for CRM is required in order to improve tolerance to antineoplastic therapy and clinical outcome.

In this situation, the introduction of a Nutritional Oncology Board (NOB) in daily practice, aimed at a multidisciplinary assessment of patients and at implementing an early nutritional therapy from oncological diagnosis onward seems to be the right path to take [6]. The NOB, sharing common experiences, goals, obstacles and unmet needs, can be an optimal fertile ground for the birth of collaborative research activities. Indeed, the NOB aims to enhance a shared pathway of care from both a clinical and an organizational point of view, and ideally to also improve awareness towards clinical nutrition [6].

5. Conclusions

Our findings highlight that an impaired nutritional status before pancreatic surgery can strongly affect many short-term postoperative outcomes. CRM is a well-known risk factor for surgery-related complications. In cancer patients, before and after pancreatic surgery, proper and appropriate recognition and management of CRM are central clinical concerns and warrant a specific and multidisciplinary (clinical nutrition, oncology, surgery) approach to improve clinical outcomes. The measurement of nutritional status supported by CT analysis of body composition parameters, especially the muscle component, should be a gold standard for preoperative assessment in order to achieve early and appropriate nutritional support.

Further studies are needed to better understand the effect of preoperative nutritional therapy on short-term clinical outcomes in patients undergoing elective pancreatic surgery.

Author Contributions: Conception and design: R.M., F.V., A.P. and R.C.C. Analysis and interpretation of data: all authors. Statistical analysis: R.D. and R.C.C. Supervision: R.M., A.P. and P.T. All authors have read and agreed to the published version of the manuscript.

Funding: This research received no external funding.

Institutional Review Board Statement: The Ethical Review Board of each Institutional Hospital approved the present study. This study was performed in line with the principles of the Declaration of Helsinki.

Informed Consent Statement: Informed consent was obtained from all subjects involved in the study.

Data Availability Statement: Data are available on request from the authors.

Conflicts of Interest: The authors declare no conflict of interest.

References

1. Mizrahi, J.D.; Surana, R.; Valle, J.W.; Shroff, R.T. Pancreatic cancer. *Lancet* **2020**, *395*, 2008–2020. [CrossRef]
2. Arnold, M.; Rutherford, M.J.; Bardot, A.; Ferlay, J.; Andersson, T.M.L.; Myklebust, T.Å. Progress in cancer survival, mortality, and incidence in seven high-income countries 1995–2014 (ICBP SURVMARK-2): A population-based study. *Lancet Oncol.* **2019**, *20*, 1493–1505. [CrossRef]
3. Sung, H.; Ferlay, J.; Siegel, R.L.; Laversanne, M.; Soerjomataram, I.; Jemal, A.; Bray, F. Global Cancer Statistics 2020: GLOBOCAN Estimates of Incidence and Mortality Worldwide for 36 Cancers in 185 Countries. *CA Cancer J. Clin.* **2021**, *71*, 209–249. [CrossRef]
4. Ducreux, M.; Cuhna, A.S.; Caramella, C.; Hollebecque, A.; Burtin, P.; Goéré, D.; Seufferlein, T.; Haustermans, K.; Van Laethem, J.L.; Conroy, T.; et al. ESMO Guidelines Committee. Cancer of the pancreas: ESMO Clinical Practice Guidelines for diagnosis, treatment and follow-up. *Ann. Oncol.* **2015**, *26* (Suppl. S5), v56–v68. [CrossRef] [PubMed]

5. Wolfgang, C.L.; Herman, J.M.; Laheru, D.A.; Klein, A.P.; Erdek, M.A.; Fishman, E.K.; Hruban, R.H. Recent progress in pancreatic cancer. *CA Cancer J. Clin.* **2013**, *63*, 318–348. [CrossRef]

6. Rovesti, G.; Valoriani, F.; Rimini, M.; Bardasi, C.; Ballarin, R.; Di Benedetto, F.; Menozzi, R.; Dominici, M.; Spallanzani, A. Clinical Implications of Malnutrition in the Management of Patients with Pancreatic Cancer: Introducing the Concept of the Nutritional Oncology Board. *Nutrients* **2021**, *13*, 3522. [CrossRef] [PubMed]

7. Schmidt, C.M.; Choi, J.; Powell, E.S.; Yiannoutsos, C.T.; Zyromski, N.J.; Nakeeb, A.; Pitt, H.A.; Wiebke, E.A.; Madura, J.A.; Lillemoe, K.D. Pancreatic fistula following pancreaticoduodenectomy: Clinical predictors and patient outcomes. *HPB Surg.* **2009**, *2009*, 404520. [CrossRef]

8. Addeo, P.; Delpero, J.R.; Paye, F.; Oussoultzoglou, E.; Fuchshuber, P.R.; Sauvanet, A.; Cunha, A.S.; Le Treut, Y.P.; Adham, M.; Mabrut, J.Y.; et al. Pancreatic fistula after a pancreaticoduodenectomy for ductal adenocarcinoma and its association with morbidity: A multicentre study of the French Surgical Association. *HPB* **2014**, *16*, 46e55. [CrossRef] [PubMed]

9. Van Heek, N.T.; Kuhlmann, K.F.; Scholten, R.J.; de Castro, S.M.; Busch, O.R.; van Gulik, T.M.; Obertop, H.; Gouma, D.J. Hospital volume and mortality after pancreatic resection—A systematic review and an evaluation of intervention in The Netherlands. *Ann. Surg.* **2005**, *242*, 781e90. [CrossRef]

10. Nimptsch, U.; Krautz, C.; Weber, G.; Mansky, T.; Grützmann, R. Nationwide inhospital mortality following pancreatic surgery in Germany is higher than anticipated. *Ann. Surg.* **2016**, *264*, 1082e90. [CrossRef] [PubMed]

11. Uzunoglu, F.G.; Reeh, M.; Vettorazzi, E.; Ruschke, T.; Hannah, P.; Nentwich, M.F.; Vashist, Y.K.; Bogoevski, D.; König, A.; Janot, M.; et al. Preoperative Pancreatic Resection (PREPARE) score: A prospective multicenter-based morbidity risk score. *Ann. Surg.* **2014**, *260*, 857e64. [CrossRef]

12. Greenblatt, D.Y.; Kelly, K.J.; Rajamanickam, V.; Wan, Y.; Hanson, T.; Rettammel, R.; Winslow, E.R.; Cho, C.S.; Weber, S.M. Preoperative factors predict perioperative morbidity and mortality after pancreaticoduodenectomy. *Ann. Surg. Oncol.* **2011**, *18*, 2126e35. [CrossRef] [PubMed]

13. DeOliveira, M.L.; Winter, J.M.; Schafer, M.; Cunningham, S.C.; Cameron, J.L.; Yeo, C.J.; Clavien, P.A. Assessment of complications after pancreatic surgery: A novel grading system applied to 633 patients undergoing pancreaticoduodenectomy. *Ann. Surg.* **2006**, *244*, 931e9. [CrossRef] [PubMed]

14. Büchler, M.W.; Wagner, M.; Schmied, B.M.; Uhl, W.; Friess, H.; Z'graggen, K. Changes in morbidity after pancreatic resection: Toward the end of completion pancreatectomy. *Arch. Surg.* **2003**, *138*, 1310e4. [CrossRef]

15. Miligkos, M.; Wächter, S.; Manoharan, J.; Maurer, E.; Bartsch, D.K. Sarcopenia and sarcopenic obesity are significantly associated with poorer overall survival in patients with pancreatic cancer: Systematic review and meta-analysis. *Int. J. Surg.* **2018**, *59*, 19–26.

16. Chan, M.Y.; Chok, K.S.H. Sarcopenia in pancreatic cancer—Effects on surgical outcomes and chemotherapy. *World J. Gastrointest Oncol.* **2019**, *11*, 527–537. [CrossRef]

17. Pierobon, E.S.; Moletta, L.; Zampieri, S.; Sartori, R.; Brazzale, A.R.; Zanchettin, G.; Serafini, S.; Capovilla, G.; Valmasoni, M.; Merigliano, S.; et al. The Prognostic Value of Low Muscle Mass in Pancreatic Cancer Patients: A Systematic Review and Meta-Analysis. *J. Clin. Med.* **2021**, *10*, 3033. [CrossRef] [PubMed]

18. Bundred, J.; Kamarajah, S.K.; Roberts, K.J. Body composition assessment and sarcopenia in patients with pancreatic cancer: A systematic review and meta-analysis. *HPB* **2019**, *21*, 1603–1612. [CrossRef]

19. Ratnayake, C.B.; Loveday, B.P.; Shrikhande, S.V.; Windsor, J.A.; Pandanaboyana, S. Impact of preoperative sarcopenia on postoperative outcomes following pancreatic resection: A systematic review and meta-analysis. *Pancreatology* **2018**, *18*, 996–1004. [CrossRef]

20. Arends, J.; Bachmann, P.; Baracos, V.; Barthelemy, N.; Bertz, H.; Bozzetti, F.; Fearon, K.; Hütterer, E.; Isenring, E.; Kaasa, S.; et al. ESPEN guidelines on nutrition in cancer patients. *Clin. Nutr.* **2017**, *36*, 11–48. [CrossRef]

21. Cederholm, T.; Jensen, G.L.; Correia, M.I.T.D.; Gonzalez, M.C.; Fukushima, R.; Higashiguchi, T.; Baptista, G.; Barazzoni, R. GLIM Core Leadership Committee; GLIM Working Group. GLIM criteria for the diagnosis of malnutrition—A consensus report from the global clinical nutrition community. *Clin. Nutr.* **2019**, *38*, 1–9. [CrossRef] [PubMed]

22. Prado, C.M.; Lieffers, J.R.; McCargar, L.J.; Reiman, T.; Sawyer, M.B.; Martin, L.; Baracos, V.E. Prevalence and clinical implications of sarcopenic obesity in patients with solid tumours of the respiratory and gastrointestinal tracts: A population-based study. *Lancet Oncol.* **2008**, *9*, 629–635. [CrossRef]

23. Gianotti, L.; Besselink, M.G.; Sandini, M.; Hackert, T.; Conlon, K.; Gerritsen, A.; Griffin, O.; Fingerhut, A.; Probst, P.; Abu Hilal, M.; et al. Nutritional support and therapy in pancreatic surgery: A position paper of the International Study Group on Pancreatic Surgery (ISGPS). *Surgery* **2018**, *164*, 1035–1048. [CrossRef]

24. Fearon, K.; Strasser, F.; Anker, S.D.; Bosaeus, I.; Bruera, E.; Fainsinger, R.L.; Jatoi, A.; Loprinzi, C.; MacDonald, N.; Mantovani, G.; et al. Definition and classification of cancer cachexia: An international consensus. *Lancet Oncol.* **2011**, *12*, 489–495. [CrossRef]

25. Cooper, C.; Burden, S.T.; Molassiotis, A. An explorative study of the views and experiences of food and weight loss in patients with operable pancreatic cancer perioperatively and following surgical intervention. *Support Care Cancer* **2015**, *23*, 1025–1033. [CrossRef] [PubMed]

26. Bozzetti, F.; Mariani, L. Perioperative nutritional support of patients undergoing pancreatic surgery in the age of ERAS. *Nutrition* **2014**, *30*, 1267–1271. [CrossRef] [PubMed]

27. Pausch, T.; Hartwig, W.; Hinz, U.; Swolana, T.; Bundy, B.D.; Hackert, T.; Grenacher, L.; Büchler, M.W.; Werner, J. Cachexia but not obesity worsens the postoperative outcome after pancreatoduodenectomy in pancreatic cancer. *Surgery* **2012**, *152*, S81. [CrossRef]

28. Kanda, M.; Fujii, T.; Kodera, Y.; Nagai, S.; Takeda, S.; Nakao, A. Nutritional predictors of postoperative outcome in pancreatic cancer. *Br. J. Surg.* **2011**, *98*, 268–274. [CrossRef]
29. Cruz-Jentoft, A.J.; Bahat, G.; Bauer, J.; Boirie, Y.; Bruyère, O.; Cederholm, T. Sarcopenia: Revised European consensus on definition and diagnosis. *Age Ageing* **2019**, *48*, 16–31. [CrossRef]
30. Fielding, R.A.; Vellas, B.; Evans, W.J.; Bhasin, S.; Morley, J.E.; Newman, A.B.; Abellan van Kan, G.; Andrieu, S.; Bauer, J.; Breuille, D.; et al. Sarcopenia: An undiagnosed condition in older adults. Current consensus definition: Prevalence, etiology, and consequences. International working group on sarcopenia. *J. Am. Med. Dir. Assoc.* **2011**, *12*, 249–256. [CrossRef]
31. Muscaritoli, M.; Anker, S.D.; Argilés, J.; Aversa, Z.; Bauer, J.M.; Biolo, G.; Boirie, Y.; Bosaeus, I.; Cederholm, T.; Costelli, P.; et al. Consensus definition of sarcopenia, cachexia and pre-cachexia: Joint document elaborated by Special Interest Groups (SIG) "cachexia-anorexia in chronic wasting diseases" and "nutrition in geriatrics". *Clin. Nutr.* **2010**, *29*, 154–159. [CrossRef] [PubMed]
32. Morley, J.E.; Abbatecola, A.M.; Argiles, J.M.; Baracos, V.; Bauer, J.; Bhasin, S.; Cederholm, T.; Coats, A.J.; Cummings, S.R.; Evans, W.J.; et al. Society on Sarcopenia, Cachexia and Wasting Disorders Trialist Workshop. Sarcopenia with limited mobility: An international consensus. *J. Am. Med. Dir. Assoc.* **2011**, *12*, 403–409. [CrossRef] [PubMed]
33. Chen, L.K.; Woo, J.; Assantachai, P.; Auyeung, T.W.; Chou, M.Y.; Iijima, K.; Jang, H.C.; Kang, L.; Kim, M.; Kim, S.; et al. Asian Working Group for Sarcopenia: 2019 Consensus Update on Sarcopenia Diagnosis and Treatment. *J. Am. Med. Dir. Assoc.* **2020**, *21*, 300–307.e2. [CrossRef]
34. Cruz-Jentoft, A.J.; Gonzalez, M.C.; Prado, C.M. Sarcopenia ≠ low muscle mass. *Eur. Geriatr. Med.* **2023**, *14*, 225–228. [CrossRef] [PubMed]
35. Probst, P.; Haller, S.; Bruckner, T.; Ulrich, A.; Strobel, O.; Hackert, T.; Diener, M.K.; Büchler, M.W.; Knebel, P. Prospective trial to evaluate the prognostic value of different nutritional assessment scores in pancreatic surgery (NURIMAS Pancreas). *Br. J. Surg.* **2017**, *104*, 1053–1062. [CrossRef]
36. Lee, J.L.; Oh, E.S.; Lee, R.W.; Finucane, T.E. Serum albumin and prealbumin in calorically restricted, nondiseased individuals: A systematic review. *Am. J. Med.* **2015**, *128*, 1023.e1–1023.e22. [CrossRef]
37. Weimann, A.; Braga, M.; Carli, F.; Higashiguchi, T.; Hübner, M.; Klek, S.; Laviano, A.; Ljungqvist, O.; Lobo, D.N.; Martindale, R.G.; et al. ESPEN practical guideline: Clinical nutrition in surgery. *Clin. Nutr.* **2021**, *40*, 4745–4761. [CrossRef]

Systematic Review

Neoadjuvant Gastric Cancer Treatment and Associated Nutritional Critical Domains for the Optimization of Care Pathways: A Systematic Review

Marta Correia [1,*,†], Ines Moreira [1,†], Sonia Cabral [2], Carolina Castro [3], Andreia Cruz [3,4], Bruno Magalhães [5,6], Lúcio Lara Santos [3,7] and Susana Couto Irving [2,†]

[1] CBQF—Centro de Biotecnologia e Química Fina—Laboratório Associado, Escola Superior de Biotecnologia, Universidade Católica Portuguesa, Rua Diogo Botelho 1327, 4169-005 Porto, Portugal
[2] Portuguese Oncology Institute of Porto (IPO-Porto)—Nutrition, 4200-072 Porto, Portugal
[3] Experimental Pathology and Therapeutics Group, Portuguese Oncology Institute of Porto (IPO-Porto), 4200-072 Porto, Portugal
[4] Medical Oncology Department, Portuguese Oncology Institute of Porto (IPO-Porto), 4200-072 Porto, Portugal
[5] School of Health, University of Trás-os-Montes e Alto Douro (UTAD), 5000-801 Vila Real, Portugal
[6] Oncology Nursing Research Unit IPO Porto Research Center (CI-IPOP), Portuguese Oncology Institute of Porto (IPO Porto), 4200-072 Porto, Portugal
[7] Surgical Oncology Department, Portuguese Oncology Institute of Porto (IPO-Porto), 4200-072 Porto, Portugal
* Correspondence: mmcorreia@ucp.pt
† These authors contributed equally to this work.

Abstract: (1) Background: Gastric cancer patients are known to be at a high risk of malnutrition, sarcopenia, and cachexia, and the latter impairs the patient's nutritional status during their clinical course and also treatment response. A clearer identification of nutrition-related critical points during neoadjuvant treatment for gastric cancer is relevant to managing patient care and predicting clinical outcomes. The aim of this systematic review was to identify and describe nutrition-related critical domains associated with clinical outcomes. (2) Methods: We performed a systematic review (PROSPERO ID:CRD42021266760); (3) Results: This review included 14 studies compiled into three critical domains: patient-related, clinical-related (disease and treatment), and healthcare-related. Body composition changes during neoadjuvant chemotherapy (NAC) accounted for the early termination of chemotherapy and reduced overall survival. Sarcopenia was confirmed to have an independent prognostic value. The role of nutritional interventions during NAC has not been fully explored. (4) Conclusions: Understanding critical domain exposures affecting nutritional status will enable better clinical approaches to optimize care plans. It may also provide an opportunity for the mitigation of poor nutritional status and sarcopenia and their deleterious clinical consequences.

Keywords: nutritional status; nutrition support; nutrition impact symptoms; sarcopenia; neoadjuvant chemotherapy; pre-operative; stomach neoplasms

Citation: Correia, M.; Moreira, I.; Cabral, S.; Castro, C.; Cruz, A.; Magalhães, B.; Santos, L.L.; Irving, S.C. Neoadjuvant Gastric Cancer Treatment and Associated Nutritional Critical Domains for the Optimization of Care Pathways: A Systematic Review. *Nutrients* **2023**, *15*, 2241. https://doi.org/10.3390/nu15102241

Academic Editor: Lauri Byerley

Received: 27 March 2023
Revised: 26 April 2023
Accepted: 4 May 2023
Published: 9 May 2023

1. Introduction

Gastric cancer is the fifth most commonly diagnosed solid tumor and one of the leading causes of cancer-related deaths worldwide [1].

Gastric cancer patients are known to be at high risk of malnutrition, sarcopenia, and cachexia [2]. Often, malnutrition can be observed at diagnosis [2], and weight loss is also commonly reported at presentation [3]. Evidence has been accumulating to strengthen the adverse influence of an impaired nutritional status on a patient's clinical course, treatment response [4], and quality of life [5].

Neoadjuvant treatment (NT) encompasses the therapeutic approaches in the immediate period leading to surgery. NT in gastric cancer only includes chemotherapy [6] with the intention to reduce tumor size, increase the possibility of a R0 resection, attempt to treat potential micrometastatic disease, and improve overall survival.

ESMO's (European Society of Medical Oncology) 2022 guideline, which has been widely adopted in Europe [7], recommends a perioperative chemotherapy regimen with a combination of platinum/fluoropyrimidine for patients with resectable gastric cancer [8]. Following on from the MAGIC [9] and the FFCD/FNCLCC trials [10], the use of ECF (epirubicin, cisplatin, and 5-fluorouracil) or CF (cisplatin and 5-FU), respectively, is common. More recently, the FLOT4-AIO trial showed an increased benefit in the use of the FLOT (fluorouracil, leucovorin, oxaliplatin, and docetaxel) scheme in the perioperative setting [9]. This approach of a fluoropyrimidine-platinum doublet or triplet before surgery is recommended for 2 to 3 months [9]. During neoadjuvancy, most patients are managed at outpatient clinics; hence, it is crucial that this population be best supported to minimize adverse symptoms while remaining in the community. Further, and as a consequence, locally advanced gastric patients have longer care continuums with the prospect of accumulating several nutritional risk exposures along the way, encompassing both disease and iatrogenic impact.

Nutritional status has been shown to strongly impair chemotherapy (CT) success, postoperative prognosis, overall and disease-specific survival (DSS), the rate of complications, and the length of hospital stay.

Thus, a clearer identification and description of nutrition-related critical points throughout neoadjuvant treatment for gastric cancer might be relevant for improving patient care and outcomes.

2. Materials and Methods

This systematic review followed the preferred reporting items for systematic reviews and meta-analyses (PRISMA) reporting guidelines. The protocol has also been registered on the International Prospective Register of Systematic Reviews (PROSPERO), the University of York Centre for Reviews and Dissemination PROSPERO, August 2021 (CRD42021266760). Available from: https://www.crd.york.ac.uk/prospero/display_record.php?ID=CRD420 21266760, accessed on 8 August 2021.

2.1. Sources and Searches

The following databases: Pubmed/Medline, US National Library of Medicine's PubMed, ISI's Web of Knowledge, Cochrane, and Scopus databases were systematically searched using the search string (((Gastric OR Stomach) AND (Cancer OR Neoplasm OR Carcinoma OR Malignancy)) AND (Neoadjuvant OR Pre-operatory) AND (Nutritional status OR Nutritional intervention OR Nutritional support OR Dietary counseling OR Oral nutritional supplements)). An example of the search strategy used can be found in File S1 (Supplementary Data).

2.2. Study Selection

Two reviewers (MC and ICM) screened the studies against the review's predefined inclusion criteria (Table 1).

The types of studies that were included in this review were randomized clinical trials (RCTs), surveys, and observational studies such as cohort and case-control studies. All disagreements were debated until a consensus was reached with the assistance of a third subsequent reviewer (MC, ICM, and SCI). Fourteen studies were selected for inclusion.

2.3. Data Extraction

Data extraction was performed independently by two reviewers (MC and ICM), using a standardized data extraction template, and following the PI/ECO format. The extraction data divergence was resolved by the third independent reviewer (SCI).

Table 1. Inclusion and exclusion criteria.

Criteria	Inclusion	Exclusion
Patients' characteristics	Human adults aged $\geq$ 18 years	$\leq$18 years, pregnant women
	Medical oncology outpatients	Patients hospitalized: wards, care in acute or intensive or critical or long-term or end of life units. Surgical patients. Palliative patients.
Disease characteristics	Histologically documented primary gastric cancer suitable for a neoadjuvant treatment approach: - locally advanced gastric cancer, - newly diagnosed - without any prior antitumor treatment, - potentially resectable disease - clinically diagnosed stage: cT2-4/cN-any/cM0 or according to reported ultrasound, endoscopy, or enhanced CT/MRI scan: cT any/cN +/cM0.	Healthy In situ disease Other early stages Metastatic settings
Outcomes	Nutrition-critical domains: Patient-related critical points Clinical-related critical points (disease and treatment) Healthcare-related critical points	
Language	English, Portuguese, Spanish, and French	
Year	2011–2021	All other years

In cases of uncertainties about the data reported, the trials' authors are contacted in order to obtain more information; if contact is not possible, a team consensus decision is made about the inclusion or exclusion of studies.

3. Results

This systematic review included 14 studies (Figure 1), two of which (14.3%) were RCTs and 11 (78.6%) were cohorts, mainly retrospectively assessed; one out of the eleven included cohorts was assessed prospectively (9.1%). More than half (57.1%) of the included studies comprised body composition analysis data using CT scans or ultrasounds (42.9%), followed by nutritional biomarkers or indices (28.6%). Lastly, only three nutrition support studies (21.4%), comprising an immunonutrition and an ERAS protocol, were eligible.

The selected studies encompassed 1910 eligible patients, with 1360 included. The population characteristics may be found in the below diagram (Figure 2).

The included study overview and findings can be found summarized in Table 2.

Subsequently, the study findings were compiled into three previously defined critical domains: patient-related (Table 3), clinical-related (disease and treatment) (Table 4), and healthcare-related (Table 5). For further definition of the critical domains it was considered that patient-related critical points would include baseline (admission for cancer care) descriptions of advanced age, comorbidities, presence sarcopenia, and/or frailty including performance status, nutritional status, body composition, and gastrointestinal or other nutrition impairing symptoms present before treatment; clinical-related (disease and treatment) would include all of the above but concerning disease characteristics, treatment induced changes and clinical outcomes; lastly, the healthcare-related domain would include descriptions of clinical care, institutional and organizational issues, such as nutritional risk screening, nutrition support, access constraints, among others deemed relevant.

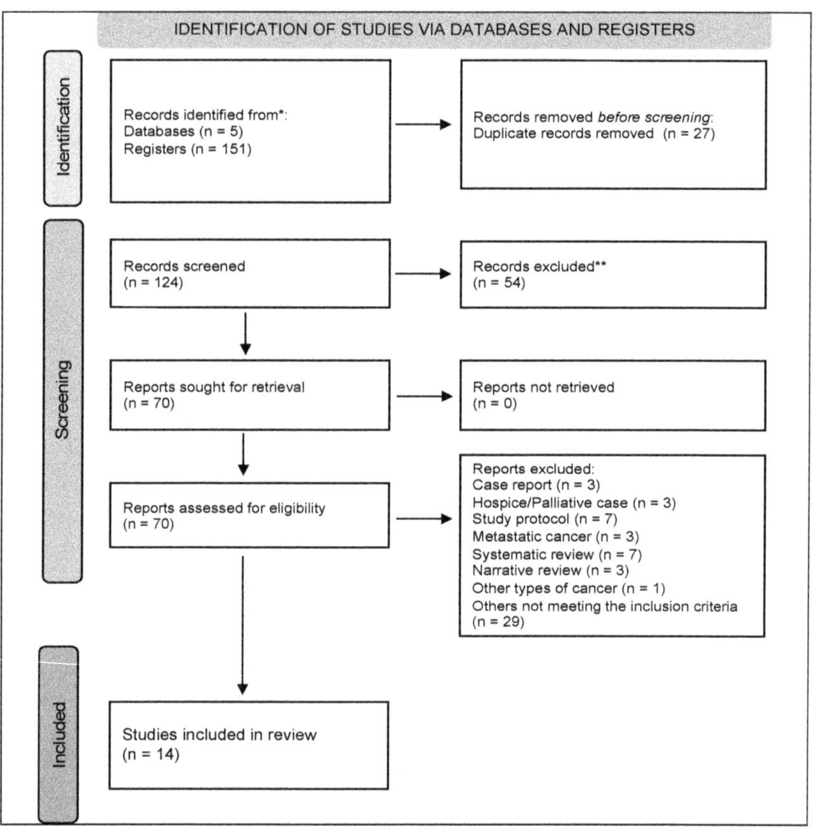

Figure 1. Flow chart of studies' selection (PRISMA) [11]. * Pubmed/Medline, US National Library of Medicine's PubMed, ISI's Web of Knowledge, Cochrane, and Scopus databases; ** records that were excluded from analysis.

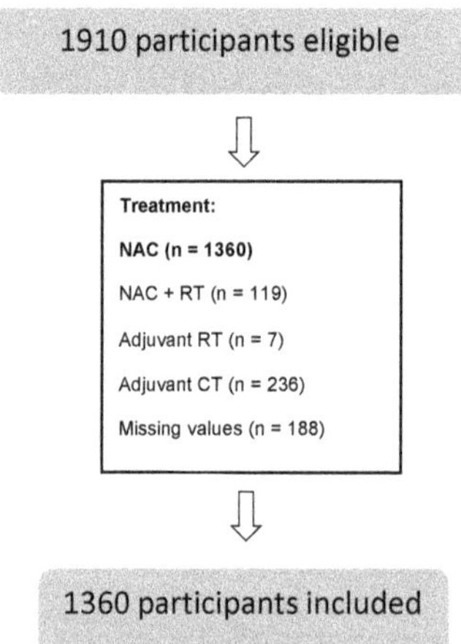

Figure 2. Population characteristics. Legend: NAC—neoadjuvant chemotherapy; CT—chemotherapy; RT—radiotherapy.

Table 2. Overview of studies and summary of findings.

Study, Country Year	Study Design	Tumor Type, Setting, and Sample Size	Study Description	Outcomes		
				Clinical Characteristics (OS, DFS, PFS, Age Comorbidities)	Treatment Complications (DLT, Completion)	Surgery-Related Events
			Body Composition Studies			
Palmela et al. [12] Portugal 2017	cohort (retrospective)	Locally advanced (LA) gastric or (GEJ) adenocarcinoma; NAC; n = 48.	CT Scan - cancer diagnosis; - completion of NAC (n = 43)	Survival reduction in sarcopenic obese patients. (median survival 6 months [95% CI = 3.9–8.5] vs. 25 months for patients who were obese and did not have sarcopenia [95% CI = 20.2–38.2]; log-rank test p = 0.000)	Higher percentage of DLT in sarcopenic/sarcopenic obese patients (non-significant trend). DLT in patients with sarcopenia (64% vs. 39%; p = 0.181) and sarcopenic obesity (80% vs. 42%; p = 0.165 Sarcopenic patients was associated with early CT termination (non-significant). (sarcopenic obesity (100% vs. 28%; p = 0.004) and sarcopenia (64% vs. 28%; p = 0.069) associated with early termination of CT; OR = 4.23; p = 0.050)	
Yamaoka et al. [13] Japan 2014	cohort (retrospective)	Gastric cancer; Open total gastrectomy (roux-en-y) - None or adjuvant CT < 6 months n = 102 - Adjuvant CT > 6 months n= 38	CT Scan - preoperatively; - postoperatively (1 year);		SMI decreased with NAC. Loss of skeletal muscle was not associated with sex, age (p > 0.05), diabetes, pathological stage, and preoperative SMI and ATI.	Loss of skeletal muscle was not associated with postoperative complications. NAC was an independent risk factor for loss of skeletal muscle.

Table 2. *Cont.*

Study, Country Year	Study Design	Tumor Type, Setting, and Sample Size	Study Description	Outcomes		
				Clinical Characteristics (OS, DFS, PFS, Age Comorbidities)	Treatment Complications (DLT, Completion)	Surgery-Related Events
Tan et al. [14] UK 2015	cohort (retrospective)	Esophagogastric cancer; NAC; $n = 89$	Combination of CT Scan, endoscopic ultrasound (EUS) and laparoscopy. Pre-treatment serum albumin levels and neutrophil-lymphocyte ratio, weight and height.	Median OS for sarcopenic patients was lower than for not sarcopenic patients. (569 days (IQ range: 357–1230 days) and for not sarcopenic 1013 days (IQ range: 496–1318 days, log-rank test, $p = 0.04$) No significant difference in OS in patients who experienced DLT compared with those that did not. (810 days [IQ range: 323–1417] vs. 859 days [IQ range: 445–1269]; $p = 0.665$)	Sarcopenic patients had lower BMI and BSA. BMI, BSA and sarcopenia were associated with DLT. (OR 2.95; 95% confidence interval, 1.23–7.09; $p = 0.015$)	
Zhang et al. [15] China 2021	cohort (retrospective)	Gastric cancer; Laparoscopic radical gastrectomy D2 lymph node dissection NAC; $n = 110$.	CT Scan Skeletal muscle, VAT and SAT: - Before NAC - After NAC (before the surgery).	Low VAT before NAC and low SAT after NAC was associated with low OS. Low VAT before and after NAC independent predictors for shorter DFS. (OR, 2.901; 95% CI, 1.205–6.983; $p = 0.018$)	Sarcopenia before NAC predicted adverse effects. Body composition and tumor pathological response were not significantly associated.	Higher BMI after NAC was associated with postoperative complications. Higher VAT was associated with higher incidence of postoperative complications.
Zhou et al. [16] China 2020	cohort (retrospective)	Gastric cancer; Radical gastrectomy; $n = 187$	Definition of gender-specific skeletal muscle/adipose cut-off values (CT Scan): BCS0 (normal) BCS1 (low skeletal muscle only) BCS2 (both low)	BCS2 group progressively shorter OS NAT was not the 3y OS independent prognostic factor after radical gastrectomy. (BCS2 HR: 3.5; 95% CI: 1.5–15.2; $p = 0.002$) were independent prognostic factor of 3 year OS; also low VAT before NAT (HR, 2.542; 95% CI; $p = 0.027$) and low SAT after NAT (HR, 2.743; 95% CI, 1.248–6.027; $p = 0.012$) were significantly associated with low OS	BCS2 group associated with lower BMI and higher NRS2002 score. ($p < 0.001$)	Body composition does not affect post-surgery complications. BCS2 group worse preoperative markers (hypoalbuminemia ($p < 0.001$), lower prealbumin, ($p < 0.001$), and IGF-1 levels ($p = 0.031$).

Table 2. *Cont.*

Study, Country Year	Study Design	Tumor Type, Setting, and Sample Size	Study Description	Outcomes		
				Clinical Characteristics (OS, DFS, PFS, Age Comorbidities)	Treatment Complications (DLT, Completion)	Surgery-Related Events
Zhang et al. [15] China 2021	cohort (retrospective)	Advanced GC Radical gastrectomy and NAC; $n = 157$	Skeletal muscle, VAT and SAT (CT Scan): - Before NAC - After NAC (before the surgery).	Marked loss of VAT, marked loss of SAT predicted shorter OS ($p = 0.022$) and DFS (Independent predictor for shorter DFS (hazards ratio = 2.67; 95% confidence interval = 1.182–6.047; $p = 0.018$) Skeletal muscle mass loss did not correlate well with nutritional status.	Marked loss of VAT and lower albumin levels not related.	
Jiang et al. [17] China 2021	cohort (retrospective)	Gastric adenocarcinoma; Radical surgery after NAC; $n = 203$	Body weight recorded at two-time points: - Before NAC - After NAC (before the surgery).	Independent risk factor for pathological response: - age (OR = 1.840, 95% CI 1.016–3.332, $p = 0.044$) - histological type	Weight loss was independent risk factor influencing NAC pathological responses: - >2.95% of body weight loss during NAC worsens CT response - maintaining weight trends (non-significant, (66.4% vs. 53.3%, $p = 0.059$): - better pathological response - higher rate of ONS usage Patients without weight loss had a higher rate of oral nutritional supplements than patients with weight loss during NAC (82.3%% vs. 70%, $\chi^2 = 4.261$, $p = 0.039$)	
Rinninela et al. [18] Italy 2021	cohort (retrospective)	Gastric adenocarcinoma; NAC; $N = 26$	CTScan Preoperative pre- and post-FLOT Lumbar SMI and adipose indices: - Before FLOT - After FLOT		BMI, SMI, and VAI variations not associated with short term outcomes: - Toxicity, delay and completion of perioperative FLOT (BMI from 24.4 kg/m^2 ± 3.7 to 22.6 kg/m^2 ± 3.1; $p < 0.0001$) - RECIST and Mandard A decrease of SMI $\geq$ 5% associates with a higher Mandard tumor-regression grade	Execution of gastrectomy not related with BMI, SMI, and VAI variations.

Table 2. *Cont.*

Study, Country Year	Study Design	Tumor Type, Setting, and Sample Size	Study Description	Outcomes		
				Clinical Characteristics (OS, DFS, PFS, Age Comorbidities)	Treatment Complications (DLT, Completion)	Surgery-Related Events
			Nutritional markers			
Jin et al. [19] China 2021	cohort (retrospective)	Gastric adenocarcinoma; NAC; $n = 272$	Serum albumin, total lymphocyte count, CONUT score. Blood samples: - within 2 weeks before the initial CT; - within 1 week before surgery; - at least 7 days after surgery (discharge)	For PFS and OS: - No prognostic significance between groups moderate/severe MN vs. normal/light MN group (pretreatment: $p = 0.482$, preoperative: $p = 0.446$; postoperative: $p = 0.464$, Kaplan–Meier with log-rank test) - worse association with high pre-treatment Hight pre-treatment CONUT score (HR, 1.618; 95% CI, 1.111–2.356; $p = 0.012$ independently associated with worse OS). Age Older age associates with high CONUT score (48.2% vs. 31.9%, $p = 0.010$) CONUT score) OS was better in pre-CT PNI-high group (3 year survival rate: 66.0% vs. 43.5%; 5 year survival rate: 55.5% vs. 25.6%, HR = 2.237, 95% CI = 1.271–3.393, $p = 0.005$), but there were no significant differences in OS between the post-CT groups (3 year survival rate: 61.5% vs. 61.9%, 5 year survival rate: 49.8% vs. 49.0%, $p = 0.775$)	CONUT-high-score associates, invasion, and lower pathological complete response rate. (HR, 1.615; 95% CI, 1.112–2.347; $p = 0.012$)	No change in the Moderate/severe MN status during NAT. Moderate/severe MN status increased postoperatively. No association between CONUT-score and postoperative complication.

Table 2. *Cont.*

Study, Country Year	Study Design	Tumor Type, Setting, and Sample Size	Study Description	Outcomes		
				Clinical Characteristics (OS, DFS, PFS, Age Comorbidities)	Treatment Complications (DLT, Completion)	Surgery-Related Events
Li et al. [20] China 2020	Cohort (prospective)	Gastric adenocarcinoma; Gastrectomy and NAC; $n = 225$	Nutritional markers (serum abumin, BMI, PNI) - pre-NAC - post-NAC		No significant differences in PNI, Alb, and mSISo after NAT ($p > 0.05$)	
Sun et al. [21] China 2016	cohort (retrospective)	Gastric cancer; NAC and radical surgery; $n = 117$	Markers for the PNI score: serum albumin, total lymphocyte count. Blood samples - 1 week before NAC - within 1 week before surgery. Patients PNI-high ($\geq$45) and PNI-low (<45).	OS Higher OS for PNI-high pre-NAC patients; No differences in OS for post-CT groups; Age Low pre-CT PNI associates with older age ($p = 0.007$ pre-CT PNI)	Anemia and lymphocytopenia associates with lower pre-NAC PNI (HR = 1.963, 95% CI = 1.101–3.499, $p = 0.022$), Pre-NAC PNI is an independent prognostic factor.	Pre-NAC PNI not associated with surgical complications ($p = 0.157$).
				Nutritional support studies		
Zhao et al. [22] China 2018	Randomized clinical trial	Adenocarcinoma of the esophagogastric junction; NAC and radiotherapy; $n = 66$	Control group: routine preoperative diet (35 kcal/kg/day) and research group: 500 mL of EN suspension # Data collected 48 h within the first hospitalization, the first day after NT and the first and 8th day after surgery		Higher BMI, serum PA, TP and ALB in trial group and a faster gastrointestinal recovery, shorter term use of drainage tubes, shorter hospital stay and less complications ($p < 0.05$)	Preoperative EN and ALB were independent risk factors for PRNS. ($p < 0.05$) Lower NRS2002 and PGSGA in the trial group ($p < 0.05$).
Claudino et al. [23] Brazil 2019	cohort (retrospective)	Gastric cancer. Subtotal or total gastrectomy. Patients who did or did not undergo NAC $n = 164$	Two groups: - immunonutrition ᵞ before surgery - conventional	No significant difference in OS rates at 6 months, 1 year, and 5 years (no significant difference in OS rates at 6 months (92.6% versus 85.0%; $p = 0.154$) 1 year (87.0% versus 78.5%; $p = 0.153$) and 5 years (69.6% versus 58.3%; $p = 0.137$). A trend for longer OS was found in immunonutrition group.	Immunonutrition group with less infectious complications (non-significant)	Immunonutrition group with less readmissions for surgical complications (non-significant) (41.1% vs. 48.1%; $p = 0.413$)

Table 2. *Cont.*

Study, Country Year	Study Design	Tumor Type, Setting, and Sample Size	Study Description	Outcomes		
				Clinical Characteristics (OS, DFS, PFS, Age Comorbidities)	Treatment Complications (DLT, Completion)	Surgery-Related Events
Zhao et al. [22] China 2018	Randomized clinical trial	Locally advanced gastric cancer; NAC; $n = 106$	$ ERAS group or standard care group.	Sarcopenic patients had lower OS than non-sarcopenic patients ($p < 0.05$). No significant differences in OS for patients who experienced DLT.	BMI and BSA were lower in sarcopenic patients and associated with DLT.	

Legend: NAC—neoadjuvant chemotherapy; LA—locally advanced; GEJ—gastroeosophageal junction; DLT—dose-limiting toxicity; NAT—neoadjuvant treatment; CT—chemotherapy; VAT—visceral adipose tissue; SAT—subcutaneous adipose tissue; DFS—disease-free survival; MN—malnutrition; BMI—body mass index; PA—prealbumin; PRNS—prognostic-related nutritional score; mSIS—modified systemic inflammation score; CT Scan—computed tomography scan; RECIST—response evaluation criteria in solid tumors; FLOT—fluorouracil plus leucovorin, oxaliplatin, and docetaxel; PGSGA—patient generated subjected global assessment; SMI—skeletal muscle index; CONUT—controlling nutritional status. # Nutrison fiber and oral nutritional supplementation (500 mL per bottle containing 500 kcal, 20 g protein, 19.45 g fat, and 61.5 g CH); 7 days before surgery apart from routine preoperative diet (35 kcal/kg/day). Both groups on Nutrison fiber within 48 h after surgery. ¥ Immune enteral diet enriched with arginine, omega-3 fatty acids, and nucleotides. $ ERAS group: sufficient preoperative patient education, normal diet until 6 h before surgery, liquid intake until 2 h before surgery, preoperative carbohydrate loading before surgery, analgesia with nonsteroidal anti-inflammatory drugs, minimization of opioid pain management, avoidance of perioperative fluid overload, no routine use of NGT, no abdominal drains, early removal of bladder catheters, liquid diet on recovery from anesthesia, semi-liquid diet on return of bowel function, tolerated liquid diet and forced ambulation on the day of the surgery; NGT placed preoperatively and remained until flatus occurred, intra-abdominal drains placed during surgery until the day before discharge, not allowed oral intake until bowel flatus gastrointestinal movement occurred, usually remained in bed for approximately 2 days after surgery. Conventional group: gastrointestinal preparation before surgery, fasting from midnight.

3.1. Patient-Related Critical Points

3.1.1. Advanced Age

Age was described as relating to neoadjuvant chemotherapy pathological response and lower blood counts. It is an independent risk factor that significantly impacts pathological response in patients older than 60 years old (OR = 1.840, 95% CI 1.016–3.332, p = 0.044) [17]. Additionally, older age was significantly associated with both a lower (p = 0.007) pre-chemotherapy prognostic nutritional index (PNI) [21] and a high (48.2% vs. 31.9%, p = 0.010) controlling nutritional status (CONUT) score [19]. Surprisingly, age did not arise as a risk factor for significant loss of skeletal muscle (p > 0.05) [13].

3.1.2. Sarcopenia (Baseline, Pre-Treatment)

Sarcopenia accounted for adverse effects during treatment, including early termination of CT and reduced survival, but also a reduced BMI and body surface area (BSA). Sarcopenia at diagnosis was prevalent in three quarters (73.1%) of patients in the Rinninela et al. study [18]. Zhang et al. [5] identified sarcopenia before NT as a significant risk factor for treatment adverse effects during univariate analyses, and, subsequently, by multivariate logistic regression analyses (OR, 2.901; 95% CI, 1.205–6.983; p = 0.018), it remained an independent predictor for overall treatment-related adverse effects [5].

Regarding sarcopenic obesity, Palmela et al. [12] showed reduced OS (overall survival) (median survival 6 months [95% CI = 3.9–8.5] vs. 25 months for patients who were obese and did not have sarcopenia [95% CI = 20.2–38.2]; log-rank test p = 0.000). In the same study, sarcopenic obesity (100% vs. 28%; p = 0.004) and sarcopenia (64% vs. 28%; p = 0.069) were also associated with early termination of chemotherapy, with none of these patients capable of completing treatment plans. As such, the odds ratio of treatment termination was higher in patients with sarcopenia compared with patients without it (OR = 4.23; p = 0.050). When the authors analyzed muscle radiation attenuation, they also found the same outcomes (higher mean vs. lower, OR = 0.20; p = 0.040) [12]. Tan et al. [14] showed a median OS for sarcopenic patients of 569 days (IQ range: 357–1230 days) and for patients who were not sarcopenic of 1013 days (IQ range: 496–1318 days) (log-rank test, p = 0.04). However, they found no significant difference in overall survival in patients who experienced DLT compared with those that did not (810 days [IQ range: 323–1417] vs. 859 days [IQ range: 445–1269]; p = 0.665).

Looking at dose-limiting toxicity (DLT), only sarcopenia (multivariate analysis) was independently associated with DLT (odds ratio, 2.95; 95% confidence interval, 1.23–7.09; p = 0.015) [14]. On the contrary, Palmela et al. [12] found a non-significant trend for a DLT in patients with sarcopenia (64% vs. 39%; p = 0.181) and sarcopenic obesity (80% vs. 42%; p = 0.165), but no corresponding significant association with subsequent treatment response [12]. On multivariate analysis, the odds of treatment termination were higher in patients with sarcopenia (odds ratio = 4.23; p = 0.050).

Sarcopenic patients also seemed to have a lower BMI and BSA when compared with those who did not have sarcopenia [22].

Only one study assessed loss of skeletal muscle related to gender or comorbidities, such as type 1 diabetes, but did not find any significant association [13].

3.1.3. BMI (Baseline, Pre-Treatment)

Baseline BMI (pre-NAC) is associated with adverse effects during treatments and overall survival (OS). Two studies showed that both underweight and overweight at baseline BMI seem significantly associated with OS and a significant risk factor for adverse effects (pre-treatment BMI < 18.5 kg/m^2; univariate analysis: HR = 2.015; p = 0.002; multivariate analysis: HR =1.456; p = 0.163) [19] and a BMI of 25 kg/m^2 (p = 0.04) [5]. Zhou et al. [16] indicated that a lower BMI in this setting was also significantly associated with low skeletal muscle mass (p < 0.001) and higher nutritional risk scores, NRS 2002 (p < 0.001). A study by Rinninela et al. also showed a decrease in the mean of the BMI with FLOT (from 24.4 kg/m^2 ± 3.7 to 22.6 kg/m^2 ± 3.1; p < 0.0001) [18].

3.1.4. Body Composition (Baseline, Pre-Treatment)

In the studies included, several associations were described between different body compositions and OS, but not all were significant. Patients with low skeletal muscle or, both, low skeletal and adipose mass had progressively shorter OS than patients with normal body composition parameters (3 year OS rates were 44.4% and 76.3%, respectively, for low skeletal muscle and adipose mass patients or for low skeletal muscle mass only vs. 88.2% for normal body composition parameters, $p < 0.001$). Low skeletal muscle mass (HR: 1.7; 95% CI: 1.2–3.7; $p < 0.001$) and low skeletal muscle and adipose mass (HR: 3.5; 95% CI: 1.5–15.2; $p = 0.002$) were independent prognostic factors of 3 year OS, namely after radical gastrectomy [16]. Other studies verified that, before NAT, the group with low visceral adipose tissue (VAT), defined as <120 cm^2, had significantly shorter OS ($p = 0.033$), as did the group with low (<99.5 cm^2) subcutaneous adipose tissue (SAT), after NAT ($p = 0.032$). In multivariate Cox regression analyses, low VAT before NAT (HR, 2542; 95% CI, $p = 0.027$) and low SAT after NAT (HR, 2.743; 95% CI, 1.248–6.027; $p = 0.012$) were significantly associated with low OS [5]. Moreover, patients with a marked loss of VAT ($\geq$35.7%) during NAT had significantly shorter OS ($p = 0.028$) compared to those with no or minor (<35.7%) VAT losses. In this study, during NAT, marked loss of adiposity (as per VAT or SAT) was considered a risk factor for long-term survival. Marked ($\geq$35.7%) VAT loss accompanied by marked SAT loss (high-risk group = NRS $\geq$ 3) independently predicted shorter OS (hazards ratio = 2.447; 95% confidence interval = 1.022–5.861; $p = 0.045$) [5]. However, Jin et al. [19] found no prognostic significance between the moderate or severe malnutrition group and the normal or light malnutrition group for OS at different times (pretreatment: $p = 0.482$; preoperative: $p = 0.446$; postoperative: $p = 0.464$, Kaplan–Meier with log-rank test).

There were no significant associations between different body compositions and progression free survival (PFS) or postoperative complications. Zhou et al. [16] found no significant differences in postoperative complications within 30 days among the different body composition groups, and Yamaoka et al. [13] found no association between postoperative complications and significant loss of skeletal muscle.

Different body compositions are related to disease-free survival (DFS). In the Zhang et al. [5] study, patients with low VAT before NT (<120 cm^2) had significantly poor DFS ($p = 0.022$), similar to those with low VAT after NT (<106 cm^2; $p = 0.025$). Multivariate analyses of DFS identified low VAT before NT (<120 cm^2; HR, 2.50; 95% CI, 1.22; $p = 0.012$) and low VAT after NT (<106 cm^2; HR, 2.51; 95% CI, 1.1725358; $p = 0.018$) as independent predictors for shorter DFS [5]. Moreover, patients with a marked loss of VAT ($\geq$35.7%) during NT had significantly shorter DFS ($p = 0.03$). Simultaneously, marked VAT loss with marked SAT loss (the high-risk group) was an independent predictor for shorter DFS (hazards ratio = 2.67; 95% confidence interval = 1.182–6.047; $p = 0.018$) [15].

In most studies, there was no significant relation between body composition and tumor pathological response, except for Rinninela et al., where a decrease higher than 5% in SMI was associated with a higher Mandard tumor regression grade [5,18], whereas Jiang et al. reported that weight loss significantly influences the pathological response to treatment [17].

3.1.5. Nutritional Markers and Indices

Regarding nutritional markers, patients with low skeletal muscle and adipose mass had a higher incidence of hypoalbuminemia ($p < 0.001$), lower prealbumin ($p < 0.001$), and lower IGF-1 levels ($p = 0.031$). Despite this, there were no significant differences in the preoperative concentrations of retinol-binding protein and transferrin [16]. Zhang et al. [15] found correlations between a marked loss of VAT and lower albumin levels ($p < 0.05$).

Associations between nutritional indices and OS are not consistent. Jin et al. [19] confirm that a high pre-treatment CONUT score (HR, 1.618; 95% CI, 1.111–2.356; $p = 0.012$) was independently associated with worse OS. According to Li et al. [20], PNI, albumin, and modified systemic inflammation score (mSIS) showed no significant difference after NT, and none of the pre-NT markers were independent prognostic factors for OS. However, OS was better in the pre-chemotherapy PNI-high group (3 year survival rate: 66.0% vs. 43.5%; 5 year survival rate: 55.5% vs. 25.6%, HR = 2.237, 95% CI = 1.271–3.393, p = 0.005), but there were no significant differences in OS between the post-chemotherapy groups (3 year survival rate: 61.5% vs. 61.9%; 5 year survival rate: 49.8% vs. 49.0%, p = 0.775) [21].

A high pre-treatment CONUT score (HR, 1.615; 95% CI, 1.112–2.347; p = 0.012) was independently associated with worse PFS [21].

Anemia and lymphocytopenia were significantly associated with a lower pre-chemotherapy PNI (p < 0.05) [21]. In the Sun et al. [21] study, pre-chemotherapy PNI was an independent prognostic factor (HR = 1.963, 95% CI = 1.101–3.499, p = 0.022), but no association was found between PNI and surgical complications (p = 0.157).

Table 3. Patient-related critical points: summary and findings.

Study and Country	Study Design	Tumor Type, Setting, and Sample Size	Study Description	Outcomes
Jiang et al. [17] China 2021	cohort (retrospective)	Gastric adenocarcinoma; Radical surgery after NAC; n = 203.	Body weight recorded at two-time points: evaluated before and after NAC (before the surgery)	Weight loss was independent risk factor influencing NAC pathological responses: - >2.95% of body weight loss during NAC worsens CT response
Jin et al. [19] China 2021	cohort (retrospective)	Gastric adenocarcinoma; NAC; n = 272.	Serum albumin, total lymphocyte count, CONUT score. Blood samples: - within 2 weeks before the initial CT; - within 1 week before surgery; - at least 7 days after surgery (discharge)	No change in the Moderate/severe MN status during NAT Moderate/severe MN status increased postoperatively MN group: worse association with high pre-treatment CONUT score Older age associates with a high CONUT score
Sun et al. [21] China 2016	cohort (retrospective)	GC; Preoperative CT and radical surgery; n = 117.	Markers for the PNI score: serum albumin, total lymphocyte count. Blood samples - 1 week before NAC - within 1 week before surgery. Patients PNI-high ($\geq$45) and PNI-low (<45).	Pre-NAC PNI not associated with surgical complications. Anemia and lymphocytopenia associates with lower pre-NAC PNI. Pre-NAC PNI is an independent prognostic factor. Higher survival for PNI-high pre-NAC patients. No differences in survival for post-CT groups. Low pre-CT PNI associates with older age.
Yamaoka et al. [13] Japan 2014	cohort (retrospective)	Primary GC; Open total gastrectomy with roux-en-y; n = 102 (none or adjuvant CT < 6 months) n= 38 (adjuvant CT > 6 months).	CT Scan - preoperatively; - postoperatively (1 year);	Loss of skeletal muscle was not associated with postoperative complications. NAC was an independent risk factor for loss of skeletal muscle. SMI decreased with NAC. Loss of skeletal muscle was not associated with sex, age, diabetes.
Zhang et al. [15] China 2021	cohort (retrospective)	GC Laparoscopic radical gastrectomy with D2 lymph node dissection followed by roux-en-y or billroth I reconstruction. NAC or CT (SOX, XELOX or FOLFOX); n = 110.	Skeletal muscle, VAT and SAT: - Evaluated before and after NAC (before the surgery).	Low VAT before NAC and low SAT after NAC was associated with low OS. Low VAT before and after NAC independent predictors for shorter DFS. Sarcopenia before NAC predicted adverse effects. Body composition and tumor pathological response were not significantly associated. Higher BMI after NAC was associated with postoperative complications. Higher VAT was associated with higher incidence of postoperative complications

Table 3. *Cont.*

Study and Country	Study Design	Tumor Type, Setting, and Sample Size	Study Description	Outcomes
Rinninela et al. [18] Italy 2021	cohort (retrospective)	Gastric adenocarcinoma; NAC; *n* = 26	Lumbar CTScan SMI and adipose indices: - Before FLOT - After FLOT	Almost $\frac{3}{4}$ of patients were sarcopenic at diagnosis
Zhang et al. [15] China 2021	cohort (retrospective)	Advanced GC (including gastroesophageal junction); Radical gastrectomy and NAC or CT. *n* = 157.	CTScan Skeletal muscle, VAT and SAT measure: - Before NAT - After NAT	Marked loss of VAT, marked loss of SAT predicted shorter OS and DFS. Skeletal muscle mass loss did not correlate well with nutritional status. Marked loss of VAT and lower albumin levels not related.
Palmela et al. Portugal 2017	cohort (retrospective)	Locally advanced adenocarcinoma from the stomach or gastroesophageal junction; NAC; *n* = 48.	CTScan - cancer diagnosis; - completion of NAC (*n* = 43)	Higher percentage of DLT in sarcopenic/sarcopenic obese patients (non-significant trend). Survival reduction in sarcopenic obese patients. Sarcopenic patients was associated with early CT termination (non-significant).
Zhou et al. [16] China 2020	cohort (retrospective)	GC Radical gastrectomy; *n* = 187.	Definition of gender-specific skeletal muscle/adipose cut-off values: BCS0 (normal) BCS1 (low skeletal muscle only) BCS2 (both low)	BCS2 group progressively shorter OS NAT was not the 3y OS independent prognostic factor after radical gastrectomy. BCS2 group associated with lower BMI and higher NRS2002 score. Body composition does not affect post-surgery complications. BCS2 group worse preoperative markers (hypoalbuminemia, lower prealbumin and IGF-1 levels).
Tan et al. [14] UK 2015	cohort (retrospective)	Oesophagogastric cancer; NAC; *n* = 89	Combination of CTScan, endoscopic ultrasound (EUS) and laparoscopy. Pre-treatment serum albumin levels, neutrophil-lymphocyte ratio, weight, height.	Median OS for sarcopenic patients was lower than for not sarcopenic patients. No significant difference in OS in patients who experienced DLT compared with those that did not. Sarcopenic patients had lower BMI and BSA. BMI, BSA and sarcopenia were associated with DLT.

Legend: NAC—neoadjuvant chemotherapy; LA—locally advanced; GC—gastric cancer; GEJ—gastroesophageal junction; DLT—dose-limiting toxicity; NAT—neoadjuvant treatment; CT—chemotherapy; VAT—visceral adipose tissue; SAT—subcutaneous adipose tissue; DFS—Disease free survival; MN—malnutrition; BMI—body mass index; PRNS—prognostic-related nutritional score; mSIS—modified systemic inflammation score; CT Scan—computed tomography scan; FLOT—fluorouracil plus leucovorin, oxaliplatin, and docetaxel; PGSGA—patient-generated subjected global assessment; ONS—oral nutritional supplements; CONUT—controlling nutritional status; PA—prealbumin; SMI—skeletal muscle index.

3.2. Clinical-Related Critical Points (Disease and Treatment)

The independent prognostic factor for 3-year OS after radical gastrectomy was tumor stage III (HR: 4.1; 95% CI: 2.1–17.8; $p < 0.001$) [16]. According to Jiang et al. [17], the independent risk factors influencing the effect of neoadjuvant chemotherapy were histological types. In the same study, clinical T stage and histological type of biopsy significantly influenced pathological response to the treatment [17].

The pathological stage was not associated with a significant loss of skeletal muscle [13]. However, Jiang et al. [17] described that those patients that did not lose weight had a better, although not significant, trend for pathological response than patients suffering from weight loss (66.4% vs. 53.3%, $p = 0.059$). Likewise, Rinninela et al. described a change in body composition (a decrease in SMI of $\geq 5\%$) and a lack of tumor-regressive changes [18].

Deep tumor invasion ($p = 0.025$) and a lower pathological complete response rate (1.2% vs. 6.6%, $p = 0.107$) were significantly associated with a higher CONUT-score [19], while Li et al. [20] found no significant difference in PNI, albumin, or mSIS after NAC.

Table 4. Clinical (disease and treatment)-related critical points: summary and findings.

Study and Country	Study Design	Tumor Type, Setting, and Sample Size	Study Description	Outcomes
Zhou et al. [16] China 2020	cohort (retrospective)	G; Radical gastrectomy; $n = 187$.	Gender-specific skeletal muscle/adipose cut-off values: BCS0 (normal) BCS1 (low skeletal muscle only) BCS2 (both low)	Body composition does not affect post-surgery complications. BCS2 group worse preoperative markers (hypoalbuminemia, lower prealbumin and IGF-1). BCS2 group progressively shorter OS. NAT was not the 3y OS independent prognostic factor after radical gastrectomy.
Yamaoka et al. [13] Japan 2014	cohort (retrospective)	Total gastrectomy with roux-en-y; $n = 102$ (none or adjuvant CT < 6 months) $n = 38$ (adjuvant CT > 6 months).	CT Scan: - preoperatively; - postoperatively (1 year);	SMI decreased with NAC (independent risk factor for loss of skeletal muscle). Loss of skeletal muscle was not associated with pathological stage, preoperative SMI and ATI. Loss of skeletal muscle was not associated with postoperative complications.
Li et al. [20] China 2020	cohort (Prospective)	Gastric adenocarcinoma; Gastrectomy and NAC $n = 225$	Nutritional markers (serum albumin, BMI, PNI): - pre-NAC - post-NAC	No significant differences in PNI, Alb, and mSISo after NAT.
Zhang et al. [15] China 2021	cohort (retrospective)	GC Laparoscopic radical gastrectomy, D2 lymph node dissection Neoadjuvant CT or CT-radiotherapy (SOX, XELOX or FOLFOX); $n = 110$	Skeletal muscle, VAT and SAT; CT Scan: - before NAT - after NAT	Sarcopenia before NAT is a significant and independent predictor for overall treatment AEs; Higher BMI after NAT was significantly correlated with postoperative complications; High VAT was significantly associated with higher incidence of postoperative complications; Low VAT before NAT and low SAT after NAT was significantly associated with low OS; Low VAT before and after NAT were independent predictors for shorter DFS; No significant association between body composition and tumor pathological response.
Rinninela et al. [18] Italy 2021	cohort (retrospective)	Gastric adenocarcinoma; NAC; $n = 26$	Lumbar CTScan SMI and adipose indices: - Before FLOT - After FLOT	BMI, SMI, and VAI variations were not associated with short outcomes: - toxicity - delay and completion of perioperative FLOT - RECIST, response - the execution of gastrectomy; A decrease in SMI $\geq$ 5% was associated with a higher Mandard tumor-regression grade Preoperative FLOT was associated with a reduction in SMI, BMI, and VAI

Table 4. *Cont.*

Study and Country	Study Design	Tumor Type, Setting, and Sample Size	Study Description	Outcomes
Jin et al. [19] China 2021	cohort (retrospective)	Gastric adenocarcinoma; NAC; $n = 272$	Serum albumin, total lymphocyte count, CONUT score. Blood samples: - within 2 weeks before the initial CT; - within 1 week before surgery; - at least 7 days after surgery (discharge)	No change in the moderate/severe MN status during NAT. Moderate/severe MN status increased postoperatively. No association between CONUT-score and postoperative complication. CONUT-high score associates: invasion and lower pathological complete response rate. For PFS and OS: no prognostic significance between MN groups.
Jiang et al. [17] China 2021	cohort (retrospective)	Gastric adenocarcinoma; Radical surgery after NAC; $n = 203$	Body weight recorded at two-time points: - before; - after NAC (before the surgery)	Weight loss was independent risk factor influencing NAC pathological responses: - >2.95% of body weight loss during NAC worsens chemotherapy response - maintaining weight trends (non-significant) better pathological response

Legend: NAC—neoadjuvant chemotherapy; NAT—neoadjuvant treatment; CT—chemotherapy; GC—gastric cancer; AEs—adverse events; VAT—visceral adipose tissue; SAT—subcutaneous adipose tissue; SMI—skeletal muscle index; OS—overall survival; PFS—progression free survival; MN—malnutrition; BMI—body mass index; CTScan—computed tomography scan; CONUT—controlling nutritional status; PRNS—prognostic-related nutritional score; PA—prealbumin; RECIST—response evaluation criteria in solid tumors; FLOT—fluorouracil plus leucovorin, oxaliplatin, and docetaxel.

3.3. Healthcare-Related Critical Points

It is known that the identification of nutritional risk by assessment tools and higher scores achieved by PG-SGA are more associated with postsurgical complications, such as anastomotic leakage and intra-abdominal infection [24]. Zhao et al. [22] found that the trial group had a higher BMI than the control group ($p < 0.005$), and on the eighth day after surgery, the rate of malnutrition according to the PG-SGA and nutritional risk according to the NRS-2002 became lower in the trial group ($p < 0.05$). This group had a faster gastrointestinal recovery, a shorter-term use of drainage tubes, a shorter hospital length of stay, fewer complications ($p < 0.05$), and higher concentrations of serum prealbumin, total proteins, and albumin ($p < 0.05$) [22].

Regarding nutritional support, the group using immunonutrition intervention had fewer infectious complications when compared with the conventional intervention group, but the differences were not statistically significant (41.1% vs. 48.1%; $p = 0.413$). Although the immunonutrition group had a lower percentage of patients who were readmitted for surgical complications than the conventional group, this difference was also not significant. Claudino et al. found no significant difference in survival rates at 6 months (92.6% versus 85.0%; $p = 0.154$), 1 year (87.0% versus 78.5%; $p = 0.153$), and 5 years (69.6% versus 58.3%; $p = 0.137$). Nevertheless, the immunonutrition patient group showed a trend for longer survival when compared with the conventional nutritional group [23].

Patients without weight loss had a higher rate of oral nutritional supplements than patients with weight loss during neoadjuvant chemotherapy (82.3% vs. 70%, $\chi^2 = 4.261$, $p = 0.039$) [17].

Table 5. Healthcare-related critical points: summary and findings.

Study and Country	Study Design	Tumor Type, Setting, and Sample Size	Study Description	Outcomes
Zhao et al. [22] China 2018	Randomized clinical trial	Adenocarcinoma of the esophagogastric junction; NAC and radiotherapy; $n = 66$	Control group: routine preoperative diet (35 kcal/kg/day) and research group: 500 mL of EN suspension [#] Data collected 48 h within the first hospitalization, the first day after NT and the first and eighth day after surgery	Higher BMI, serum PA, TP and ALB in trial group and a faster gastrointestinal recovery, shorter term use of drainage tubes, shorter hospital stay and less complications. Preoperative EN and ALB were independent risk factors for PRNS. Lower NRS2002 and PGSGA in the trial group
Claudino et al. [23] Brazil 2019	cohort (retrospective)	Stomach cancer; Patients who did or did not undergo NAC and who did undergo subtotal or total gastrectomy; $n = 164$.	The patients were divided into 2 groups: the immunonutrition group (received immune-modulatory diet oral or enteral, polymeric, hyperproteic diet, enriched with arginine, omega-3 fatty acids and nucleotides total 600 mL/d and 600 kcal/d for 5 to 7 days before surgery with at least 80% adherence) and conventional group	- immunonutrition group had less infectious complications compared with the conventional group, and had a lower percentage of patients who were readmitted for surgical complications than the conventional group, although differences were not significant; - immunonutrition group showed a trend for longer survival compared with the conventional nutrition group. - no significant difference in survival rates at 6 months or 1 year.
Zhao et al. [22] China 2018	Randomized clinical trial	Locally advanced gastric cancer; NAC; $n = 106$.	Patients were randomly assigned to the [$] ERAS or standard care group.	- serum PA, TP, and ALB concentrations were higher in the ERAS group than in the standard group.
Jiang et al. [17] China 2021	cohort (retrospective)	Gastric adenocarcinoma; Radical surgery after NAC; $n = 203$.	Body weight was recorded at the starting of NAC and before surgery, but after the last NAC. Patients with declining body weight during NAC were classified as weight loss group and patients who maintained/increased their weight during NAC were classified as no weight loss group.	Maintaining weight trends (non-significant): >higher rate of ONS usage.

Legend: NAC—neoadjuvant chemotherapy; LA—locally advanced; GC—gastric cancer; GEJ—gastroesophageal junction; DLT—dose-limiting toxicity; NAT—neoadjuvant treatment; CT—chemotherapy; VAT—visceral adipose tissue; SAT—subcutaneous adipose tissue; DFS—disease free; MN—malnutrition; BMI—body mass index; PA—prealbumin; PRNS—prognostic-related nutritional score; mSIS—modified systemic inflammation score; CT Scan—computed tomography scan; PGSGA—patient-generated subjective global assessment; ONS—oral nutritional supplements; CONUT—controlling nutritional status; SMI—skeletal muscle index. [#] Nutrison fiber and oral nutritional supplementation (500 mL per bottle containing 500 kcal, 20 g protein, 19.45 g fat, and 61.5 g CH); 7 days before surgery apart from routine preoperative diet (35 kcal/kg/day). Both groups on Nutrison fiber within 48 h after surgery. [$] ERAS group: sufficient preoperative patient education, normal diet until 6 h before surgery, liquid intake until 2 h before surgery, preoperative carbohydrate loading before surgery, analgesia with nonsteroidal anti-inflammatory drugs, minimization of opioid pain management, avoidance of perioperative fluid overload, no routine use of NGT, no abdominal drains, early removal of bladder catheters, liquid diet on recovery from anesthesia, semi-liquid diet on return of bowel function, tolerated liquid diet and forced ambulation on the day of the surgery; NGT placed preoperatively and remained until flatus occurred, intra-abdominal drains placed during surgery until the day before discharge, not allowed oral intake until bowel flatus gastrointestinal movement occurred, usually remained in bed for approximately 2 days after surgery. # Conventional group: gastrointestinal preparation before surgery, fasting from midnight, NGT placed preoperatively and remained until flatus occurred, intra-abdominal drains placed during surgery until the day before discharge, not allowed oral intake until bowel flatus gastrointestinal movement occurred, usually remained in bed for approximately 2 days after surgery.

4. Discussion

Gastric cancer (GC) is one of the most significant malignancies worldwide, with an annual burden prediction of ~1.8 million new cases and ~1.3 million deaths by 2040 [25]. Preoperative nutritional status is known to affect prognosis, OS, and DFS rates in surgical patients [26]. Indeed, the presence of MN in patients with radical surgical resections contributes to an increased incidence of postoperative complications and extended hospitalization [27].

It has been shown that NAC improves the overall therapeutic effects in locally advanced GC patients and does not increase the incidence of surgical complications. Additionally, undergoing GC surgery without previous NAC might significantly decrease the chance of effective reduction and radical resection [28]. NAC has been established because it confers clinical benefits over surgery [9], and it seems to be capable of enhancing immunological status, ameliorating GC patients' postoperative prognosis. Nevertheless, these widely adopted treatment proposals (e.g., FLOT) are also known to be frequently associated with a variety of gastrointestinal adverse effects, including anorexia, nausea, vomiting, stomatitis, and diarrhea, which can lead to a further deterioration of a patient's nutritional status, especially because these frequently present an already high risk of MN [29]. Furthermore, nutritional-related problems are one of the leading causes of hospital readmissions. Commonly, patients are not able to meet nutritional needs because of inadequate intake due to intolerance to oral and/or enteral feedings, typically manifested by nausea, vomiting, and/or early satiety [18]. For all these reasons, this review attempted to identify nutrition-related critical points during GC neoadjuvant management and their associations with clinical outcomes, as described in the selected literature.

Fourteen studies were analyzed, with 1360 patients included. Most studies were related to body composition and nutritional indexes. The results can be categorized as patient- and clinical- (disease- and treatment-) related ones. This review found considerably fewer concerning healthcare-related critical points, besides the application of nutritional risk identification tools.

Sarcopenia was predominantly considered a significant risk factor for adverse effects or the worst outcomes during treatment [5]. In addition, lower BMI and BSA relate to DLT and seem to lead to early treatment termination [14]. Interestingly, and still concerning the relationship of BSA with DLT, sarcopenic obesity was indeed associated with early treatment termination and reduced survival [12].

In these studies, GC patients' clinical outcomes, including OS, were shown to be closely related to many nutritional parameters, such as body weight. In fact, a lower BMI was associated with a poor OS [19], while a higher BMI seems to also be a significant risk factor for adverse effects during treatments [5]. Importantly, patients who lose weight during NAC seem to be at higher risk of worse CT effects. CT adverse effects, such as nausea, vomiting, and dysgeusia, may compromise food intake, which in turn could exacerbate weight loss. This weight loss is often sharp and marked and may contribute to the loss of skeletal muscle and to MN, which might account for the description of a low BMI being related to a poorer OS. Even though BMI signifies a relationship between weight and height and cannot describe body compartments. Furthermore, NAC trajectories are long, and the timing of some of the body composition analyses might not capture the dynamic nature of the body composition variations throughout treatment.

Adding on, GC patients, who simultaneously present with a high BMI and sarcopenia, had a higher BSA but low muscle mass [14]. This is an important consideration, as it is now established that patients with low muscle mass during CT treatments will have higher toxicity and more treatment interruptions. When compared with patients with normal muscle mass, sarcopenic obesity seems capable of shaping low OS [12]. Visceral adipose tissue, strongly linked with inflammation, is shown to have a higher risk of relapse in several cancer types. Here, DFS is also associated with low VAT, both before and after NT [12]. Indeed, adiposity levels are known to be associated with both increased cancer incidence and progression in multiple tumor types, and obesity is estimated to contribute to up to 20% of cancer-related deaths [30]. Adipose tissue mechanistically disrupts physiological homeostasis, but the underlying relationships between obesity and cancer are still poorly understood.

Concerning a patient's pathological response following treatment, this was also associated with weight loss, even though body composition did not seem to be. In addition, patients with low skeletal muscle and adipose mass had a higher incidence of hypoalbuminemia and low IGF-1 levels.

Regarding postoperative complications (within 30 days), Zhou et al. failed to show a significant association with body composition. Nonetheless, a higher BMI with a high VAT after NAC was significantly correlated with postoperative and treatment complications [15,16].

In relation to nutritional interventions, immunonutrition did not seem to have a significant association with complications or survival rates. On the other hand, patients with nutritional support strategies, such as oral nutritional supplements, were shown to have better weight stability throughout the proposed treatments [22,23].

Age has been found to be associated with physiological changes influencing drug pharmacokinetics, thus affecting cancer therapies [31]. In this review, one study related age to pathological response [17], showing a better pathological response in older patients than in younger ones. This could imply a more aggressive gastric cancer in younger patients and, hence, a poorer clinical response. Interestingly, older patients present lower CONUT and PNI scores, indicative of lower serum albumin and lymphocyte counts. Ageing also carries the risk of an impaired immune and hematologic system, potentially making elderly patients more vulnerable to infections and, in turn, more susceptible to earlier treatment termination [31].

Yamaoka et al. found that age was not a risk factor associated with a significant loss of skeletal muscle after total gastrectomy, even though it is expected that a higher percentage of muscle wasting occurs in the elderly over 65 years of age [13].

This systematic review tried to clarify the exposures and critical determinants that may be impacting GC patients' nutritional status during neoadjunvancy, and our findings seem to reinforce the importance of body composition throughout the course of NT. GC is known to be accompanied by MN, altered metabolism, and cancer-associated cachexia, with a significant impact on the patient's nutritional status, muscle compartments, function, and OS [32]. GC patients will then be exposed to the burden of persistent inflammation and metabolic deregulation, along with decreased food intake due to anorexia, nausea, and digestive impairments such as epigastric pain and early satiety. Many of these symptoms endured since clinical presentation and/or diagnosis, if unabated, will potentially be made worse by the prolonged multimodal treatment, which, in turn, might aggravate any involuntary weight loss or sarcopenia [26,33]. Although current guidelines already recommend screening and the systematic identification of nutritional risk as the first step for the nutritional care process of cancer patients, and as sarcopenia's independent prognostic value becomes more established, body composition assessment could emerge as a broader tool to support clinical decision making in patients with GC, namely dose and toxicity management [34].

Regardless, the exact role of nutritional support during NAC has yet to be fully explored. Even though the evidence shows that nutritional support in the immediate perioperative period with immune-nutrient-enriched formulas seems to reduce surgical complications, little is known about the type of nutritional interventions during NAC [26,33].

In addition, it is urgent to better comprehend the role of nutritional support in stabilizing and reversing sarcopenia and its role during cancer-associated body composition changes, specifically throughout NAC.

Most of the studies found and included in this review had a retrospective design and recruited a small sample size (single center). Many have also identified the following limitations: heterogeneous clinical data, inconsistencies in the prescribed treatment plan, time to follow up, and diverse cut-off values (File S2—Supplementary Data). This review has also identified a lack of studies documenting wider aspects that might influence nutritional status, such as healthcare and organizational critical points. Patients with NT proposals will be exposed for longer to treatments and hospital visits, and this might be even more concerning for those having to accommodate farther travel to reach reference centers. More is needed to better understand the nutritional status implications of these prolonged care continuum exposures and subsequent clinical outcomes.

Finally, NT is a period that normally encompasses several weeks and could undoubtedly represent an opportunity to identify, manage, and tackle nutritional-related issues that seem to be associated with several clinical outcomes and to provide the best supportive measures for GC patients.

5. Conclusions

Neoadjuvant chemotherapy in gastric cancer patients has the potential to contribute to an increase in catabolic stress, nutritional impact symptoms, malnutrition, and sarcopenia. Pursuing a better understanding of the exposure to critical domains affecting nutritional status risk and their determinants will enable proactive clinical approaches and optimized care plans by deploying appropriate and timely nutrition support so that there is an opportunity to mitigate poor nutritional status and sarcopenia alongside their deleterious clinical consequences.

Supplementary Materials: The following supporting information can be downloaded at: https://www.mdpi.com/article/10.3390/nu15102241/s1, File S1: Supplementary data: search strategy example; File S2: Limitations of the included studies.

Author Contributions: Conceptualization, M.C. and S.C.I.; Methodology, M.C., I.M. and S.C.I.; Software, I.M.; Formal Analysis, M.C., I.M. and S.C.I.; Investigation, all authors; Resources, M.C. and S.C.; Data Curation, M.C. and I.M.; Writing—Original Draft Preparation, M.C., I.M. and S.C.I.; Writing—Review and Editing, all authors; Supervision, M.C., S.C.I., L.L.S. and B.M.; Project Funding Acquisition, M.C. All authors have read and agreed to the published version of the manuscript.

Funding: This work was supported by national funds from FCT—Fundação para a Ciência e a Tecnologia through project UIDB/50016/2020.

Institutional Review Board Statement: Not applicable.

Informed Consent Statement: Not applicable.

Data Availability Statement: Data is unavailable due to privacy or ethical restrictions.

Acknowledgments: The authors would like to thank Filomena Gomes for her contribution. The authors would like to thank Paula Alves for her support.

Conflicts of Interest: The authors declare no conflict of interest.

References

1. Siegel, R.L.; Miller, K.D.; Fuchs, H.E.; Jemal, A. Cancer Statistics, 2022. *CA Cancer J. Clin.* **2022**, *72*, 7–33. [CrossRef] [PubMed]
2. Arends, J. Struggling with Nutrition in Patients with Advanced Cancer: Nutrition and Nourishment—Focusing on Metabolism and Supportive Care. *Ann. Oncol.* **2018**, *29* (Suppl. 2), ii27–ii34. [CrossRef] [PubMed]
3. Mansoor, W.; Roeland, E.J.; Chaudhry, A.; Liepa, A.M.; Wei, R.; Knoderer, H.; Abada, P.; Chatterjee, A.; Klempner, S.J. Early Weight Loss as a Prognostic Factor in Patients with Advanced Gastric Cancer: Analyses from REGARD, RAINBOW, and RAINFALL Phase III Studies. *Oncologist* **2021**, *26*, e1538–e1547. [CrossRef] [PubMed]
4. Fukuda, Y.; Yamamoto, K.; Hirao, M.; Nishikawa, K.; Maeda, S.; Haraguchi, N.; Miyake, M.; Hama, N.; Miyamoto, A.; Ikeda, M.; et al. Prevalence of Malnutrition Among Gastric Cancer Patients Undergoing Gastrectomy and Optimal Preoperative Nutritional Support for Preventing Surgical Site Infections. *Ann. Surg. Oncol.* **2015**, *22*, 778–785. [CrossRef] [PubMed]
5. Zhang, Y.; Li, Z.; Jiang, L.; Xue, Z.; Ma, Z.; Kang, W.; Ye, X.; Liu, Y.; Jin, Z.; Yu, J. Impact of Body Composition on Clinical Outcomes in People with Gastric Cancer Undergoing Radical Gastrectomy after Neoadjuvant Treatment. *Nutrition* **2021**, *85*, 111135. [CrossRef] [PubMed]
6. Byrd, D.R.; Brierley, J.D.; Baker, T.P.; Sullivan, D.C.; Gress, D.M. Current and Future Cancer Staging after Neoadjuvant Treatment for Solid Tumors. *CA Cancer J. Clin.* **2021**, *71*, 140–148. [CrossRef]
7. Lordick, F.; Carneiro, F.; Cascinu, S.; Fleitas, T.; Haustermans, K.; Piessen, G.; Vogel, A.; Smyth, E.C. Gastric Cancer: ESMO Clinical Practice Guideline for Diagnosis, Treatment and Follow-Up. *Ann. Oncol.* **2022**, *33*, 1005–1020. [CrossRef]
8. Smyth, E.C.; Nilsson, M.; Grabsch, H.I.; van Grieken, N.C.; Lordick, F. Gastric Cancer. *Lancet* **2020**, *396*, 635–648. [CrossRef]

9.   Cunningham, D.; Allum, W.; Stenning, S.; Thompson, J.; Van de Velde, C.; Nicolson, M.; Scarffe, H.; Lofts, F.; Falk, S.; Iveson, T.; et al. Perioperative Chemotherapy versus Surgery Alone for Resectable Gastroesophageal Cancer. *N. Engl. J. Med.* **2006**, *355*, 11–20. [CrossRef]

10.   Ychou, M.; Boige, V.; Pignon, J.P.; Conroy, T.; Bouché, O.; Lebreton, G.; Ducourtieux, M.; Bedenne, L.; Fabre, J.M.; Saint-Aubert, B.; et al. Perioperative Chemotherapy Compared with Surgery Alone for Resectable Gastroesophageal Adenocarcinoma: An FNCLCC and FFCD Multicenter Phase III Trial. *J. Clin. Oncol.* **2011**, *29*, 1715–1721. [CrossRef]

11.   Page, M.J.; McKenzie, J.E.; Bossuyt, P.; Boutron, I.; Hoffmann, T.C.; Mulrow, C.D.; Shamseer, L.; Tetzlaff, J.M.; Akl, E.; Brennan, S.E.; et al. The Prisma 2020 Statement: An Updated Guideline for Reporting Systematic Reviews. *Med. Flum.* **2021**, *57*, 444–465. [CrossRef]

12.   Palmela, C.; Velho, S.; Agostinho, L.; Branco, F.; Santos, M.; Santos, M.P.C.; Oliveira, M.H.; Strecht, J.; Maio, R.; Cravo, M.; et al. Body Composition as a Prognostic Factor of Neoadjuvant Chemotherapy Toxicity and Outcome in Patients with Locally Advanced Gastric Cancer. *J. Gastric Cancer* **2017**, *17*, 74–87. [CrossRef] [PubMed]

13.   Yamaoka, Y.; Fujitani, K.; Tsujinaka, T.; Yamamoto, K.; Hirao, M.; Sekimoto, M. Skeletal Muscle Loss after Total Gastrectomy, Exacerbated by Adjuvant Chemotherapy. *Gastric Cancer* **2015**, *18*, 382–389. [CrossRef] [PubMed]

14.   Tan, B.H.L.; Brammer, K.; Randhawa, N.; Welch, N.T.; Parsons, S.L.; James, E.J.; Catton, J.A. Sarcopenia Is Associated with Toxicity in Patients Undergoing Neo-Adjuvant Chemotherapy for Oesophago-Gastric Cancer. *Eur. J. Surg. Oncol.* **2015**, *41*, 333–338. [CrossRef]

15.   Zhang, Y.; Li, Z.; Jiang, L.; Xue, Z.; Ma, Z.; Kang, W.; Ye, X.; Liu, Y.; Jin, Z.; Yu, J. Marked Loss of Adipose Tissue during Neoadjuvant Therapy as a Predictor for Poor Prognosis in Patients with Gastric Cancer: A Retrospective Cohort Study. *J. Hum. Nutr. Diet.* **2021**, *34*, 585–594. [CrossRef]

16.   Zhou, D.; Zhang, Y.; Gao, X.; Yang, J.; Li, G.; Wang, X. Long-Term Outcome in Gastric Cancer Patients with Different Body Composition Score Assessed via Computed Tomography. *J. Investig. Surg.* **2021**, *34*, 875–882. [CrossRef]

17.   Jiang, L.; Ma, Z.; Ye, X.; Kang, W.; Yu, J. Clinicopathological Factors Affecting the Effect of Neoadjuvant Chemotherapy in Patients with Gastric Cancer. *World J. Surg. Oncol.* **2021**, *19*, 44. [CrossRef]

18.   Rinninella, E.; Cintoni, M.; Raoul, P.; Pozzo, C.; Strippoli, A.; Bria, E.; Tortora, G.; Gasbarrini, A.; Mele, M.C. Muscle Mass, Assessed at Diagnosis by L3-CT Scan as a Prognostic Marker of Clinical Outcomes in Patients with Gastric Cancer: A Systematic Review and Meta-Analysis. *Clin. Nutr.* **2020**, *39*, 2045–2054. [CrossRef]

19.   Jin, H.; Zhu, K.; Wang, W. The Predictive Values of Pretreatment Controlling Nutritional Status (Conut) Score in Estimating Shortand Long-Term Outcomes for Patients with Gastric Cancer Treated with Neoadjuvant Chemotherapy and Curative Gastrectomy. *J. Gastric Cancer* **2021**, *21*, 155–168. [CrossRef]

20.   Li, Z.; Li, S.; Ying, X.; Zhang, L.; Shan, F.; Jia, Y.; Ji, J. The Clinical Value and Usage of Inflammatory and Nutritional Markers in Survival Prediction for Gastric Cancer Patients with Neoadjuvant Chemotherapy and D2 Lymphadenectomy. *Gastric Cancer* **2020**, *23*, 540–549. [CrossRef]

21.   Sun, J.; Wang, D.; Mei, Y.; Jin, H.; Zhu, K.; Liu, X.; Zhang, Q.; Yu, J. Value of the Prognostic Nutritional Index in Advanced Gastric Cancer Treated with Preoperative Chemotherapy. *J. Surg. Res.* **2017**, *209*, 37–44. [CrossRef] [PubMed]

22.   Zhao, Q.; Li, Y.; Yu, B.; Yang, P.; Fan, L.; Tan, B.; Tian, Y. Effects of Preoperative Enteral Nutrition on Postoperative Recent Nutritional Status in Patients with Siewert II and III Adenocarcinoma of Esophagogastric Junction after Neoadjuvant Chemoradiotherapy. *Nutr. Cancer* **2018**, *70*, 895–903. [CrossRef] [PubMed]

23.   Claudino, M.M.; Lopes, J.R.; Rodrigues, V.D.; de Pinho, N.B.; Martucci, R.B. Postoperative Complication Rate and Survival of Patients with Gastric Cancer Undergoing Immunonutrition: A Retrospective Study. *Nutrition* **2020**, *70*, 110590. [CrossRef] [PubMed]

24.   Yang, D.; Zheng, Z.; Zhao, Y.; Zhang, T.; Liu, Y.; Xu, X. Patient-Generated Subjective Global Assessment versus Nutritional Risk Screening 2002 for Gastric Cancer in Chinese Patients. *Futur. Oncol.* **2019**, *16*, 4475–4483. [CrossRef] [PubMed]

25.   Morgan, E.; Arnold, M.; Camargo, M.C.; Gini, A.; Kunzmann, A.T.; Matsuda, T.; Meheus, F.; Verhoeven, R.H.A.; Vignat, J.; Laversanne, M.; et al. The Current and Future Incidence and Mortality of Gastric Cancer in 185 Countries, 2020–2040: A Population-Based Modelling Study. *eClinicalMedicine* **2022**, *47*, 101404. [CrossRef]

26.   Rosania, R.; Chiapponi, C.; Malfertheiner, P.; Venerito, M. Nutrition in Patients with Gastric Cancer: An Update. *Gastrointest. Tumors* **2015**, *2*, 178–187. [CrossRef]

27.   Guo, W.; Ou, G.; Li, X.; Huang, J.; Liu, J.; Wei, H. Screening of the Nutritional Risk of Patients with Gastric Carcinoma before Operation by NRS 2002 and Its Relationship with Postoperative Results. *J. Gastroenterol. Hepatol.* **2010**, *25*, 800–803. [CrossRef]

28.   Cho, H.; Nakamura, J.; Asaumi, Y.; Yabusaki, H.; Sakon, M.; Takasu, N.; Kobayashi, T.; Aoki, T.; Shiraishi, O.; Kishimoto, H.; et al. Long-Term Survival Outcomes of Advanced Gastric Cancer Patients Who Achieved a Pathological Complete Response with Neoadjuvant Chemotherapy: A Systematic Review of the Literature. *Ann. Surg. Oncol.* **2015**, *22*, 787–792. [CrossRef]

29.   Deftereos, I.; Yeung, J.M.C.; Arslan, J.; Carter, V.M.; Isenring, E.; Kiss, N. Assessment of Nutritional Status and Nutrition Impact Symptoms in Patients Undergoing Resection for Upper Gastrointestinal Cancer: Results from the Multi-Centre Nourish Point Prevalence Study. *Nutrients* **2021**, *13*, 3349. [CrossRef]

30.   Quail, D.F.; Dannenberg, A.J. The Obese Adipose Tissue Microenvironment in Cancer Development and Progression. *Nat. Rev. Endocrinol.* **2019**, *15*, 139–154. [CrossRef]

31.   Joharatnam-Hogan, N.; Shiu, K.K.; Khan, K. Challenges in the Treatment of Gastric Cancer in the Older Patient. *Cancer Treat. Rev.* **2020**, *85*, 101980. [CrossRef] [PubMed]

32.   Xia, L.; Zhao, R.; Wan, Q.; Wu, Y.; Zhou, Y.; Wang, Y.; Cui, Y.; Shen, X.; Wu, X. Sarcopenia and Adverse Health-Related Outcomes: An Umbrella Review of Meta-Analyses of Observational Studies. *Cancer Med.* **2020**, *9*, 7964–7978. [CrossRef]

33. Kamarajah, S.K.; Bundred, J.; Tan, B.H.L. Body Composition Assessment and Sarcopenia in Patients with Gastric Cancer: A Systematic Review and Meta-Analysis. *Gastric Cancer* **2019**, *22*, 10–22. [CrossRef] [PubMed]
34. Muscaritoli, M.; Arends, J.; Bachmann, P.; Baracos, V.; Barthelemy, N.; Bertz, H.; Bozzetti, F.; Hütterer, E.; Isenring, E.; Kaasa, S.; et al. ESPEN Practical Guideline: Clinical Nutrition in Cancer. *Clin. Nutr.* **2021**, *40*, 2898–2913. [CrossRef] [PubMed]

Article

Obesity Risk of Pediatric Central Nervous System Tumor Survivors: A Cross-Sectional Study

Rebekah L. Wilson [1,2], Jacqueline Soja [3], Alexandra G. Yunker [1,4], Hajime Uno [1,2], Erin Gordon [5], Tabitha Cooney [1,2,3,†] and Christina M. Dieli-Conwright [1,2,*,†]

[1] Division of Population Sciences, Department of Medical Oncology, Dana-Farber Cancer Institute, Boston, MA 02215, USA; rebekahl_wilson@dfci.harvard.edu (R.L.W.)

[2] Department of Medicine, Harvard Medical School, Boston, MA 02215, USA

[3] Boston Children's Cancer and Blood Disorders Center, Boston, MA 02215, USA

[4] Department of Nutrition, Harvard T.H. Chan School of Public Health, Boston, MA 02215, USA

[5] Boston Children's Hospital, Boston, MA 02215, USA

*   Correspondence: christinam_dieli-conwright@dfci.harvard.edu

†   These authors contributed equally to this work.

Citation: Wilson, R.L.; Soja, J.; Yunker, A.G.; Uno, H.; Gordon, E.; Cooney, T.; Dieli-Conwright, C.M. Obesity Risk of Pediatric Central Nervous System Tumor Survivors: A Cross-Sectional Study. *Nutrients* 2023, 15, 2269. https://doi.org/10.3390/nu15102269

Academic Editors: Fernando Mendes, Diana Martins and Nuno Borges

Received: 22 February 2023
Revised: 29 April 2023
Accepted: 9 May 2023
Published: 11 May 2023

Abstract: Adult survivors of pediatric central nervous system (CNS) tumors are at the highest risk for morbidity and late mortality among all childhood cancers due to a high burden of chronic conditions, and environmental and lifestyle factors. This study aims to epidemiologically characterize young adult survivors of pediatric CNS tumors using body mass index (BMI) to assess risk factors for obesity. Using a cross-sectional design, young adults (18–39 years) previously treated for pediatric CNS tumors and followed in a survivorship clinic during 2016–2021 were examined. Demographic, BMI, and diagnosis information were extracted from medical records of the most recent clinic visit. Data were assessed using a two-sample t-test, Fisher's exact test, and multivariable logistical regression. 198 survivors (53% female, 84.3% White) with a BMI status of underweight (4.0%), healthy weight (40.9%), overweight (26.8%), obesity (20.2%), and severe obesity (8.1%) were examined. Male sex (OR, 2.414; 95% CI, 1.321 to 4.414), older age at follow-up (OR, 1.103; 95% CI, 1.037 to 1.173), and craniopharyngioma diagnosis (OR, 5.764; 95% CI, 1.197 to 27.751) were identified as significant ($p < 0.05$) obesity-related ($\geq$25.0 kg/m^2) risk factors. The majority of patients were overweight or obese. As such, universal screening efforts with more precise determinants of body composition than BMI, risk stratification, and targeted lifestyle interventions are warranted during survivorship care.

Keywords: nutrition status; pediatric central nervous system tumors; obesity risk; young adults; risk stratification

1. Introduction

Central nervous system (CNS) tumors are the second most common childhood cancer with 75% of children diagnosed surviving $\geq$5 years [1]. With this high long-term survival rate comes a need for clinical care with a focus on late effects, which can be just as detrimental to a survivor's health as the original cancer diagnosis. Specifically, among pediatric cancer survivors, pediatric CNS tumor survivors are at the highest risk for new chronic medical conditions (e.g., cardiovascular disease, endocrinopathies), poor health-related quality of life, and late mortality, with a cumulative mortality rate of more than 27% by 45 years old [2–6]. Obesity, which has a prevalence of 28.8–42.0% in pediatric CNS tumor survivors [7–9], is a known modifiable risk factor for secondary malignancies and higher cancer relapse in the pediatric cancer population [9–11]. Additionally, survivors of CNS tumors are at increased risk of obesity development, potentially as a result of hypothalamic insult, metabolic changes, or reduced physical activity levels as a result of the tumor and/or its treatment [12–14]. At the other end of the spectrum, pediatric cancer survivors with an underweight status have an increased likelihood of reporting adverse health and major medical conditions, which may also contribute to early mortality [8].

Monitoring of the nutritional status of adult survivors of pediatric CNS tumors is encouraged within the survivorship guidelines [15]; however, identification of key risk factors of poor nutritional status among pediatric CNS tumor survivors that need the most attention, how nutritional status should be assessed, and updated comparison studies with the general population where ongoing societal increases in obesity levels continue to trend [16] are not well examined. The literature examining the nutritional status within the pediatric cancer population is sparse with limited work completed within the past 5–10 years and the majority of the research conducted in hematological cancers [13,17]. A systematic review examining the prevalence of malnutrition in pediatric cancer patients identified 29 studies targeting hematological cancers, 13 examining solid tumors, two on brain tumors, eight targeting those with International Classification of Childhood Cancer (3rd edition), and six including a mixed cohort [13]. As such, the CNS tumor survivor population is drastically underrepresented within malnutrition-related research, despite obesity being a known late effect within this population.

As adults, more than 45% of survivors of pediatric CNS tumors are likely to experience at least one or more poor health outcomes (e.g., poor general health, adverse mental health, functional impairment, activity limitations) compared to less than 20% of non-cancer siblings [18]. Furthermore, poor health outcomes are further exacerbated by the presence of malnutrition, either underweight or obesity, a risk factor that can be modified [13,18]. As such, to know who, if, how, and when to intervene with regard to improving nutritional status, it is important to understand the current scope of malnutrition within the pediatric CNS tumor survivor population. While the limitations of BMI have become increasingly known and scrutinized for its inability to distinguish between body composition (i.e., fat mass vs. muscle mass), differences amongst ethnic populations, and inability to provide any indicator of metabolic health, it continues to be a simple screening tool to help identify those at risk for potential health and/or nutritional complications. Additionally, current diagnostic criteria for malnutrition (both under- and over-nutrition) continue to use BMI as an indicator of overall nutrition status. Therefore, the purpose of this observational study was to epidemiologically characterize young adult survivors of pediatric CNS tumors using BMI to assess risk factors of obesity.

2. Materials and Methods

2.1. Study Design and Population

Data collected for this retrospective, cross-sectional study were obtained from medical records of young adults seen through the Stop & Shop Family Pediatric Neuro-Oncology Outcomes Clinic at Dana-Farber/Boston Children's Cancer and Blood Disorders Center. The clinic provides multi-disciplinary, long-term follow-up to survivors of pediatric CNS tumors through young adulthood. Patients were eligible for this analysis if they were aged 18–39 years at the time of their most recent visit, previously diagnosed with a CNS tumor at 18 years or younger, and were seen in the survivorship clinic between 2016 and 2021. Due to the minimal risk of the study to patients, a consent waiver was granted.

2.2. Data Collection

Data obtained from medical records included: (1) height (cm) and weight (kg) data from most recent survivorship appointment, these were used to calculate BMI, (2) age at diagnosis and follow-up appointment, (3) sex (male, female), (4) race (White, Black, Asian, other or multiple races) and ethnicity (Hispanic/Latino yes/no), (5) insurance (Medicaid/Mass health, private, other), (6) tumor histology (low-grade glioma, embryonal, craniopharyngioma, other e.g., ependymoma, choroid plexus tumor, germ cell tumor), (7) tumor location (posterior fossa, hypothalamus/optic pathway, supratentorial, cervi-comedullary, spinal cord), (8) treatment (surgery, cranial radiotherapy exposure, chemotherapy), (9) presence of neurodevelopmental and/or endocrine disorder, and (10) current stimulant medication use (methylphenidate, amphetamine/dextroamphetamine).

2.3. Definitions

Age at diagnosis and follow-up. Age at diagnosis was taken from the date of magnetic resonance imaging showing tumor presence. If an exact date was not available, then the first of the month was entered. Age at follow-up was taken from the most recent visit to the survivorship clinic between 2016 and 2021 when both weight and height variables were available.

Underweight, healthy weight, overweight, obesity, and severe obesity. BMI was calculated by dividing weight (kg) by height squared (kg/m^2) that was recorded at the most recent survivorship clinic appointment. Underweight was defined as a BMI < 18.5 kg/m^2, healthy weight 18.5–24.9 kg/m^2, overweight 25.0–29.9 kg/m^2, obesity (class I) 30.0–34.9 kg/m^2, and severe obesity (combined class II and III) $\geq$ 35.0 kg/m^2 [19].

Treatment. Receipt of surgery included those who had a biopsy, partial/subtotal resection, and/or a total resection. Those who only had a shunt procedure were not classified as having had a tumor-related surgery. Only those patients who received cranial radiation were identified as being exposed to radiotherapy. Finally, any persons who received chemotherapy were identified as exposed to chemotherapy.

Endocrine disorder. An endocrine disorder was defined as any condition relating to the hypothalamic-pituitary system including growth hormone deficiency, thyroid stimulating hormone deficiency, adrenocorticotropic hormone deficiency, gonadotropic releasing hormone deficiency, diabetes insipidus, central precocious puberty, disorders of the thyroid and gonads.

Neurodevelopment disorder. A neurodevelopment disorder was defined as any condition that influences brain function and alters neurological development including autism, attention deficit hyperactivity disorder, speech, motor, and learning disorders, and intellectual disability.

Stimulant use. Stimulant use was defined by whether patients were currently taking methylphenidate or amphetamine/dextroamphetamine at the time of their most recent survivorship clinic visit.

2.4. Statistical Analysis

Statistical analyses were conducted using IBM SPSS version 28 (SPSS Inc., IBM, Armonk, NY, USA). The normality of distribution was assessed using the Kolmogorov–Smirnov test. Data are presented as mean ± standard deviation (SD), median and interquartile range [IQR], or number (percentage). The two-sample *t*-test, or the Wilcoxon rank sum test for the non-normally distributed variables, was used to assess between-group differences for continuous variables and Fisher's exact test for categorical data. Forward stepwise logistic regression analysis was used to identify risk factors for overweight and obesity diagnosis. Patients that were underweight or healthy weight were grouped together as the reference group and compared to those classed as overweight, obese, and severely obese given that a BMI $\geq$ 25.0 kg/m^2 is associated with a higher risk of mortality [20]. The result is presented as odd ratios (ORs) and 95% confidence intervals (CIs). Independent variables to be included in the forward stepwise variable selection procedure were selected based on the significance of the bivariate association between the variables and the weight group, in addition to variables deemed clinically relevant (e.g., an endocrine disorder, location, and histology of tumor). In an exploratory analysis, due to a small number of underweight survivors, only bivariate associations were examined for differences in the variables between underweight and those not underweight. Tests were two-tailed and statistical significance was set at $p < 0.05$.

3. Results

3.1. Patient Characteristics

We identified 198 (53% female) survivors of childhood CNS tumors that met the inclusion criteria (Figure 1). The cohort was separated into five BMI categories where 4% were identified as underweight, 40.9% as healthy weight, 26.8% as overweight, 20.3% as obese, and 8.1% as severely obese (Table 1). Overall, patients predominantly identified as White (84.3%), non-Hispanic/Latino (66.7%), and were diagnosed at a median age of 8 (4–12) years. At the time of the most recent survivorship clinic appointment, patients were a median age of 24 (20–28) years and had been followed for a mean of 16.5 ± 6.4 years. Tumors were commonly located in the posterior fossa (39.4%), with low-grade glioma being the most prevalent primary tumor histology (50.5%). The majority of patients received surgery (90.4%) with just over half of patients exposed to chemotherapy (51.5%) and cranial radiotherapy (52.5%). Finally, 41.9% were identified as having an endocrine disorder, 23.2% as a neurodevelopmental disorder, and 12.6% were taking stimulants at the time of their most recent survivorship clinic appointment.

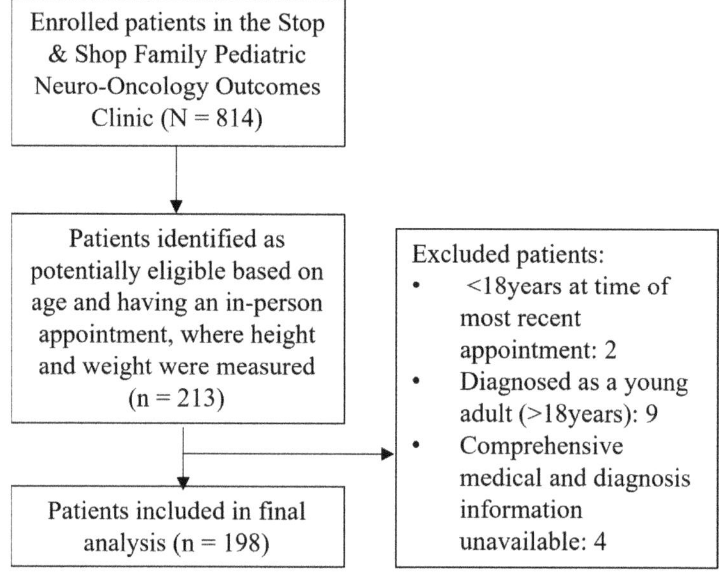

Figure 1. CONSORT diagram.

Table 1. Characteristics of young adult survivors of pediatric CNS tumors with a bivariate analysis comparing overweight, obesity, and severe obesity versus those without this status; similarly comparing underweight status with those without this status.

Characteristic	Total (N = 198; 100%)	Underweight (n = 8; 4.0%)	Healthy weight (n = 81; 40.9%)	Overweight (n = 53; 26.8%)	Obesity (n = 40; 20.2%)	Severe obesity (n = 16; 8.1%)	p-Value Overweight or Obese vs. Not	Underweight vs. Not
Sex N (%)								
Male	93 (47.0)	-	32 (39.5)	34 (64.2)	20 (50.0)	7 (43.8)	*0.006*	*0.007*
Female	105 (53.0)	8 (100.0)	49 (60.5)	19 (35.8)	20 (50.0)	9 (56.3)		
Age Median [IQR]								
At diagnosis, years	8 [4.0–12.0]	6 [3.5–14.5]	8 [5.0–13.0]	8 [3.5–12.0]	10 [5.0–12.8]	5.5 [3.3–9.0]	0.860	0.743
At follow-up, years	24.0 [20.0–28.0]	20.5 [19.3–25.5]	23.0 [20.0–27.0]	24.0 [20.0–28.0]	27.0 [20.3–32.8]	25.0 [23.3–31.8]	*0.002*	0.181
Follow-up time, years (mean ± SD)	16.5 ± 6.4	14.2 ± 6.3	15.1 ± 6.2	16.8 ± 5.6	17.9 ± 7.3	19.7 ± 6.2	*0.005*	0.323
Race N (%)								
White	167 (84.3)	4 (50.0)	70 (86.4)	47 (88.7)	31 (77.5)	15 (93.8)		
Black/African American	8 (4.0)	3 (37.5)	3 (3.7)	1 (1.9)	1 (2.5)	-	0.698	*0.022*
Asian	2 (1.0)	-	-	2 (3.8)	-	-		
Other or multiple races	7 (3.5)	1 (12.5)	2 (2.5)	-	4 (10.0)	-		
Unknown	14 (7.1)	-	6 (7.4)	3 (5.7)	4 (10.0)	1 (6.3)		
Ethnicity N (%)								
Hispanic/Latino	14 (7.1)	-	6 (7.4)	2 (3.8)	4 (10.0)	2 (12.5)		
Not Hispanic/Latino	132 (66.7)	8 (100.0)	46 (56.8)	37 (69.8)	28 (70.0)	13 (81.3)	1.000	0.229
Unknown	52 (26.3)	-	29 (35.8)	14 (26.4)	8 (20.0)	1 (6.3)		
Insurance N (%)								
Medicaid/Mass Health	78 (39.4)	4 (50.0)	24 (29.6)	21 (39.6)	21 (52.5)	8 (50.0)		
Private	116 (58.6)	4 (50.0)	55 (67.9)	31 (58.5)	18 (45.0)	8 (50.0)	0.109	0.760
Other	4 (2.0)	-	2 (2.5)	1 (1.9)	1 (2.5)	-		
Tumor histology at primary diagnosis N (%)								
Low-grade glioma	100 (50.5)	6 (75.0)	44 (54.3)	25 (47.2)	19 (47.5)	6 (37.5)		
Embryonal tumor	40 (20.2)	-	18 (22.2)	11 (20.8)	9 (22.5)	2 (12.5)	0.087	0.054
Craniopharyngioma	14 (7.1)	1 (12.5)	1 (1.2)	2 (3.8)	5 (12.5)	5 (31.3)		
Other	44 (22.2)	1 (12.5)	18 (22.2)	15 (28.3)	7 (17.5)	3 (18.8)		

Table 1. *Cont.*

Characteristic	Total (N = 198; 100%)	Underweight (n = 8; 4.0%)	Healthy weight (n = 81; 40.9%)	Overweight (n = 53; 26.8%)	Obesity (n = 40; 20.2%)	Severe obesity (n = 16; 8.1%)	*p*-Value	
							Overweight or Obese vs. Not	Underweight vs. Not
Primary tumor location N (%)								
Posterior fossa	78 (39.4)	3 (37.5)	33 (40.7)	22 (41.5)	16 (40.0)	4 (25.0)	0.295	0.054
Hypothalamus/optic pathway	44 (22.2)	1 (12.5)	13 (16.0)	9 (17.0)	13 (32.5)	8 (50.0)		
Supratentorial	62 (31.3)	2 (25.0)	30 (37.0)	16 (30.2)	11 (27.5)	3 (18.8)		
Cervicomedullary	5 (2.5)	2 (25.0)	1 (1.2)	1 (1.9)	-	1 (6.3)		
Spinal cord	8 (4.0)	-	4 (4.9)	4 (7.5)	-	-		
Tumor treatment N (%)								
Surgery	179 (90.4)	7 (87.5)	74 (91.4)	48 (90.6)	36 (90.0)	14 (87.5)	1.000	0.561
Chemotherapy	102 (51.5)	2 (25.0)	48 (59.3)	25 (47.2)	19 (47.5)	8 (50.0)	0.255	0.160
Radiotherapy	104 (52.5)	1 (12.5)	36 (44.4)	29 (54.7)	26 (65.0)	12 (75.0)	*0.007*	*0.028*
Presence of neurodevelopmental disorder N (%)	46 (23.2)	1 (12.5)	16 (19.8)	14 (26.4)	12 (30.0)	3 (18.8)	0.239	0.684
Presence of endocrine disorder N (%)	83 (41.9)	2 (25.0)	29 (35.8)	21 (39.6)	22 (55.0)	9 (56.3)	0.083	0.472
Stimulant use N (%)	25 (12.6)	1 (12.5)	12 (14.8)	7 (13.2)	3 (7.5)	2 (12.5)	0.521	1.000

3.2. Risk of Overweight and Obesity at Follow-Up

At the most recent survivorship appointment, 109 (55.1%) survivors had overweight (n = 53, 26.8%), obesity (n = 40, 20.2%), or severe obesity (n = 16, 8.1%), after a mean follow-up time of 16.8 ± 5.6 years, 17.9 ± 7.3 years, and 19.7 ± 6.2 years respectively. A significant difference in the prevalence of patients with overweight and obesity was found between females and males (p = 0.006) (Figure 2). On multivariate analyses, male sex (OR, 2.414; 95% CI, 1.321 to 4.414), older age at follow-up (OR, 1.103; 95% CI, 1.037 to 1.173), and diagnosis of craniopharyngioma (OR, 5.764; 95% CI, 1.197 to 27.751) were associated with overweight or obesity at last follow-up (Table 2). Twelve (85.7%) of the craniopharyngioma survivors had overweight (n = 2, 14.3%), obesity (n = 5, 35.7%), or severe obesity (n = 5, 35.7%) at the last follow-up. In sub-hoc analysis, which excluded those with a craniopharyngioma diagnosis (n = 184), male sex (OR, 2.312; 95% CI, 1.256 to 4.256, p = 0.007) and older age at follow-up (OR, 1.101; 95% CI, 1.034 to 1.172, p = 0.003) were still significant risk factors for overweight/obesity.

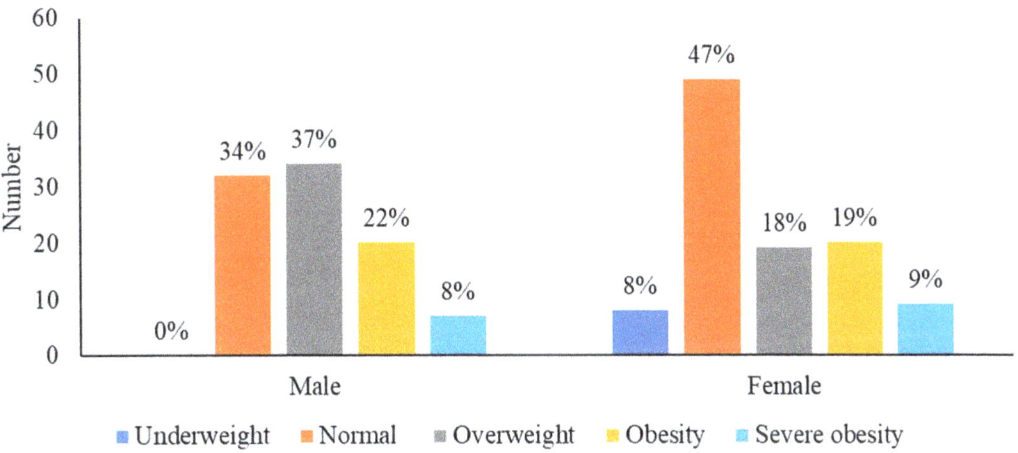

Figure 2. Distribution of sex across the five BMI groups.

Table 2. Regression analysis assessing risk factors for developing overweight, obesity, or severe obesity during survivorship care.

Variable	Odds Ratio	95% CI	p Value
Males	2.414	1.321–4.414	*0.004*
Age at follow up	1.103	1.037–1.173	*0.002*
Craniopharyngioma	5.764	1.197–27.751	*0.029*

3.3. Underweight at Follow-Up

At the most recent follow-up, 8 (4.0%) survivors were underweight, after a mean follow-up of 14.2 ± 6.3 years. A significant difference in underweight prevalence between females and males was found (p = 0.007) (Figure 2). On bivariate analysis, female sex (100.0% vs. 51.1%, p = 0.007) and non-white race (50.0% vs. 14.2%, p = 0.022) were significantly associated with underweight status at follow-up when compared to survivors with healthy weight, overweight, obesity, or severe obesity. One (7.1%) had a tumor in the hypothalamic/optic pathway, which was a craniopharyngioma. Underweight survivors received cranial radiotherapy less frequently when compared with survivors with healthy weight, overweight, obesity, or severe obesity (12.5% vs. 54.2%, p = 0.028) (Table 1).

4. Discussion

We evaluated the BMI status of young adult survivors of pediatric CNS tumors followed in a pediatric neuro-oncology survivorship clinic. There were three important findings: (1) over half (55.1%) of the population analyzed were identified to have overweight, obesity, or severe obesity, (2) those who were male sex, older age at follow-up, and had a craniopharyngioma diagnosis were more likely to have overweight, obesity, or severe obesity, and (3) female sex, non-white race, and exposure to radiotherapy were more likely among patients with an underweight status, compared with non-underweight.

Presence of obesity in childhood cancer survivors is associated with an increased risk of relapse, obesity-related comorbidity development (e.g., type two diabetes, metabolic syndrome), obesity-related cancer development (e.g., colorectal, kidney) as well as decreased overall survival [11]. Within the examined cohort, 55.1% were overweight, obese, or severely obese. Compared to other studies examining survivors of pediatric CNS tumors with shorter follow-up durations, the present study found a higher prevalence of obesity rates (55.1% vs. 28.8–42.0%) [7–9]. Of note, Wilson et al. [21] examined 158 survivors of pediatric CNS tumors with an older median age at follow-up (32.4 years) and reported an overweight and obese prevalence of 66.4%. Given the present analysis determined age at a follow-up appointment to be a significant risk factor for obesity presence, it is likely the rates of overweight/obesity continue to climb throughout survivorship years, similar to the general population [14,22]. Furthermore, the proportion of obesity (28.3%; BMI > 30 kg/m^2) in our cohort is comparable to the general U.S. young adult population of 32.7% (20–25 years) (assessed 1976–2018) [16]. These findings would suggest similar rates of obesity-related cardiometabolic comorbidities between age-matched pediatric CNS tumor survivors and non-cancer controls; however, this is not the case. For example, significant relative risks between 2.0 and 359.7 have been reported for survivors of pediatric CNS tumors in developing endocrine and cardiovascular-related conditions within five years post-diagnosis when compared to siblings [23]. This discrepancy in cardiometabolic comorbidities, but no difference in BMI status, would suggest that the tumor and its treatment may impact the adiposity levels and metabolic function of survivors. Such impact has been highlighted in a systematic review by Wang et al. [24] where despite similar BMI distributions, body fat percent was 4.1% greater among survivors of CNS tumors compared to non-cancer controls. While BMI is associated with adiposity and metabolic dysregulation at the population level, it does not always reflect accurate obesity or generalizable nutrition status at the individual level. Therefore, a further examination into the body composition and metabolic health of survivors of pediatric CNS tumors is required to better understand the potential mechanisms involved in their increased risk of obesity-driven comorbidities [24].

Within this analysis we found male sex to be a significant risk factor associated with overweight, obesity, or severe obesity during survivorship. This is an interesting finding as the majority of the previous research has identified females to be at a higher risk [11,25–27]. A systematic review comparing the prevalence of overweight and obesity between survivors of pediatric brain tumors and non-cancer controls reported male cancer survivors to be at higher odds of an overweight status compared to female cancer survivors, but the risk was similar between both sexes when examining an obesity status [24]. Similar to the systematic review, our study did not restrict tumor type, unlike Lek, Prentice [25] who only examined suprasellar brain tumors, which may partially explain the discrepancy in findings. Additionally, our study exclusively targeted survivors ≥ 18 years of age, as such pubertal state in previous studies may also explain sex-associated risk discrepancies. Mechanistic insight into why one sex over the other may be more vulnerable to developing obesity during survivorship is unclear. However, while not in line with the findings of the current analysis, Lek et al. [25] proposed that hyperleptinemia induced by injury to the hypothalamus may play a role in that females typically have higher leptin levels than males for the same degree of fat mass and as such females may develop leptin resistance and have a greater loss of appetite suppression [28]. Given the conflicting evidence, further analysis into whether females or males are at a higher risk of obesity as a young adult survivor of a CNS tumor, as well as mechanistic insight into why this may be, is needed.

Older age at most recent survivorship appointment and a craniopharyngioma diagnosis were also identified as obesity-related risk factors, which is in line with age-related trends in the general population [29,30] and previous pediatric CNS tumor studies [9,11,14,17,22]. The location of the craniopharyngioma tumor and damage to the hypothalamus because of the tumor and cancer-related therapies can lead to the development of "hypothalamic obesity," a syndrome in which lifestyle interventions are rarely effective [31]. Nonetheless, physical activity may still be critical in obesity management, where a lack of physical activity has also been noted as a potential key component of obesity development among patients with craniopharyngioma [32]. Further analysis is required to identify how much physical inactivity contributes to obesity development, in addition to why physical activity is lower among survivors of both craniopharyngioma and other CNS tumors. Refined understanding will assist in targeted interventions to manage and modify obesity status.

An underweight status is uncommon among survivors of pediatric CNS tumors in Western countries [9,33]. This is reflected in the current analysis with only 4.0% of the assessed population with an underweight status. However, being underweight is still associated with comorbidity development (e.g., frailty), or a higher risk of mortality [11,34]. Consequently, we did a bi-variate analysis between those with underweight versus not underweight statuses. Because of the small sample size of survivors with underweight (n = 8), caution is warranted with this analysis; however, female sex, non-white race, and no exposure to radiotherapy are potential candidates as risk factors for developing underweight as a young adult survivor of pediatric CNS tumors.

This study has several strengths. The cohort analyzed represents a typical distribution of tumor types, locations, and treatments from a unique clinical setup that is dedicated to enhancing survivorship quality among young adult survivors of CNS tumors. Additionally, compared to similar studies [9,25,35], our study describes the longest follow-up time (16.5 ± 6.4 years) providing a better epidemiological description of BMI status in young adult survivors of pediatric CNS tumors. However, there are several limitations. This was a cross-sectional study of a single institution analysis with the majority of participants identifying as White, therefore, the generalizability of this study to populations not racialized as White is unclear. Additionally, while our study targeted survivors, we need to consider survivor bias where those who were underweight may have died shortly after diagnosis. While BMI is often used as an easy, inexpensive variable to gauge the nutrition status of a patient or population, the distribution of body composition (i.e., fat and muscle mass) provides more accurate information regarding metabolic health and mechanistic understanding of obesity development within the CNS tumor population. Additionally, CNS tumor patients are known to be at risk of short stature. BMI relies on height across a normal distribution, which may result in the misclassification of obesity, and further emphasizes that body composition assessment (e.g., fat and muscle tissue) may be a useful measure when considering obesity management strategies [26,36].

5. Conclusions

Within the current analysis, the majority of patients were overweight, obese, or severely obese, which is comparable to the non-cancer young adult U.S. population. Our refined understanding of the impact of disease and treatment exposure in this population will allow for more strategic identification of potential survivors in need of obesity management. Within this analysis, we identified male sex, age at follow-up, and craniopharyngioma diagnosis as key risk factors for obesity development. While these may help guide clinicians to identify at-risk survivors, the ideal intervention to prevent or modify obesity is still unclear in this population. While some forms of obesity within the CNS tumor population may respond to lifestyle-based interventions, those with hypothalamic obesity may not respond to such interventions [11,22]. To date, there are no combined nutrition and exercise-based interventions within the pediatric CNS tumor population, either on or off treatment [37].

Future studies are required to examine the benefit of such interventions to establish the best prescription to manage not only obesity status but adiposity and metabolic function, among long-term survivors of CNS tumors.

Author Contributions: Conceptualization, T.C. and E.G.; Methodology, T.C.; Software, R.L.W.; Validation, T.C., R.L.W. and C.M.D.-C.; Formal Analysis, R.L.W.; Investigation, R.L.W., A.G.Y. and J.S.; Resources, T.C.; Data Curation, R.L.W., J.S. and A.G.Y.; Writing—Original Draft Preparation, R.L.W.; Writing—Review & Editing, R.L.W., A.G.Y., J.S., E.G., H.U., T.C. and C.M.D.-C.; Visualization, T.C. and E.G.; Supervision, T.C. and C.M.D.-C.; Project Administration, T.C., R.L.W., C.M.D.-C., T.C. and C.M.D.-C. equally contributed to leading the development of this study design, as such they are indicated as co-last authors. All authors have read and agreed to the published version of the manuscript.

Funding: This research received no external funding.

Institutional Review Board Statement: Ethical review and approval were waived for this study due to minimal risk to patients.

Informed Consent Statement: Patient consent was waived due to minimal risk to patients and data obtained from medical records retrospectively.

Data Availability Statement: Available on request.

Conflicts of Interest: The authors declare no conflict of interest.

References

1. National Cancer Institute Surveillance Research Program. Cancer Stat Facts: Childhood Brain and Other Nervous System Cancer (Ages 0–19). 2022. Available online: https://seer.cancer.gov/statfacts/html/childbrain.html (accessed on 21 January 2023).
2. Mertens, A.C.; Yasui, Y.; Neglia, J.P.; Potter, J.D.; Nesbit, M.E., Jr.; Ruccione, K.; Smithson, W.A.; Robison, L.L. Late mortality experience in five-year survivors of childhood and adolescent cancer: The Childhood Cancer Survivor Study. *J. Clin. Oncol.* **2001**, *19*, 3163–3172. [CrossRef] [PubMed]
3. Taylor, A.J.; Little, M.P.; Winter, D.L.; Sugden, E.; Ellison, D.W.; Stiller, C.A.; Stovall, M.; Frobisher, C.; Lancashire, E.R.; Reulen, R.C.; et al. Population-based risks of CNS tumors in survivors of childhood cancer: The British Childhood Cancer Survivor Study. *J. Clin. Oncol.* **2010**, *28*, 5287–5293. [CrossRef] [PubMed]
4. Dama, E.; Pastore, G.; Mosso, M.L.; Ferrante, D.; Maule, M.M.; Magnani, C.; Merletti, F. Late deaths among five-year survivors of childhood cancer. A population-based study in Piedmont Region, Italy. *Haematologica* **2006**, *91*, 1084–1091. [PubMed]
5. Armstrong, G.T.; Liu, Q.; Yasui, Y.; Huang, S.; Ness, K.K.; Leisenring, W.; Hudson, M.M.; Donaldson, S.S.; King, A.A.; Stovall, M.; et al. Long-term outcomes among adult survivors of childhood central nervous system malignancies in the Childhood Cancer Survivor Study. *J. Natl. Cancer Inst.* **2009**, *101*, 946–958. [CrossRef]
6. Stavinoha, P.L.; Askins, M.A.; Powell, S.K.; Pillay Smiley, N.; Robert, R.S. Neurocognitive and Psychosocial Outcomes in Pediatric Brain Tumor Survivors. *Bioengineering* **2018**, *5*, 73. [CrossRef]
7. Warner, E.L.; Fluchel, M.; Wright, J.; Sweeney, C.; Boucher, K.M.; Fraser, A.; Smith, K.R.; Stroup, A.M.; Kinney, A.Y.; Kirchhoff, A.C. A population-based study of childhood cancer survivors' body mass index. *J. Cancer Epidemiol.* **2014**, *2014*, 531958. [CrossRef]
8. Meacham, L.R.; Gurney, J.G.; Mertens, A.C.; Ness, K.K.; Sklar, C.A.; Robison, L.L.; Oeffinger, K.C. Body mass index in long-term adult survivors of childhood cancer: A report of the Childhood Cancer Survivor Study. *Cancer* **2005**, *103*, 1730–1739. [CrossRef]
9. van Schaik, J.; van Roessel, I.; Schouten-van Meeteren, N.; van Iersel, L.; Clement, S.C.; Boot, A.M.; Claahsen-van der Grinten, H.L.; Fiocco, M.; Janssens, G.O.; van Vuurden, D.G.; et al. High Prevalence of Weight Gain in Childhood Brain Tumor Survivors and Its Association With Hypothalamic-Pituitary Dysfunction. *J. Clin. Oncol.* **2021**, *39*, 1264–1273. [CrossRef]
10. Moke, D.J.; Hamilton, A.S.; Chehab, L.; Deapen, D.; Freyer, D.R. Obesity and Risk for Second Malignant Neoplasms in Childhood Cancer Survivors: A Case-Control Study Utilizing the California Cancer Registry. *Cancer Epidemiol. Biomark. Prev.* **2019**, *28*, 1612–1620. [CrossRef]
11. Co-Reyes, E.; Li, R.; Huh, W.; Chandra, J. Malnutrition and obesity in pediatric oncology patients: Causes, consequences, and interventions. *Pediatr. Blood Cancer* **2012**, *59*, 1160–1167. [CrossRef]
12. Brouwer, C.A.; Gietema, J.A.; Kamps, W.A.; de Vries, E.G.; Postma, A. Changes in body composition after childhood cancer treatment: Impact on future health status—A review. *Crit. Rev. Oncol./Hematol.* **2007**, *63*, 32–46. [CrossRef] [PubMed]
13. Iniesta, R.R.; Paciarotti, I.; Brougham, M.F.; McKenzie, J.M.; Wilson, D.C. Effects of pediatric cancer and its treatment on nutritional status: A systematic review. *Nutr. Rev.* **2015**, *73*, 276–295. [CrossRef] [PubMed]

14. Müller, H.L.; Emser, A.; Faldum, A.; Bruhnken, G.; Etavard-Gorris, N.; Gebhardt, U.; Oeverink, R.; Kolb, R.; Sörensen, N. Longitudinal study on growth and body mass index before and after diagnosis of childhood craniopharyngioma. *J. Clin. Endocrinol. Metab.* **2004**, *89*, 3298–3305. [CrossRef] [PubMed]

15. Children's Oncology Group. *Long-Term Follow-Up Guidelines for Survivors of Childhood, Adolescent, and Young Adult Cancers. Version 5.0*; Children's Oncology Group: Monrovia, CA, USA, 2018; Available online: http://www.survivor-shipguidelines.org/ (accessed on 21 January 2023).

16. Ellison-Barnes, A.; Johnson, S.; Gudzune, K. Trends in Obesity Prevalence Among Adults Aged 18 through 25 Years, 1976–2018. *JAMA* **2021**, *326*, 2073–2074. [CrossRef] [PubMed]

17. Prasad, M.; Arora, B.; Chinnaswamy, G.; Vora, T.; Narula, G.; Banavali, S.; Kurkure, P. Nutritional status in survivors of childhood cancer: Experience from Tata Memorial Hospital, Mumbai. *Indian J. Cancer* **2015**, *52*, 219–223. [CrossRef]

18. Hudson, M.M.; Oeffinger, K.C.; Jones, K.; Brinkman, T.M.; Krull, K.R.; Mulrooney, D.A.; Mertens, A.; Castellino, S.M.; Casillas, J.; Gurney, J.G.; et al. Age-dependent changes in health status in the Childhood Cancer Survivor cohort. *J. Clin. Oncol.* **2015**, *33*, 479–491. [CrossRef]

19. Centers for Disease Control and Prevention. Defining Adult Overweight & Obesity. 2022. Available online: https://www.cdc.gov/obesity/basics/adult-defining.html (accessed on 21 January 2023).

20. The Global BMI Mortality Collaboration; Di Angelantonio, E.; Bhupathiraju, S.N.; Wormser, D.; Gao, P.; Kaptoge, S.; Berrington de Gonzalez, A.; Cairns, B.J.; Huxley, R.; Jackson, C.L.; et al. Body-mass index and all-cause mortality: Individual-participant-data meta-analysis of 239 prospective studies in four continents. *Lancet* **2016**, *388*, 776–786. [CrossRef]

21. Wilson, C.L.; Liu, W.; Yang, J.J.; Kang, G.; Ojha, R.P.; Neale, G.A.; Srivastava, D.K.; Gurney, J.G.; Hudson, M.M.; Robison, L.L.; et al. Genetic and clinical factors associated with obesity among adult survivors of childhood cancer: A report from the St. Jude Lifetime Cohort. *Cancer* **2015**, *121*, 2262–2270. [CrossRef]

22. Lustig, R.H.; Post, S.R.; Srivannaboon, K.; Rose, S.R.; Danish, R.K.; Burghen, G.A.; Xiong, X.; Wu, S.; Merchant, T.E. Risk factors for the development of obesity in children surviving brain tumors. *J. Clin. Endocrinol. Metab.* **2003**, *88*, 611–616. [CrossRef]

23. Gurney, J.G.; Kadan-Lottick, N.S.; Packer, R.J.; Neglia, J.P.; Sklar, C.A.; Punyko, J.A.; Stovall, M.; Yasui, Y.; Nicholson, H.S.; Wolden, S.; et al. Endocrine and cardiovascular late effects among adult survivors of childhood brain tumors: Childhood Cancer Survivor Study. *Cancer* **2003**, *97*, 663–673. [CrossRef]

24. Wang, K.W.; Fleming, A.; Johnston, D.L.; Zelcer, S.M.; Rassekh, S.R.; Ladhani, S.; Socha, A.; Shinuda, J.; Jaber, S.; Burrow, S.; et al. Overweight, obesity and adiposity in survivors of childhood brain tumours: A systematic review and meta-analysis. *Clin. Obes.* **2018**, *8*, 55–67. [CrossRef] [PubMed]

25. Lek, N.; Prentice, P.; Williams, R.M.; Ong, K.K.; Burke, G.A.; Acerini, C.L. Risk factors for obesity in childhood survivors of suprasellar brain tumours: A retrospective study. *Acta Paediatr.* **2010**, *99*, 1522–1526. [CrossRef] [PubMed]

26. Gurney, J.G.; Ness, K.K.; Stovall, M.; Wolden, S.; Punyko, J.A.; Neglia, J.P.; Mertens, A.C.; Packer, R.J.; Robison, L.L.; Sklar, C.A. Final height and body mass index among adult survivors of childhood brain cancer: Childhood cancer survivor study. *J. Clin. Endocrinol. Metab.* **2003**, *88*, 4731–4739. [CrossRef] [PubMed]

27. Gance-Cleveland, B.; Linton, A.; Arbet, J.; Stiller, D.; Sylvain, G. Predictors of Overweight and Obesity in Childhood Cancer Survivors. *J. Pediatr. Oncol. Nurs.* **2020**, *37*, 154–162. [CrossRef] [PubMed]

28. Kennedy, A.; Gettys, T.W.; Watson, P.; Wallace, P.; Ganaway, E.; Pan, Q.; Garvey, W.T. The metabolic significance of leptin in humans: Gender-based differences in relationship to adiposity, insulin sensitivity, and energy expenditure. *J. Clin. Endocrinol. Metab.* **1997**, *82*, 1293–1300. [CrossRef]

29. Stierman, B.; Afful, J.; Carroll, M.D.; Chen, T.-C.; Davy, O.; Fink, S.; Fryar, C.D.; Gu, Q.; Hales, C.M.; Hughes, J.P. National Health and Nutrition Examination Survey 2017–March 2020 Prepandemic Data Files Development of Files and Prevalence Estimates for Selected Health Outcomes. In *National Health Statistics Reports*; National Center for Health Statistics (NCHS): Hyattsville, MD, USA, 2021.

30. Mizuno, T.; Shu, I.W.; Makimura, H.; Mobbs, C. Obesity over the life course. *Sci. Aging Knowl. Environ.* **2004**, *2004*, re4. [CrossRef] [PubMed]

31. Lustig, R.H. Hypothalamic obesity after craniopharyngioma: Mechanisms, diagnosis, and treatment. *Front. Endocrinol.* **2011**, *2*, 60. [CrossRef] [PubMed]

32. Harz, K.J.; Müller, H.L.; Waldeck, E.; Pudel, V.; Roth, C. Obesity in patients with craniopharyngioma: Assessment of food intake and movement counts indicating physical activity. *J. Clin. Endocrinol. Metab.* **2003**, *88*, 5227–5231. [CrossRef]

33. Cohen, J.; Collins, L.; Gregerson, L.; Chandra, J.; Cohn, R.J. Nutritional concerns of survivors of childhood cancer: A "First World" perspective. *Pediatr. Blood Cancer* **2020**, *67* (Suppl. S3), e28193. [CrossRef]

34. Hayek, S.; Gibson, T.M.; Leisenring, W.M.; Guida, J.L.; Gramatges, M.M.; Lupo, P.J.; Howell, R.M.; Oeffinger, K.C.; Bhatia, S.; Edelstein, K.; et al. Prevalence and Predictors of Frailty in Childhood Cancer Survivors and Siblings: A Report From the Childhood Cancer Survivor Study. *J. Clin. Oncol.* **2020**, *38*, 232–247. [CrossRef]

35. van Roessel, I.; van Schaik, J.; Meeteren, A.; Boot, A.M.; der Grinten, H.L.C.; Clement, S.C.; van Iersel, L.; Han, K.S.; van Trotsenburg, A.S.P.; Vandertop, W.P.; et al. Body mass index at diagnosis of a childhood brain tumor; a reflection of hypothalamic-pituitary dysfunction or lifestyle? *Support. Care Cancer* **2022**, *30*, 6093–6102. [CrossRef] [PubMed]

36. Karlage, R.E.; Wilson, C.L.; Zhang, N.; Kaste, S.; Green, D.M.; Armstrong, G.T.; Robison, L.L.; Chemaitilly, W.; Srivastava, D.K.; Hudson, M.M.; et al. Validity of anthropometric measurements for characterizing obesity among adult survivors of childhood cancer: A report from the St. Jude Lifetime Cohort Study. *Cancer* **2015**, *121*, 2036–2043. [CrossRef] [PubMed]
37. Runco, D.V.; Yoon, L.; Grooss, S.A.; Wong, C.K. Nutrition & Exercise Interventions in Pediatric Patients with Brain Tumors: A Narrative Review. *J. Natl. Cancer Inst. Monogr.* **2019**, *2019*, 163–168. [CrossRef] [PubMed]

Article

Identification of ^{18}F-FDG PET/CT Parameters Associated with Weight Loss in Patients with Esophageal Cancer

Thierry Galvez [1,2], Ikrame Berkane [1], Simon Thézenas [3], Marie-Claude Eberlé [1], Nicolas Flori [4,5], Sophie Guillemard [1], Alina Diana Ilonca [1], Lore Santoro [1], Pierre-Olivier Kotzki [1,5], Pierre Senesse [4,5] and Emmanuel Deshayes [1,5,*]

[1] Department of Nuclear Medicine, Institut du Cancer de Montpellier, Université de Montpellier, 34298 Montpellier, Cedex 5, France

[2] Department of Endocrinology, Diabetes and Nutrition, CHU de Montpellier, Université de Montpellier, 34295 Montpellier, France

[3] Biometry Unit, Institut du Cancer de Montpellier, Université de Montpellier, 34298 Montpellier, Cedex 5, France

[4] Department of Clinical Nutrition and Gastroenterology, Institut du Cancer de Montpellier, Université de Montpellier, 34298 Montpellier, Cedex 5, France

[5] Institut de Recherche en Cancérologie de Montpellier (IRCM), INSERM U1194, Université de Montpellier, 34298 Montpellier, France

* Correspondence: emmanuel.deshayes@icm.unicancer.fr; Tel.: +33-4-67-61-31-90

Citation: Galvez, T.; Berkane, I.; Thézenas, S.; Eberlé, M.-C.; Flori, N.; Guillemard, S.; Ilonca, A.D.; Santoro, L.; Kotzki, P.-O.; Senesse, P.; et al. Identification of ^{18}F-FDG PET/CT Parameters Associated with Weight Loss in Patients with Esophageal Cancer. *Nutrients* **2023**, *15*, 3042. https://doi.org/10.3390/nu15133042

Academic Editors: Harvey J. Murff, Nuno Borges, Fernando Mendes and Diana Martins

Received: 8 May 2023
Revised: 16 June 2023
Accepted: 1 July 2023
Published: 5 July 2023

Abstract: ^{18}F-FDG PET-CT is routinely performed as part of the initial staging of numerous cancers. Other than having descriptive, predictive and prognostic values for tumors, ^{18}F-FDG PET-CT provides full-body data, which could inform on concurrent pathophysiological processes such as malnutrition. To test this hypothesis, we measured the ^{18}F-FDG uptake in several organs and evaluated their association with weight loss in patients at diagnosis of esophageal cancer. Forty-eight patients were included in this retrospective monocentric study. ^{18}F-FDG uptake quantification was performed in the brain, the liver, the spleen, bone marrow, muscle and the esophageal tumor itself and was compared between patients with different amounts of weight loss. We found that Total Lesion Glycolysis (TLG) and peak Standardized Uptake Values (SUV_{peak}) measured in the brain correlated with the amount of weight loss: TLG was, on average, higher in patients who had lost more than 5% of their usual weight, whereas brain SUV_{peak} were, on average, lower in patients who had lost more than 10% of their weight. Higher TLG and lower brain SUV_{peak} were associated with worse OS in the univariate analysis. This study reports a new and significant association between ^{18}F-FDG uptake in the brain and initial weight loss in patients with esophageal cancer.

Keywords: malnutrition; ^{18}F-FDG PET/CT; weight loss; brain metabolism; esophageal cancer

1. Introduction

In oncology, 2-deoxy-2-(^{18}F)-fluoro-D-glucose positron emission tomography coupled to computed tomography (^{18}F-FDG PET-CT) is widely used in order to detect regional and distant tumor spread as part of the initial tumor staging or treatment evaluation [1]. Specific metrics based on ^{18}F-FDG uptake at the tumor level are also recognized as independent prognostic factors in several cancers, including esophageal cancer [2–4]. However, ^{18}F-FDG uptake by neoplastic or non-neoplastic tissues, though measurable, has only rarely been used to explore concurrent pathophysiological processes, such as malnutrition [5,6]. Cancer-associated malnutrition is a severe systemic metabolic condition, with a high incidence in patients with esophageal cancers [7,8]. It is defined by involuntary WL, low body mass index (BMI) or reduced muscle mass in the context of reduced food intake, reduced nutrient absorption or active disease [9]. It impairs quality of life, associates with worse tolerance to treatment and negatively impacts overall survival (OS) [10–12]. How malnutrition impacts

^{18}F-FDG uptake of tumors and healthy organs is largely unknown [13,14]. Likewise, to what extent ^{18}F-FDG uptake could inform the pathophysiological changes occurring in malnourished patients is also unknown.

In this study, using data from routinely performed ^{18}F-FDG PET-CT at the initial staging of esophageal cancer, we retrospectively and systematically assessed the association of ^{18}F-FDG uptake values in the brain, the liver, the spleen, bone marrow, muscle and the esophageal tumor itself with weight loss.

2. Methods

2.1. Patient Selection

Patients aged 18 years old or above diagnosed with esophageal cancer (squamous cell carcinoma or adenocarcinoma) that underwent an ^{18}F-FDG PET-CT scan for initial staging (before any treatment) between January 2014 and June 2019 were eligible for inclusion. The study was approved by the Institutional Review Board (ART-2021-02). Patients were excluded if anti-diabetic medications were listed in their medical files, if their capillary blood glucose concentration at the time of the ^{18}F-FDG injection was below 65 mg/mL or above 135 mg/mL, if the time between tracer injection and imaging was under 55 min or above 75 min, if their brain had not been scanned, if they presented with brain, liver or spleen metastases or if CT and PET images were misaligned at visual inspection.

2.2. Imaging Data Acquisition and Processing

Patients were asked not to ingest anything other than plain water and to avoid intense physical activity for 6 h before the injection of ^{18}F-FDG (3.5 MBq/kg). Their venous blood glucose level was measured before injection using a glucometer. Image acquisition from skull to mid-thigh was performed on the same Discovery PET/CT 690 scanner (GE Healthcare, Waukesha, WI, USA). Non-contrast CT scans of patients in the supine position were acquired, followed by 3D PET imaging. Data were corrected for geometrical response and detector efficiency, dead time, random coincidences, scatter and attenuation, as recommended in [15], and reconstructed into matrices of 256 × 256 pixels. Our PET/CT imaging facility was accredited for tumor imaging by the European Association of Nuclear Medicine Research Ltd.

2.3. Quantification of ^{18}F-FDG Uptake

The quantification of ^{18}F-FDG uptake was retrospectively performed. Spherical volumes of interest (VOI) were manually positioned over relevant organs using CT images and OsiriX MD software (version 7.5): over the right lobe of the liver (19.2 cm^3), over the spleen (5.2 cm^3), inside the brain (centered on the putamen) and inside the left iliac tuberosity in order to measure tracer uptake in the bone marrow. The putamen is an easily recognizable brain structure, which we used to reproducibly center the brain VOI. This warranted consistent measurements. SUV_{peak} were computed within these VOI using OsiriX MD. SUV_{peak} correspond to the average value within a 1 cm^3 sphere positioned around the highest voxel value (SUV_{max}) [15–17]. SUV_{peak} were proposed to be more robust than SUV_{max}, especially in low-count conditions, as was the case for most organs in this study [18]. The esophageal tumor was circumscribed within a large spherical VOI. The Metabolic Tumor Volume (MTV) was defined as the volume inside the 3D isocontour at 41% of the maximum pixel value (as recommended in [15]) and the Total Lesion Glycolysis (TLG) as MTV multiplied by mean voxel SUV (SUV_{mean}) within the MTV. For skeletal muscle, the mean SUV (SUV_{mean}) of a 2D, manually drawn region of interest (ROI) delineating the cross-sectional area of skeletal muscle at the third lumbar vertebra was chosen. This region has been shown to be representative of whole-body muscle mass [19]. When indicated, SUVs were normalized to lean body mass (LBM) according to James' and Janmahasatian's predictive equations [20–22]) and referred to as SUL_{James} or SUL_{Janma}, respectively. Similarly, when indicated, brain SUV_{peak} were also normalized to blood glucose concentrations at the time of the ^{18}F-FDG injection as $SUV_{glu} = SUV_{peak} \times$ (blood glucose in mg/dL)/100) [23].

2.4. Clinical Data Collection and Nutritional Assessment

Clinical parameters and imaging conditions were obtained from patients' medical records, PET/CT reports and associated DICOM files. The reference weight (weight[ref]) was defined as the patient-reported usual stable weight. Weight loss (WL) was defined as: WL = $(weight[PET] - weight[ref])/weight[ref]$ with weight[PET] defined as the weight on the day of PET/CT. WL was categorized according to two thresholds: WL $\geq$ 5% and WL $\geq$ 10%. The reference weight was obtained from nutritional reports systematically filed for all patients by dieticians or physicians during consultation.

2.5. Statistical Analysis

Categorical variables are expressed as numbers in the indicated category and (%), with continuous variables as median and (range). Group differences between quantitative variables were tested using the non-parametric Kruskal-Wallis test by ranks or Pearson's chi-square test for categorical variables. In order to examine the optimal cut-off values for SUVs, Receiver Operating Characteristic (ROC) curves were assessed with WL $\geq$ 10% as the reference. The cut-off value corresponding to the highest predictive value, which maximized the Youden index, was chosen. Overall survival (OS) was defined as the time between diagnosis and death or last follow-up (censored data). OS was estimated using the Kaplan-Meier estimator. The log-rank test was performed to assess differences between groups. Patients alive without event were censored at the last news date. The median follow-up was estimated according to «reverse Kaplan-Meier method» and presented with 95% confidence intervals (CIs). Multivariate analyses were carried out using logistic regressions or Cox's proportional hazards regressions, with a stepwise selection procedure on covariables with $p < 0.1$ (dichotomized at median value) in univariate analyses. We added 3 more variables of interest, i.e., Glycemia, TLG and MTV, which were not automatically selected as categorical variables for the multivariate logistic regression, but were associated with WL $\geq$ 10% ($p < 0.1$) as continuous variables. Odds ratio (OR) and hazard ratios (HR) are presented with 95% CIs. All p values reported were two-sided and the significance level was set to 5% ($p < 0.05$) and indicated by *. Statistical analysis was performed using the STATA 16.1 software (Stata Corporation, College Station, TX, USA).

3. Results

3.1. Demographic and Nutritional Characteristics of Patients

Two hundred and eighteen patients with esophageal cancer underwent an initial ^{18}F-FDG PET-CT scan in our institute between January 2014 and June 2019. One hundred and fifty-three patients were excluded because the PET-CT scans did not encompass their brain or showed improper alignment between the PET and CT images in the brain area; thirteen patients were excluded because of known diabetes or blood glucose outside of the 65–135 mg/L range; four patients were excluded because the time between the ^{18}F-FDG injection and imaging was outside the predefined range. Forty-eight patients were selected for the study. The median usual BMI was 27.2 kg·m^{-2} before the onset of initial symptoms. In comparison to the usual weight, the median WL was 7% on the PET scan day. Thirty-two persons (67%) lost 5% or more of their usual weight and eighteen (37.5%) lost 10% or more of their usual weight. Values of BMI and glycemia before PET imaging were significantly different between patients who lost 10% or more of their initial weight compared to the rest of the cohort (Table 1).

Table 1. Distribution of patient characteristics according to weight loss. Fractional weight difference is between the usual weight, stable weight and the weight measured just before PET imaging. Median and (range) are indicated. p is according to Kruskal-Wallis test by rank for continuous variables and to Pearson's chi-square test for categorical variables.

Variable	All	% WL $\geq$ 5	% WL < 5	p	% WL $\geq$ 10	% WL < 10	p
Number of patients (%)	48	32 (67)	16 (33)		18 (37.5)	30 (62.5)	

Table 1. *Cont.*

Number of females (%)	8 (16.7)	5 (16)	3 (19)		3 (17)	5 (17)	
Age at diagnosis, years	64 (36:88)	61.5 (48.0:82.0)	67.0 (36.0:88.0)	0.251	61.0 (54.0:82.0)	66.5 (36.0:88.0)	0.273
Usual BMI, kg·m^{-2}	27.2 (17.0:40.8)	27.6 (17.0:40.8)	25.7 (20.0:40.3)	0.718	27.2 (17.0:33.3)	27.0 (20.0:40.8)	0.647
BMI on PET scan day, kg·m^{-2}	24.6 (14.1:38.5)	24.3 (14.1:36.9)	25.4 (19.4:38.5)	0.088	22.1 (14.1:28.7)	25.6 (19.4:38.5)	0.004 *
Fractional weight difference	−0.07 (−0.42:08)	−0.11 (−0.4:−0.1)	0.00 (−0.0:0.1)		−0.14 (−0.4:−0.1)	−0.04 (−0.1:0.1)	<0.001 *
Number of Histological type (%) Squamous cell carcinoma	28 (58)	17 (53)	11 (69)	0.300	11 (61)	17 (57)	0.762
Adenocarcinoma	20 (42)	15 (47)	5 (31)		7 (39)	13 (43)	
Patients with distant metastasis	7	5	2	0.772	4	3	0.245
History of former cancer (%)	14 (29)	6 (19)	8 (50)	0.042 *	4 (22)	10 (33)	0.412
Time tracer injection–PET acquisition, min.	63.0 (55.0:73.0)	64.0 (55.0:73.0)	59.0 (55.0:68.0)	0.024 *	62.5 (55.0:71.0)	63.0 (55.0:73.0)	0.958
Glycemia before PET, mg/dL	99.5 (65:134)	101.0 (65.0:134.0)	97.5 (84.0:131.0)	0.550	103.0 (90.0:134.0)	97.5 (65.0:131.0)	0.023 *

* indicated $p < 0.05$.

3.2. TLG and Brain SUV$_{peak}$ Associated with WL $\geq$ 10%

The SUV$_{peak}$ measured in the brain, the liver, the spleen, bone marrow, muscle and primary tumor were compared between patients presenting with WL $\geq$ 5% versus <5% on one hand, and WL $\geq$ 10% versus <10% on the other hand. When a cut-off of 5% WL was chosen, no significant difference was observed between SUV$_{peak}$ from any organs (Table 2, columns 3–5 and Figure 1A). Yet, when a cut-off value of 10% WL was used, the brain SUV$_{peak}$ were significantly lower ($p < 0.001$, Kruskal-Wallis test) in patients who lost 10% or more of their usual weight compared to other patients (Table 2, columns 6–8 and Figure 1D). The Spearman correlation coefficient between the brain SUV$_{peak}$ and weight difference was −0.44, $p = 0.0015$ (Supplementary Figure S1A). Representative ^{18}F-FDG PET-CT images from two patients presenting with different levels of brain ^{18}F-FDG uptake are shown (Figure 2).

Table 2. Distribution of ^{18}F-FDG uptake values in specified organs according to weight loss (WL). MTV, Metabolic Tumor Volume; TLG, Tumor Lesion Glycolysis. Median and (range) are indicated. p is according to Kruskal-Wallis test by rank.

		All	% WL $\geq$ 5	% WL < 5	p	% WL $\geq$ 10	% WL < 10	p
Brain	SUV$_{peak}$	8.8 (3.4:13.6)	8.3 (3.4:13.6)	9.7 (6.4:12.2)	0.189	7.2 (3.4:10.3)	10.1 (6.4:13.6)	<0.001 *
Liver	SUV$_{peak}$	2.8 (1.4:4.5)	2.8 (1.4:4.5)	2.0 (1.6:2.4)	0.710	2.8 (1.4:3.4)	2.8 (2.2:4.5)	0.148
Spleen	SUV$_{peak}$	2.3 (1.4: 3.9)	2.4 (1.9:2.9)	1.7 (1.2:2.1)	0.670	2.2 (1.4:2.7)	2.4 (1.8:3.9)	0.170
Bone marrow	SUV$_{peak}$	1.7 (1.0:3.5)	1.7 (1.0:3.5)	1.6 (1.0:2.4)	0.678	1.7 (1.0:2.4)	1.7 (1.0:3.5)	0.307
Muscle at L3	SUV$_{mean}$	0.7 (0.5:1.6)	0.7 (0.5:0.9)	0.7 (0.5:1.6)	0.623	0.7 (0.6:0.9)	0.7 (0.5:1.6)	0.221
Primary tumor	SUV$_{peak}$	11.4 (1.8:28.7)	12.1 (1.8:28.7)	10.6 (3.7:22.4)	0.431	14.2 (3.9:28.7)	11.1 (1.8:22.4)	0.394
	MTV (cm3)	11.4 (0.5:69.6)	15.4 (0.5:69.6)	5.7 (1.9:47.0)	0.003 *	20.3 (4.6:61.1)	9.6 (0.5:69.6)	0.013 *
	TLG	125.8 (3.0:677.8)	171.3 (3.0:677.8)	63.7 (11.1:269.9)	0.005 *	230.1 (24.6:677.8)	99.9 (3.0:295.4)	0.005 *

* indicated $p < 0.05$.

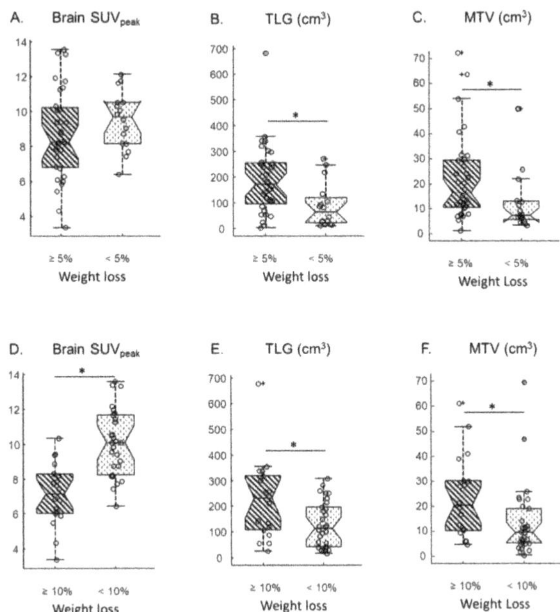

Figure 1. Comparison of brain SUV_{peak} (**A**), Total Lesion Glycolysis (TLG) (**B**) and Metabolic Tumour Volume (MTV) (**C**) from patients who lost 5% or more of their initial weight (diagonal lines) versus patients who lost less than 10% of their initial weight (dots). Comparison of brain SUV_{peak} (**D**), TLG (**E**) and MTV (**F**) from patients who lost 10% or more of their initial weight (diagonal lines) versus patients who lost less than 10% of their initial weight (dots). Circles represent data points. Central horizontal marks correspond to medians; bottom and top edges of the box indicate the 25th and 75th percentiles, respectively; notches correspond to limits of 95% CI. Whiskers extend to the most extreme data value that is not beyond $+/-2.7\sigma$. Crosses correspond to data points beyond whiskers. * indicates $p < 0.05$.

No significant difference was observed in SUV_{peak} for the spleen, bone marrow muscle or primary tumor using a WL cut-off of 10% (Table 2). For primary tumors, the median MTV and TLG were both significantly higher in patients that met either the 5% WL cut-off or the 10% WL (Figure 1B,C,E,F). The Spearman correlation coefficient between TLG and weight difference was -0.48, $p < 0.001$ (Supplementary Figure S1B). There was no significant correlation between TLG and brain SUV_{peak} (Spearman correlation coefficient of -0.23, $p > 0.1$, Supplementary Figure S1C).

Patient BMI and glycemia were potential confounding factors as both are known to influence SUVs [21,23–27] and distributions of both parameters were significantly different between patients who lost 10% or more of their usual weight and the others (Table 1). However, in a multivariate logistic regression model to predict WL $\geq$ 10%, adjusted for BMI, glycemia, brain SUV_{peak}, MTV and TLG, only brain SUV_{peak} and TLG remained significant predictors of WL $\geq$ 10% (Table 3). Moreover, using brain SUV_{peak} normalized by lean body weight (i.e., SUL) or by glycemia (i.e., SUV_{glu}) did not change the results: brain SUL or brain SUV_{glu} were lower in patients who lost 10% or more of their usual weight (Supplementary Table S1).

A cut-off value of 7.32 for brain SUV_{peak} determined with the analysis of the ROC curve (AUC 0.863) was able to predict WL $\geq$ 10% with a high specificity of 0.97, but with a low sensitivity 0.53. Similar trends were observed with ROC analysis of TLG but AUC was lower; i.e., 0.743 (Table 4).

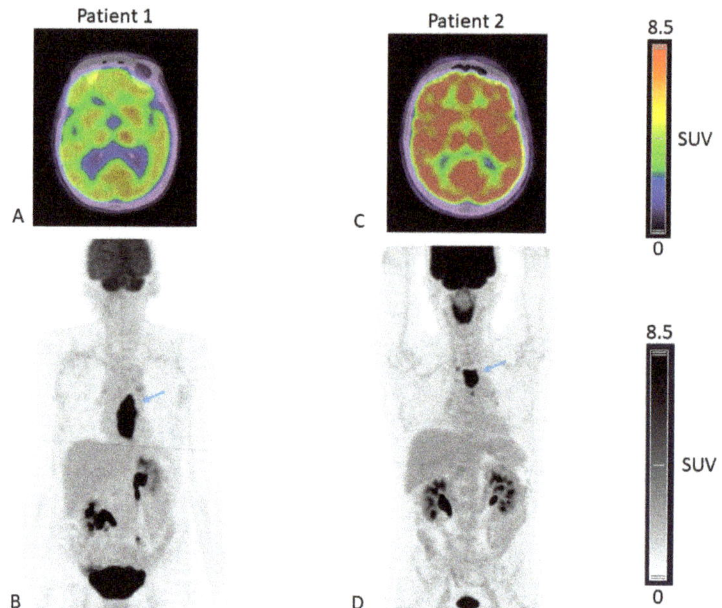

Figure 2. ^{18}F-FDG-fused PET/CT axial slices passing through the brain (**A,C**) and Maximal Intensity Projection, anterior view (**B,D**), of patient 1 presenting with low brain ^{18}F-FDG uptake (brain SUV$_{peak}$ = 5.89) and a body weight loss of 10% compared to usual weight and patient 2 presenting with higher brain ^{18}F-FDG uptake (brain SUV$_{peak}$ = 10.17) and a body weight loss of 8% compared to usual weight. Arrows depicting oesophageal tumors in both patients.

Table 3. Multivariate logistic regression predicting weight loss $\geq$ 10%. Covariables with $p < 0.1$ in univariate analysis were chosen as adjustment variables, i.e., age at diagnosis (categorical), BMI on TEP scan day (categorical), Brain SUV$_{peak}$ (categorical), Spleen SUV$_{peak}$ (categorical), Liver SUV$_{peak}$ (categorical), Glycemia before PET (mg/dL) (continuous), TLG (continuous), MTV (continuous).

		Odds Ratio	$p > |z|$	95% CI
TLG		1.004	0.031	1.000–1.009
Brain SUV$_{peak}$				
	<8.82 (median)	1		
	$\geq$8.82	0.098	<0.001	0.028–0.346

Table 4. ROC analysis of PET variables associated with weight loss. AUC, area under the ROC curve.

Variables	AUC	Optimal Cut-Point Value	# Patients above Cut-Point	AUC	Optimal Cut-Point Value	# Patients above Cut-Point
		WL $\geq$ 5%			WL $\geq$ 10%	
Brain SUV$_{peak}$		NA		0.863	7.32	10
MTV	0.763	12.03	21	0.717	40.78	5
TLG	0.748	107.86	27	0.743	291.1	7

3.3. TLG and Brain SUV$_{peak}$ Associated with Survival

The median follow-up period was 28.7 months. Using Kaplan Meier analysis and groups split at the median value, none of the PET variables significantly affected OS. Only the presence of distant metastasis, BMI on the day of the PET scan and WL $\geq$ 10% were prognostic factors (Table 5, Figure 3A). When cut-off values determined by ROC analysis to predict WL were used (Table 5), both brain SUV_{peak} and TLG were significant prognostic factors (Table 5, Figure 3B,C).

Table 5. Univariate Kaplan Meier analysis of overall survival. Cut-offs defining groups of patients are median unless otherwise specified; for variables predictive of 10%WL in multivariate logistic regression, cut-offs were determined by ROC analysis. Median are indicated in Table 1, column 2. HR: Hazard Ratio; CI: Confidence Interval.

Variables	Cut-Off	HR	95% CI	p
Glycemia before PET (mg/dL)	99.5 (median)	1.82	[0.801–4.12]	0.137
Sex (female)	yes/no	1.82	[0.546–6.07]	0.218
Age at diagnosis	64 (median)	0.902	[0.405–2.01]	0.798
Distant metastasis	yes/no	3.77	[0.871–16.3]	0.002 *
History of former cancer	yes/no	0.975	[0.406–2.34]	0.954
Usual BMI, $kg{\cdot}m^{-2}$	27.2 (median)	1.21	[0.543–2.71]	0.631
BMI (on the day of PETscan), $kg{\cdot}m^{-2}$	24.6 (median)	0.409	[0.181–0.924]	0.0268 *
Weight loss	≥5% vs. <5%	2.19	[0.98–4.94]	0.083
	≥10% vs. <10%	4.17	[1.52–11.5]	5.88×10^{-5} *
Brain SUV_{peak}	8.8 (median)	0.634	[0.274–1.47]	0.241
	≤7.32 vs. >7.32	0.31	[0.0785–1.22]	0.0065 *
Liver SUV_{peak}	2.8 (median)	1.08	[0.484–2.43]	0.844
Spleen SUV_{peak}	2.3 (median)	1.16	[0.52–2.57]	0.722
Bone marrow SUV_{peak}	1.7 (median)	1.11	[0.496–2.48]	0.798
Muscle SUV_{mean}	0.7 (median)	1.29	[0.581–2.88]	0.528
Primary tumor SUV_{peak}	11.4 (median)	0.954	[0.428–2.13]	0.908
MTV	11.4 (median)	2.15	[0.96–4.82]	0.0616
TLG	125.8 (median)	1.68	[0.736–3.85]	0.192
	≤291 vs. >291	2.89	[0.568–14.7]	0.038 *

* indicated $p < 0.05$.

In a cox multivariate model (Log likelihood = −67.31) including all variables with $p < 0.1$ in cox univariate analysis (i.e., distant metastasis, BMI on the day of the PET scan, WL ≥ 10%, brain SUV_{peak}, MTV and TLG), only the presence of distant metastasis and WL ≥ 10% were associated with overall survival (HR = 2.97, p = 0.046 and HR = 4.35, p = 0.015 respectively), indicating that brain SUV_{peak}, and TLG are not independent prognostic factors.

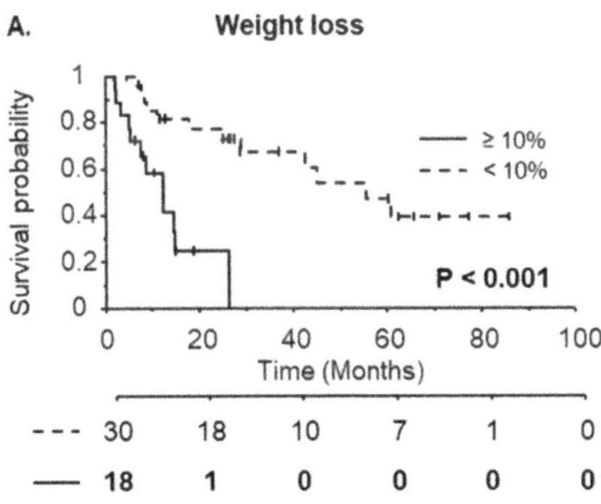

Figure 3. *Cont.*

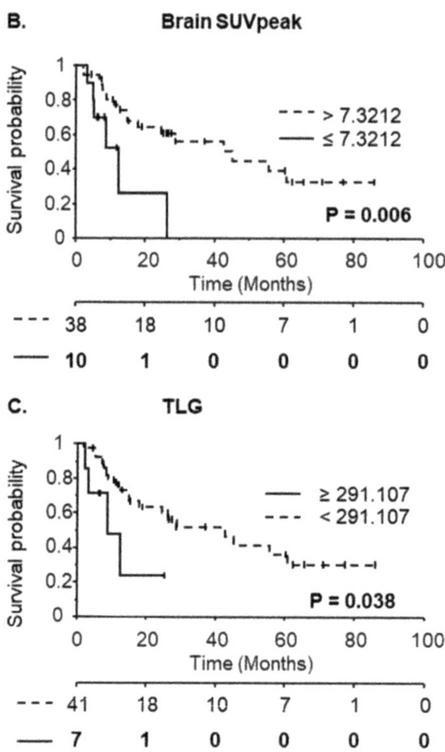

Figure 3. Kaplan-Meier curves with respect to weight loss $\geq$ 10% (**A**), brain SUV_{peak} (**B**) or Total Lesion Glycolysis (TLG) (**C**). The number at risk is indicated below the x-axis. For brain SUV_{peak} and TLG, groups of patients were defined according to cut-off values determined by the ROC analysis.

4. Discussion

Using data obtained from routine ^{18}F-FDG PET-CT, we showed that, in patients at diagnosis of esophageal cancer, WL correlated with high TLG but also with low brain SUV_{peak}. In addition, in the univariate analysis, both TLG and brain SUV_{peak} were pre-therapeutic prognostic factors in these patients, possibly in connection with weight loss. In this group of patients, weight loss did not associate with SUV_{peak} measured in the liver, the spleen, bone marrow or muscle. This specificity of the brain results cannot be simply explained by the higher amplitude of the signal observed in the brain, as tumor SUVs are similarly high and do not differ between WL categories.

WL $\geq$ 5% within the past 6 months or WL $\geq$ 10% beyond 6 months defines malnutrition in the context of cancer [9]. Lower brain SUVs are specifically associated with more pronounced WL, i.e., $\geq$10%, whereas higher TLG is associated with both high and more moderate WL; i.e., $\geq$5%. A recent study has shown a significant association between esophageal tumor SUV_{max} and weight loss. Although we found a similar association between TLG and weight loss, SUV_{peak} were not significantly associated with weight loss in our group of patients [13]. To our knowledge, this is the first clinical report of the association of a routine ^{18}F-FDG uptake measurement in the brain with malnutrition and survival in patients just diagnosed with esophageal cancer. It was made possible because of the unique and systematic survey and filing of patients' weight history by dedicated dieticians in our clinical center [24]. In a cachexia-inducing murine model of adenocarcinoma, brain uptake was significantly higher in cachexic mice compared to the group of non-cachexic mice; the reason for this discrepancy with our results is unclear, but it may be explained by the inherent limitations of the preclinical model when compared to the patients [25].

The association of weight loss with high TLG may be explained by the higher metabolic burden of a high-volume tumor, as well as by the larger hindrance of such a tumor on the esophagus, hence limiting food intake. It is more difficult to explain the correlation between weight loss and brain SUV_{peak}. The SUV of a given organ is key as the brain depends on its intrinsic metabolic properties, but also on several other parameters [26]. Systemic changes in body composition, especially changes in the fraction of fat mass, may indeed affect the biodistribution of the tracer. For instance, liver, blood and spleen SUVs are overestimated in obese persons compared to non-obese persons [22,27,28]. The normalization of SUVs by lean mass was introduced in order to circumvent this effect. The opposite phenomenon, i.e., the underestimation of tissue SUVs, may explain our results in undernourished patients who might have lost more fat than lean mass. However, lower SUVs were specifically observed in the brains of undernourished persons and not in other tissues. After normalization to lean mass estimated by predictive equations [20,21], brain SUVs were still lower in patients who lost $\geq 10\%$ of their initial weight (Supplementary Table S1). One cannot exclude that ^{18}F-FDG uptake in voluminous tumors may reduce the amount of ^{18}F-FDG available for uptake in the brain, but, in this case, a similar trend should have been observed in other tissues [29]. Moreover, we did not find any correlation between TLG and brain SUV_{peak} (Supplementary Figure S1C).

The blood glucose concentration may also affect SUVs, as endogenous glucose competes with the tracer and brain SUVs are known to be highly sensitive to glycemia [30–32]. In our group of patients, the blood glucose concentration was slightly higher in patients who lost 10% or more of their weight compared to patients who lost less than 10% of their weight (Table 1), indicating glycemia to be a possible confounding factor. However, brain SUVs reduction persisted in multivariate analysis after adjustment for glycemia, sex or age. In addition, when corrected for blood glucose [22], brain SUV_{glu} were still significantly lower in patients who lost 10% or more of their weight (Supplementary Table S1).

The pathophysiology underlying the reduced ^{18}F-FDG uptake in the brain of undernourished patients is unknown. The brain relies almost exclusively on glucose as an energy source and reduced cerebral glucose metabolism may be an adaptive mechanism to reduced nutrient availability. In agreement with this hypothesis, starvation was shown to be associated with decreased glucose consumption, specifically in the brain [33–35], and glycolytic flux and phosphofructokinase activity were significantly reduced in the neurons of starved mice [36]. Instead of glucose, neurons have been proposed to use ketone bodies as complementary fuel, which may decrease brain glucose uptake [36–40].

A decrease in brain SUV_{peak} was only observed with WLs $\geq 10\%$, which corresponds to stage 2/severe malnutrition [9]. Severe malnutrition may indeed correspond to extreme metabolic states, e.g., starvation and ketogenesis (cf above), which are associated with brain hypometabolism. A lesser weight loss may not trigger such a metabolic switch.

Several medical conditions, especially in neurology and psychiatry, have been shown to be associated with changes in ^{18}F-FDG uptake in the brain. Alzheimer's disease is associated with low ^{18}F-FDG uptake in specific regions of the brain depending on the severity and the duration of the disease [41]. ^{18}F-FDG uptake is also lower in the frontal cortex of schizophrenia patients [42] or in the thalami of patients with delirium [43]. It is unknown whether and how these observations relate to the lower ^{18}F-FDG uptake described here, but the prevalence of mood disorders is high among patients with esophageal cancer, impacting their quality of life and pain perception [44,45].

As shown by others and confirmed in this study, WL is a strong prognostic factor in esophageal cancer. In our population, brain SUV_{peak} and TLG were also pre-therapeutic prognostic factors when cut-off values predictive of WL were used in the univariate analysis. TLG has already been identified as a prognostic factor, but not brain SUV_{peak} [46]. Their prognostic value is lost in multivariate Cox models, suggesting that brain SUV_{peak} or TLG are not independent prognostic factors and affect survival because of their association with other factors; e.g., WL.

This study has several limitations: it was retrospective, performed at a single clinical center and, above all, included a small number of patients, mostly as a consequence of the exclusion criteria requiring the inclusion of the brain in the full body scan. Additional work on a larger cohort will be necessary to confirm our results. Moreover, though statistically significant, data supporting the prognostic value of brain SUV_{peak} and TLG relied on a small number of patients, especially within the group with the lowest OS (Figure 2), and must be confirmed with a larger group of patients. The clinical significance of our findings is not yet clear. Although brain SUV_{peak} are indicative of severe weight loss, they are obviously not a substitute for the clinical approach; i.e., taking the patient's actual weight and history. However, low brain SUV_{peak} could trigger nutritional assessment if it has not been carried out at the time of the PET scan.

This work revealed a so far unnoticed association between malnutrition and routine ^{18}F-FDG uptake measurements in the tumors and, more surprisingly, brains of patients diagnosed with esophageal cancer. It may open up new avenues of research aimed at understanding the systemic consequences of malnutrition, especially on the central nervous system and its cognitive and behavioral functions.

Supplementary Materials: The following supporting information can be downloaded at: https://www.mdpi.com/article/10.3390/nu15133042/s1, Figure S1: Correlation between brain SUVpeak and fractional weight difference (A). Correlation between Total Lesion Glycolysis (TLG) and fractional weight difference (B). Correlations beween brain SUVpeak and TLG (C). Spearman correlation coefficient and *p*-value of observing the null hypothesis. Table S1: Brain ^{18}FDG uptake values normalized to lean body mass or blood glucose according to weight loss.

Author Contributions: Project supervision: E.D.; project conceptualization: E.D. and T.G.; Expertise and guidance on nutritional data collection and analysis: P.S. and N.F.; images acquisition and collection: E.D., M.-C.E., S.G., A.D.I., P.-O.K. and L.S.; Provision of technical resources: P.-O.K.; Data collection and curation from patient files (biometric and imaging data): I.B. and T.G.; Data analysis: T.G. and S.T.; Supervision and validation of statistical analysis: S.T.; Writing (original draft): T.G.; Writing (review and editing): all authors. All authors have read and agreed to the published version of the manuscript.

Funding: This research received no external funding.

Institutional Review Board Statement: The study was conducted in accordance with the Declaration of Helsinki, and approved by the Institut du Cancer de Montpellier (ART-2021-02).

Informed Consent Statement: Patient consent was waived due to because of the retrospective nature of the study and the analysis of anonymous clinical data.

Data Availability Statement: The datasets analyzed during the current study are available from the corresponding author on reasonable request.

Acknowledgments: We thank all the dieticians who collected biometric data: Stéphanie Arnac, Bérénice Clavie, Anne Fallières, Laure Francioni, Arnaud Vaillé and Sophie Zaessinger. We thank also Heloïse Lecornu for support and advice regarding nutrional data interpretation.

Conflicts of Interest: The authors declare no conflict of interest.

References

1. Salaün, P.-Y.; Abgral, R.; Malard, O.; Querellou-Lefranc, S.; Quere, G.; Wartski, M.; Coriat, R.; Hindie, E.; Taieb, D.; Tabarin, A.; et al. Good Clinical Practice Recommendations for the Use of PET/CT in Oncology. *Eur. J. Nucl. Med. Mol. Imaging* **2020**, *47*, 28–50. [CrossRef] [PubMed]
2. Butof, R.; Hofheinz, F.; Zophel, K.; Stadelmann, T.; Schmollack, J.; Jentsch, C.; Lock, S.; Kotzerke, J.; Baumann, M.; van den Hoff, J. Prognostic Value of Pretherapeutic Tumor-to-Blood Standardized Uptake Ratio in Patients with Esophageal Carcinoma. *J. Nucl. Med.* **2015**, *56*, 1150–1156. [CrossRef] [PubMed]
3. Hofheinz, F.; Li, Y.; Steffen, I.G.; Lin, Q.; Lili, C.; Hua, W.; van den Hoff, J.; Zschaeck, S. Confirmation of the Prognostic Value of Pretherapeutic Tumor SUR and MTV in Patients with Esophageal Squamous Cell Carcinoma. *Eur. J. Nucl. Med. Mol. Imaging* **2019**, *46*, 1485–1494. [CrossRef] [PubMed]

4. Lemarignier, C.; Di Fiore, F.; Marre, C.; Hapdey, S.; Modzelewski, R.; Gouel, P.; Michel, P.; Dubray, B.; Vera, P. Pretreatment Metabolic Tumour Volume Is Predictive of Disease-Free Survival and Overall Survival in Patients with Oesophageal Squamous Cell Carcinoma. *Eur. J. Nucl. Med. Mol. Imaging* **2014**, *41*, 2008–2016. [CrossRef] [PubMed]
5. Otomi, Y.; Otsuka, H.; Terazawa, K.; Kondo, M.; Arai, Y.; Yamanaka, M.; Otomo, M.; Abe, T.; Shinya, T.; Harada, M. A Reduced Liver 18F-FDG Uptake May Be Related to Hypoalbuminemia in Patients with Malnutrition. *Ann. Nucl. Med.* **2019**, *33*, 689–696. [CrossRef]
6. Jiang, Y.; Wu, H.; Zhao, Y.; Cui, Y.; Dai, J.; Huang, S.; Li, C.; Mao, H.; Ju, S.; Peng, X.-G. Abnormal [18F]FDG Uptake in Liver and Adipose Tissue: A Potential Imaging Biomarker for Cancer-Associated Cachexia. *Eur. Radiol.* **2023**, *33*, 2561–2573. [CrossRef]
7. Anandavadivelan, P.; Lagergren, P. Cachexia in Patients with Oesophageal Cancer. *Nat. Rev. Clin. Oncol.* **2016**, *13*, 185–198. [CrossRef]
8. Bozzetti, F. SCRINIO Working Group Screening the Nutritional Status in Oncology: A Preliminary Report on 1,000 Outpatients. *Support. Care Cancer* **2009**, *17*, 279–284. [CrossRef]
9. Cederholm, T.; Jensen, G.L.; Correia, M.I.T.D.; Gonzalez, M.C.; Fukushima, R.; Higashiguchi, T.; Baptista, G.; Barazzoni, R.; Blaauw, R.; Coats, A.J.S.; et al. GLIM Criteria for the Diagnosis of Malnutrition—A Consensus Report from the Global Clinical Nutrition Community. *J. Cachexia Sarcopenia Muscle* **2019**, *10*, 207–217. [CrossRef]
10. Dewys, W.D.; Begg, C.; Lavin, P.T.; Band, P.R.; Bennett, J.M.; Bertino, J.R.; Cohen, M.H.; Douglass, H.O.; Engstrom, P.F.; Ezdinli, E.Z.; et al. Prognostic Effect of Weight Loss Prior Tochemotherapy in Cancer Patients. *Am. J. Med.* **1980**, *69*, 491–497. [CrossRef]
11. Pan, Y.-P.; Kuo, H.-C.; Hsu, T.-Y.; Lin, J.-Y.; Chou, W.-C.; Lai, C.-H.; Chang, P.-H.; Yeh, K.-Y. Body Mass Index-Adjusted Body Weight Loss Grading Predicts Overall Survival in Esophageal Squamous Cell Carcinoma Patients. *Nutr. Cancer* **2020**, *73*, 1130–1137. [CrossRef]
12. Di Fiore, F.; Lecleire, S.; Pop, D.; Rigal, O.; Hamidou, H.; Paillot, B.; Ducrotté, P.; Lerebours, E.; Michel, P. Baseline Nutritional Status Is Predictive of Response to Treatment and Survival in Patients Treated by Definitive Chemoradiotherapy for a Locally Advanced Esophageal Cancer. *Am. J. Gastroenterol.* **2007**, *102*, 2557–2563. [CrossRef]
13. Olaechea, S.; Gannavarapu, B.S.; Gilmore, A.; Alvarez, C.; Iyengar, P.; Infante, R. The Influence of Tumour Fluorodeoxyglucose Avidity and Cachexia Development on Patient Survival in Oesophageal or Gastroesophageal Junction Cancer. *JCSM Clin. Rep.* **2021**, *6*, 128–136. [CrossRef]
14. Nakamoto, R.; Okuyama, C.; Ishizu, K.; Higashi, T.; Takahashi, M.; Kusano, K.; Kagawa, S.; Yamauchi, H. Diffusely Decreased Liver Uptake on FDG PET and Cancer-Associated Cachexia With Reduced Survival. *Clin. Nucl. Med.* **2019**, *44*, 634–642. [CrossRef]
15. Boellaard, R.; Delgado-Bolton, R.; Oyen, W.J.G.; Giammarile, F.; Tatsch, K.; Eschner, W.; Verzijlbergen, F.J.; Barrington, S.F.; Pike, L.C.; Weber, W.A.; et al. FDG PET/CT: EANM Procedure Guidelines for Tumour Imaging: Version 2.0. *Eur. J. Nucl. Med. Mol. Imaging* **2015**, *42*, 328–354. [CrossRef]
16. Wahl, R.L.; Jacene, H.; Kasamon, Y.; Lodge, M.A. From RECIST to PERCIST: Evolving Considerations for PET Response Criteria in Solid Tumors. *J. Nucl. Med.* **2009**, *50*, 122S–150S. [CrossRef]
17. O, J.H.; Lodge, M.A.; Wahl, R.L. Practical PERCIST: A Simplified Guide to PET Response Criteria in Solid Tumors 1.0. *Radiology* **2016**, *280*, 576–584. [CrossRef]
18. Lodge, M.A.; Chaudhry, M.A.; Wahl, R.L. Noise Considerations for PET Quantification Using Maximum and Peak Standardized Uptake Value. *J. Nucl. Med.* **2012**, *53*, 1041–1047. [CrossRef]
19. Mourtzakis, M.; Prado, C.M.M.; Lieffers, J.R.; Reiman, T.; McCargar, L.J.; Baracos, V.E. A Practical and Precise Approach to Quantification of Body Composition in Cancer Patients Using Computed Tomography Images Acquired during Routine Care. *Appl. Physiol. Nutr. Metab.* **2008**, *33*, 997–1006. [CrossRef]
20. James, W.P.T. Research on Obesity. *Nutr. Bull.* **1977**, *4*, 187–190. [CrossRef]
21. Janmahasatian, S.; Duffull, S.B.; Ash, S.; Ward, L.C.; Byrne, N.M.; Green, B. Quantification of Lean Bodyweight. *Clin. Pharm.* **2005**, *44*, 1051–1065. [CrossRef] [PubMed]
22. Tahari, A.K.; Chien, D.; Azadi, J.R.; Wahl, R.L. Optimum Lean Body Formulation for Correction of Standardized Uptake Value in PET Imaging. *J. Nucl. Med.* **2014**, *55*, 1481–1484. [CrossRef] [PubMed]
23. Lindholm, P.; Minn, H.; Joensuu, H. FDG Uptake in Cancer A PET Study. *J. Nucl. Med.* **1993**, *34*, 1–6. [PubMed]
24. Zasadny, K.R.; Wahl, R.L. Standardized Uptake Values of Normal Tissues at PET with 2-[Fluorine-18]-Fluoro-2-Deoxy-D-Glucose: Variations with Body Weight and a Method for Correction. *Radiology* **1993**, *189*, 847–850. [CrossRef] [PubMed]
25. Claeys, J.; Mertens, K.; D'Asseler, Y.; Goethals, I. Normoglycemic Plasma Glucose Levels Affect F-18 FDG Uptake in the Brain. *Ann. Nucl. Med.* **2010**, *24*, 501–505. [CrossRef]
26. Viglianti, B.L.; Wong, K.K.; Wimer, S.M.; Parameswaran, A.; Nan, B.; Ky, C.; Townsend, D.M.; Rubello, D.; Frey, K.A.; Gross, M.D. Effect of Hyperglycemia on Brain and Liver 18 F-FDG Standardized Uptake Value (FDG SUV) Measured by Quantitative Positron Emission Tomography (PET) Imaging. *Biomed. Pharmacother.* **2017**, *88*, 1038–1045. [CrossRef]
27. Sprinz, C.; Zanon, M.; Altmayer, S.; Watte, G.; Irion, K.; Marchiori, E.; Hochhegger, B. Effects of Blood Glucose Level on 18F Fluorodeoxyglucose (18F-FDG) Uptake for PET/CT in Normal Organs: An Analysis on 5623 Patients. *Sci. Rep.* **2018**, *8*, 2126–2131. [CrossRef]
28. Sarikaya, I.; Albatineh, A.; Sarikaya, A. Re-Visiting SUV-Weight and SUV-Lean Body Mass in FDG PET Studies. *J. Nucl. Med. Technol.* **2019**, *48*, 233–249. [CrossRef]

29. Senesse, P.; Isambert, A.; Janiszewski, C.; Fiore, S.; Flori, N.; Poujol, S.; Arroyo, E.; Courraud, J.; Guillaumon, V.; Mathieu-Daudé, H.; et al. Management of Cancer Cachexia and Guidelines Implementation in a Comprehensive Cancer Center: A Physician-Led Cancer Nutrition Program Adapted to the Practices of a Country. *J. Pain Symptom Manag.* **2017**, *54*, 387–393.e3. [CrossRef]
30. Penet, M.-F.; Gadiya, M.M.; Krishnamachary, B.; Nimmagadda, S.; Pomper, M.G.; Artemov, D.; Bhujwalla, Z.M. Metabolic Signatures Imaged in Cancer-Induced Cachexia. *Cancer Res.* **2011**, *71*, 6948–6956. [CrossRef]
31. Huang, S.-C. Anatomy of SUV. *Nucl. Med. Biol.* **2000**, *27*, 643–646. [CrossRef]
32. Viglianti, B.L.; Wale, D.J.; Wong, K.K.; Johnson, T.D.; Ky, C.; Frey, K.A.; Gross, M.D. Effects of Tumor Burden on Reference Tissue Standardized Uptake for PET Imaging: Modification of PERCIST Criteria. *Radiology* **2018**, *287*, 993–1002. [CrossRef]
33. Hasselbalch, S.G.; Knudsen, G.M.; Jakobsen, J.; Hageman, L.P.; Holm, S.; Paulson, O.B. Brain Metabolism during Short-Term Starvation in Humans. *J. Cereb. Blood Flow Metab.* **1994**, *14*, 125–131. [CrossRef]
34. Owen, O.E.; Morgan, A.P.; Kemp, H.G.; Sullivan, J.M.; Herrera, M.G.; Cahill, G.F. Brain Metabolism during Fasting. *J. Clin. Investig.* **1967**, *46*, 1589–1595. [CrossRef]
35. Redies, C.; Hoffer, L.J.; Beil, C.; Marliss, E.B.; Evans, A.C.; Lariviere, F.; Marrett, S.; Meyer, E.; Diksic, M.; Gjedde, A.; et al. Generalized Decrease in Brain Glucose Metabolism during Fasting in Humans Studied by PET. *Am. J. Physiol.-Endocrinol. Metab.* **1989**, *256*, E805–E810. [CrossRef]
36. Buschiazzo, A.; Cossu, V.; Bauckneht, M.; Orengo, A.; Piccioli, P.; Emionite, L.; Bianchi, G.; Grillo, F.; Rocchi, A.; Di Giulio, F.; et al. Effect of Starvation on Brain Glucose Metabolism and 18F-2-Fluoro-2-Deoxyglucose Uptake: An Experimental in-Vivo and Ex-Vivo Study. *EJNMMI Res.* **2018**, *8*, 44. [CrossRef]
37. Courchesne-Loyer, A.; Croteau, E.; Castellano, C.-A.; St-Pierre, V.; Hennebelle, M.; Cunnane, S.C. Inverse Relationship between Brain Glucose and Ketone Metabolism in Adults during Short-Term Moderate Dietary Ketosis: A Dual Tracer Quantitative Positron Emission Tomography Study. *J. Cereb. Blood Flow Metab.* **2017**, *37*, 2485–2493. [CrossRef]
38. LaManna, J.C.; Salem, N.; Puchowicz, M.; Erokwu, B.; Koppaka, S.; Flask, C.; Lee, Z. Ketones Suppress Brain Glucose Consumption. *Adv. Exp. Med. Biol.* **2009**, *645*, 301–306. [CrossRef]
39. Zhang, Y.; Kuang, Y.; Xu, K.; Harris, D.; Lee, Z.; LaManna, J.; Puchowicz, M.A. Ketosis Proportionately Spares Glucose Utilization in Brain. *J. Cereb. Blood Flow Metab.* **2013**, *33*, 1307–1311. [CrossRef]
40. Svart, M.; Gormsen, L.C.; Hansen, J.; Zeidler, D.; Gejl, M.; Vang, K.; Aanerud, J.; Moeller, N. Regional Cerebral Effects of Ketone Body Infusion with 3-Hydroxybutyrate in Humans: Reduced Glucose Uptake, Unchanged Oxygen Consumption and Increased Blood Flow by Positron Emission Tomography. A Randomized, Controlled Trial. *PLoS ONE* **2018**, *13*, e0190556. [CrossRef]
41. Mosconi, L.; McHugh, P.F. FDG- and Amyloid-PET in Alzheimer's Disease: Is the Whole Greater than the Sum of the Parts? *Q. J. Nucl. Med. Mol. Imaging* **2011**, *55*, 250–264. [PubMed]
42. Townsend, L.; Pillinger, T.; Selvaggi, P.; Veronese, M.; Turkheimer, F.; Howes, O. Brain Glucose Metabolism in Schizophrenia: A Systematic Review and Meta-Analysis of 18FDG-PET Studies in Schizophrenia. *Psychol. Med.* **2022**, 1–18. [CrossRef] [PubMed]
43. Nitchingham, A.; Pereira, J.V.-B.; Wegner, E.A.; Oxenham, V.; Close, J.; Caplan, G.A. Regional Cerebral Hypometabolism on 18F-FDG PET/CT Scan in Delirium Is Independent of Acute Illness and Dementia. *Alzheimer's Dement.* **2023**, *19*, 97–106. [CrossRef] [PubMed]
44. Bergquist, H.; Ruth, M.; Hammerlid, E. Psychiatric Morbidity among Patients with Cancer of the Esophagus or the Gastro-Esophageal Junction: A Prospective, Longitudinal Evaluation. *Dis. Esophagus.* **2007**, *20*, 523–529. [CrossRef]
45. Cao, Y.; Zhao, Q.-D.; Hu, L.-J.; Sun, Z.-Q.; Sun, S.-P.; Yun, W.-W.; Yuan, Y.-G. Theory of Mind Deficits in Patients with Esophageal Cancer Combined with Depression. *World J. Gastroenterol.* **2013**, *19*, 2969–2973. [CrossRef]
46. Han, S.; Kim, Y.J.; Woo, S.; Suh, C.H.; Lee, J.J. Prognostic Value of Volumetric Parameters of Pretreatment 18F-FDG PET/CT in Esophageal Cancer: A Systematic Review and Meta-Analysis. *Clin. Nucl. Med.* **2018**, *43*, 887–894. [CrossRef]

Article

Pretreatment Modified Glasgow Prognostic Score for Predicting Prognosis and Survival in Elderly Patients with Gastric Cancer Treated with Perioperative FLOT

Ebru Melekoglu [1,*], Ertugrul Bayram [2], Saban Secmeler [3], Burak Mete [4] and Berksoy Sahin [2]

[1] Department of Nutrition and Dietetics, Faculty of Health Sciences, Cukurova University, Adana 01250, Turkey
[2] Department of Medical Oncology, Faculty of Medicine, Cukurova University, Adana 01250, Turkey; ertugrulbayram84@gmail.com (E.B.); berksoys@hotmail.com (B.S.)
[3] Department of Medical Oncology, Bahcelievler Medicalpark Hospital, Altinbas University, Istanbul 34180, Turkey; drsabansecmeler@hotmail.com
[4] Department of Public Health, Faculty of Medicine, Cukurova University, Adana 01250, Turkey; burakmete2008@gmail.com
* Correspondence: emelekoglu33@gmail.com

Abstract: The adverse effects of chemotherapy are more apparent in elderly patients and lead to worse prognosis and mortality. Identifying immunonutritional risk factors is of great importance in terms of treatment effectiveness, prognosis, and mortality in geriatric oncology. The modified Glasgow prognostic score (mGPS) is an immunonutritional index based on serum CRP and albumin levels. In this study, we aimed to investigate the role of mGPS in predicting prognosis and survival in elderly patients with gastric cancer receiving perioperative FLOT treatment. We retrospectively enrolled 71 patients aged over 65 years and grouped them according to their pretreatment mGPS score. Kaplan-Meier and Cox regression analysis showed overall survival was significantly worse in the mGPS 1 and mGPS 2 groups than in the mGPS 0 group ($p = 0.005$ and $p < 0.001$, respectively). Compared to the mGPS 0 group, the mGPS 1 group had a 6.25 times greater risk of death (95% CI: 1.61–24.28, $p = 0.008$), and the mGPS 2 group had a 6.59 times greater risk of death (95% CI: 2.08–20.85, $p = 0.001$). High BMI was identified as a significant risk factor for being in the mGPS 2 group (OR: 1.20, 95% CI: 1.018–1.425, $p = 0.030$). In conclusion, elevated pretreatment mGPS was associated with poor overall survival in elderly patients with gastric cancer treated with perioperative FLOT therapy. As such, pretreatment mGPS can be a simple and useful tool to predict mortality in this specific patient group.

Keywords: modified Glasgow prognostic score; gastric cancer; neoadjuvant chemotherapy; FLOT; elderly; prognosis; survival

Citation: Melekoglu, E.; Bayram, E.; Secmeler, S.; Mete, B.; Sahin, B. Pretreatment Modified Glasgow Prognostic Score for Predicting Prognosis and Survival in Elderly Patients with Gastric Cancer Treated with Perioperative FLOT. *Nutrients* **2023**, *15*, 4156. https://doi.org/10.3390/nu15194156

Academic Editor: Antonio Colecchia

Received: 26 August 2023
Revised: 15 September 2023
Accepted: 20 September 2023
Published: 26 September 2023

1. Introduction

Gastric cancer ranks as the fifth most prevalent cancer and the third leading cause of cancer-related deaths globally, with an annual incidence exceeding one million cases [1]. 65% of gastric cancer cases are diagnosed at locally advanced or advanced stages and have a poor prognosis [2]. In the early stage, the main treatment is surgery, and multimodal treatments, including adjuvant, neoadjuvant chemotherapy, and radiotherapy treatments, improve survival rates. Previously, neoadjuvant chemotherapy has been shown to provide 5-year overall survival (OS) rates ranging from 36% to 38% in cases of early-stage gastric cancer [3,4]. The FLOT regimen as neoadjuvant therapy in operable gastric cancers was demonstrated by the Medical Research Council Adjuvant Gastric Infusional Chemotherapy (MAGIC) study that 5-Fluorourasil (5-FU), leucovorin, oxaliplatin, and docetaxel were effective [3], and later became the standard treatment as it showed a survival advantage over other chemotherapy regimens with the FLOT4-Arbeitsgemeinschaft Internistische

Onkologie (AIO) study [5]. In these studies, the FLOT regimen has been shown to be effective and tolerable in geriatric patients with an Eastern Cooperative Oncology Group (ECOG) score of 0 or 1, which constitutes approximately 20–30% of all patient groups [3,5].

There are several predictive prognostic factors to identify high-risk gastric cancer, including age, gender, pretreatment weight, primary tumor site, tumor size, number of positive and negative lymph nodes resected, negative surgical margins, depth of invasion, and lymphovascular invasion [6–8]. Inflammation, an important cause of cancer development and progression, is also associated with poor prognosis [9]. Inflammatory markers such as lymphocytes, neutrophils, or inflammation-related C-reactive protein (CRP) and albumin have prognostic significance in cancer patients and have also been used as mortality indicators [10,11]. In addition, many inflammatory indices such as platelet/lymphocyte ratio (PLR), CRP/albumin ratio (CAR) [12], inflammatory prognostic index (IPI) [13], prognostic nutritional index (PNI) [14], and controlling nutritional status score (CONUT) [15] have been used as prognostic and mortality markers in gastric cancer patients. Despite all these prognostic indicators, novel markers are needed as the response to treatment and clinical course of patients differ. Recently, the modified Glasgow prognostic score (mGPS) in cancer patients has been used as a new indicator to determine the prognosis and survival of patients [16–18]. The prognostic value of mGPS was also confirmed in gastric cancer. However, its prognostic value in patients receiving neoadjuvant treatment is unknown [19]. However, there are no sufficient clinical studies and real-time patient experience in the literature regarding indicators that predict prognosis and survival in elderly patients with operable gastric cancer treated with the perioperative FLOT regimen. There is also a need for a prognostic tool to guide the decision-making process regarding the allocation of neoadjuvant therapy in this patient population. In this study, we aimed to investigate the importance of mGPS on prognosis and mortality in elderly patients with locally advanced gastric cancer treated with perioperative FLOT therapy.

2. Materials and Methods

2.1. Patients and Datasets

This study retrospectively enrolled 71 patients with gastric cancer aged over 65 years who were treated with perioperative FLOT at Cukurova University Balcalı Hospital from January 2013 to July 2023. We included patients over 65 years of age with locally advanced (stage 2 and 3) gastric and gastric esophageal junctional cancer who accepted preoperative chemotherapy treatment, were compatible with treatment, and had an ECOG performance score of 0 or 1. The criteria for exclusion of patients are as follows: (i) under 65 years of age, patients with (ii) early stage (stage 1) gastric cancer, (iii) metastatic disease and previous surgery, (iv) liver and kidney failure, stage 3–4 heart failure and stage 3–4 chronic obstructive pulmonary disease (COPD), (v) a low ECOG performance score (ECOG $\geq$ 2), (vi) missing data on the biochemical and pathological findings shown in Table 1, and (vii) patients who could not adapt to treatment (mental health problems, hypersensitivity to drugs, etc.).

The perioperative FLOT regimen required the following drugs to be administered biweekly for four cycles in both the preoperative and postoperative periods; "Docetaxel at a dose of 50 mg/m^2 (1 h), oxaliplatin at a dose of 85 mg/m^2 (2 h), and folinic acid at a dose of 200 mg/m^2 (2 h) by intravenous infusion on the first day, followed by a 24-h intravenous infusion of 2600 mg/m^2 of 5-fluorouracil".

Table 1. The characteristics of elderly patients with gastric cancer treated with perioperative FLOT (Mean $\pm$ SD or n [%]).

Variables	Total (n = 71)	mGPS 0 (n = 41)	mGPS 1 (n = 15)	mGPS 2 (n = 15)	p-Value
Age (years)	68.9 $\pm$ 4.16	69.7 $\pm$ 4.29	67.7 $\pm$ 4.18	68.0 $\pm$ 3.48	0.117
Gender (male/female)	50/21	29/12	11/4	10/5	0.982
BMI (kg/m^2)	24.6 $\pm$ 3.82	23.7 $\pm$ 3.27	25.2 $\pm$ 3.34	26.4 $\pm$ 4.99	0.064
ECOG-PS (0/1)	60/11	33/8	13/2	14/1	0.525
Tumor subtype					
Adenocarcinoma	53 [74.6]	31 [75.6]	12 [80]	10 [66.7]	
Mucinous adenocarcinoma	2 [2.8]	2 [4.9]	0	0	0.567
Signet-ring cell carcinoma	16 [22.5]	8 [19.5]	3 [20]	5 [33.3]	
Tumor location					
Cardia	29 [40.8]	16 [39.0]	10 [66.7]	3 [20]	
Corpus	14 [19.7]	8 [19.5]	1 [6.7]	5 [33.4]	
Antrum	21 [29.6]	14 [34.1]	3 [20]	4 [26.7]	0.126
Esophagogastric junction	7 [9.9]	3 [7.3]	1 [6.7]	3 [20]	
Lauren classification (Intestinal/Diffuse)	63/8	35/6	15/0	13/2	0.250
T Stage (T1/T2/T3/T4a/T4b)	3/10/27/28/3 [4.2/14.1/38/39.5/4.2]	2/8/17/12/2 [4.9/19.5/41.4/29.3/4.9]	1/2/5/6/1 [6.7/13.3/33.3/40/6.7]	-/-/5/10/- [-/-/33.3/66.7/-]	0.374
N Stage (N0/N1/N2/N3a/N3b)	7/21/22/18/3 [9.9/29.6/31/25.3/4.2]	5/16/12/8/- [12.2/39/29.3/19.5/-]	2/3/5/3/2 [13.3/20/33.3/20/13.3]	-/2/5/7/1 [-/13.3/33.3/46.7/6.7]	0.038
Radiologic response					
Complete response	15 [21.1]	12 [29.3]	2 [13.3]	1 [6.7]	
Partial response	30 [42.3]	20 [48.8]	5 [33.3]	5 [33.3]	
Stable disease	21 [29.6]	9 [22.0]	7 [46.7]	5 [33.3]	0.123
Progressive disease	5 [7.0]	-	1 [6.7]	4 [26.7]	
Pathologic response					
Complete response	24 [33.8]	16 [39.0]	4 [26.7]	1 [6.7]	
Residual disease	47 [66.2]	25 [61]	11 [73.4]	14 [93.3]	0.486
Neutrophil count (/L)	4.97 $\pm$ 1.89	4.70 $\pm$ 1.87	4.79 $\pm$ 1.78	4.97 $\pm$ 1.67	0.996
Lymphocyte count (/L)	1.94 $\pm$ 1.15	1.71 $\pm$ 0.49	1.79 $\pm$ 0.46	1.97 $\pm$ 0.95	0.807
Hemoglobin (g/dL)	12.1 $\pm$ 2.03	11.7 $\pm$ 1.87	12.1 $\pm$ 2.87	12.2 $\pm$ 1.52	0.730
CRP (mg/dL)	10.0 (0.34–156.0)	4.5 (2.0–10.0) [a]	15.0 (12.0–30.0) [b]	31.0 (16.0–65.0) [b]	<0.001
Albumin (g/dL)	3.59 $\pm$ 0.47	3.68 $\pm$ 0.39 [a]	3.83 $\pm$ 0.18 [a]	2.97 $\pm$ 0.31 [b]	<0.001
Preoperative CEA (ng/mL)	4.0 (0.33–74.0)	3.49 (0.51–9.30) [a]	2.85 (0.33–74.0) [a,b]	6.1 (1.59–42.0) [b]	0.027
Preoperative CA 19-9 (U/mL)	11 (0.3–2307.0)	5.20 (0.3–276.0) [a]	13.5 (0.4–2307.0) [a,b]	60.0 (0.5–1792.0) [b]	0.004
PNI	36.2 $\pm$ 4.78	37.3 $\pm$ 3.99 [a]	38.7 $\pm$ 2.87 [a,b]	30.5 $\pm$ 4.07 [b]	<0.001
GNRI	52.0 $\pm$ 7.23	50.5 $\pm$ 6.32	53.5 $\pm$ 6.42	54.7 $\pm$ 9.47	0.139

Values with different superscripts ([a, b]) in the same row indicate significant differences as a result of post hoc analysis. mGPS, modified Glasgow prognostic score; SD, standard deviation; BMI, body mass index; ECOG-PS, Eastern Cooperative Oncology Group (ECOG) Performance Status; CRP, C-reactive protein; CEA, carcinoembryonic antigen; CA 19-9, cancer antigen 19-9; PNI, prognostic nutritional index; GNRI, geriatric nutritional risk index.

Biochemical, radiological, and pathological findings of the patients were collected from medical records. The Union for International Cancer Control (UICC) tumor node metastasis (TNM) classification was used for the classification of clinicopathological factors [20]. Complications after neoadjuvant FLOT therapy and preoperatively were classified according to Clavien-Dindo (CD) grade [21]. Tumor markers, including carcinoembryonic antigen (CEA) and cancer antigen 19-9 (CA19-9), were evaluated before preoperative chemotherapy and surgery, and radiological responses to FLOT treatment were assessed with positron emission tomography and computed tomography (PET-CT). Clinical response was also assessed by pre- and postoperative endoscopic examination. Pathological response was evaluated by performing pathological examination of tissue samples taken after surgery.

2.2. Immunonutritional Indexes

We used mGPS as a prognostic factor, calculated based on serum albumin and CRP levels and categorized as 0, 1, and 2;

- A score of 0 for CRP serum levels within the normal range ($\leq$10 mg/L),
- A score of 1 for high CRP serum levels (>10 mg/L) and serum albumin levels within the normal range ($\geq$3.5 g/dL),
- A score of 2 in the presence of both high CRP serum levels (>10 mg/L) and hypoalbuminemia (<3.5 g/dL) [22].

In this study, we also calculated two additional immunonutritional indices, including PNI and geriatric nutritional risk index (GNRI), according to the following formulas:

- PNI = serum albumin (g/L) + 5 × total lymphocyte counts (10^9/L) [23].
- GNRI = 1.487 × albumin (g/L) + 41.7 × current body weight (kg)/ideal body weight (kg). Ideal body weight = height^2 (m^2) × 22 kg/m^2 [24].

2.3. Patient Follow-Up

Survival analysis included the time from the date of diagnosis until death caused by any reason. Data from two of the 71 included patients were lost to follow-up and could not be included in the survival analysis. The follow-up period for the mGPS 0 (n = 41), mGPS 1 (n = 14), and mGPS 2 (n = 14) groups was 60 months, 15 months, and 38 months, respectively.

2.4. Ethical Approval

Ethical approval was received from the Cukurova University Faculty of Medicine Non-Interventional Clinical Research Ethics Committee with decision number 54 on 7 April 2023.

2.5. Statistical Analysis

All statistical analyses were performed using Statistical Package for the Social Sciences (SPSS) software (IBM SPSS Statistics for Windows, Version 22.0. Armonk, NY, USA) and Jamovi 2.3.28 statistical software (The Jamovi project, Sydney, Australia). The normality of all variables was tested with Shapiro Wilk) test. Data are presented as mean, standard deviation, frequency, and percentage. Data were analyzed by Student's t-test, Mann Whitney U test, Kruskal Wallis test, one-way analysis of variance (ANOVA), Pearson Chi-square analysis, Cox regression analysis, Kaplan Meier survival analysis, and multinomial logistic regression analysis. Statistical significance was assigned at a p-value of less than 0.05.

3. Results

We identified 71 patients with gastric cancer, aged 65 years or older, who were treated with perioperative FLOT. The mean and standard deviation for the age and BMI of the study population were 68.9 ± 4.16 years and 24.6 ± 3.82 kg/m^2, respectively. Complete clinical follow-up for survival analysis was available for 69 patients. Patients were divided into three groups according to their pretreatment mGPS: mGPS 0 (n = 41), mGPS 1 (n = 14), and mGPS 2 (n = 14). The follow-up period was 60 months for the mGPS 0, 15 months for the mGPS 1, and 38 months for the mGPS 2 group. The general characteristics of patients across the mGPS are presented in Table 1. Compared with those in the mGPS 0 group, patients in the mGPS 1 and mGPS 2 groups had higher serum CRP levels ($p < 0.001$). Also, patients in the mGPS 2 group had lower albumin levels compared to mGPS 0 and mGPS 1 group ($p < 0.001$). Additionally, a significant difference was observed between the mGPS 1 and mGPS 2 groups in terms of preoperative CEA and CA 19-9 levels ($p = 0.004$). While there was no difference between mGPS 0 and mGPS 1, and mGPS 1 and mGPS 2 groups, patients with the highest mGPS values had lower PNI scores than those with the lowest ($p < 0.001$). No other significant difference was found between the mGPS groups in terms of general characteristics ($p \geq 0.05$).

As a result of the 60-month (5-year) survival follow-up of these patients, there were 5 deaths in the mGPS 0 group, four deaths in the mGPS 1 group, and seven deaths in the mGPS 2 group (Figure 1, Table 2). According to Cox regression analysis, the survival times of mGPS 1 and mGPS 2 groups were significantly shorter than mGPS 0. Compared to the mGPS 0 group, the risk of death in the mGPS 1 group was 6.25 times higher (95% CI: 1.61–24.28, $p = 0.008$) and 6.59 times higher in the mGPS 2 group (95% CI: 2.08–20.85, $p = 0.001$) (Table 2).

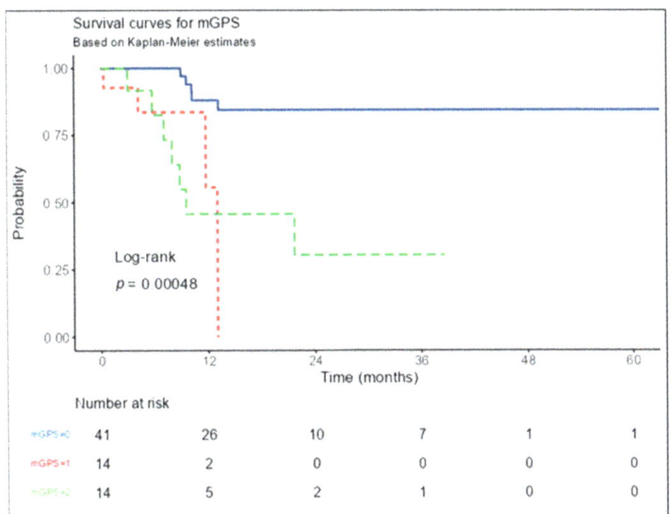

Figure 1. Survival analysis of patients according to mGPS group. Survival analysis was performed on 69 patients (2 patients were lost to follow-up).

Table 2. Median Survival Table: Levels for mGPS.

Levels	Numbers of Event		Time (Months)	Cox Table-mGPS	Pairwise Comparisons mGPS	
	Records	Events	Median	HR (95% CI)	Level	*p*-Value
mGPS 0	41	5	NA	-	1-0	0.005
mGPS 1	14	4	13.00	6.25 (1.61–24.28), $p = 0.008$	2-0	<0.001
mGPS 2	14	7	9.53	6.59 (2.08–20.85), $p = 0.001$	2-1	1.000

p-value adjustment method: Bonferroni. mGPS, modified Glasgow prognostic score; HR, hazard ratio; CI, confidence interval. Survival analysis was performed on 69 patients (2 patients were lost to follow-up).

In the mGPS 0 group, the survival rates of the patients were 88.0% at the end of the 1st year and 84.4% at the end of the 5th year. The follow-up period of the mGPS 1 group was 15 months, and the survival rate at the end of the first year was 55.7%. While the 1-year survival rate of the mGPS 2 group was 45.8%, this rate was found to be 30.6% at the end of the 3rd year. The survival rate of the mGPS 0 group was statistically significantly higher compared to the mGPS 2 group ($p < 0.001$) (Table 3). The follow-up period was 60 months in the mGPS 0 group and 38 months in the mGPS 2 group.

Table 3. 1-year, 3-year, and 5-year survival rates of patients according to mGPS group.

Levels	Time (Months)	Number at Risk	Number of Events	Survival	Lower	Upper
	12	26	4	88.0%	77.6%	99.8%
	24	10	1	84.4%	72.7%	98.0%
mGPS 0	36	7	0	84.4%	72.7%	98.0%
	48	1	0	84.4%	72.7%	98.0%
	60	1	0	84.4%	72.7%	98.0%
mGPS 1	12	2	3	55.7%	24.1%	100.0%
	12	5	6	45.8%	24.1%	87.2%
mGPS 2	24	2	1	30.6%	10.9%	85.3%
	36	1	0	30.6%	10.9%	85.3%

mGPS, modified Glasgow prognostic score.

As a result of the multinominal logistic regression analysis created to estimate the effect of BMI on the mGPS group, while the change in BMI did not affect the risk of being in the mGPS 1 group, it was found that each unit increase in BMI increased the risk of being in the mGPS 2 group by 1.20 times (20%) (OR: 1.20, 95% CI: 1.018 to 1.425, p = 0.030) (Table 4).

Table 4. Multinomial logistic regression analysis of the association between BMI and mGPS.

mGPS	Predictor	Estimate	SE	Z Score	p-Value	OR	95% CI
1-0	Intercept	−3.803	2.1284	−1.79	0.074	0.02229	3.44–1.445
	BMI	0.115	0.0854	1.34	0.179	1.12149	0.949–1.326
2-0	Intercept	−5.706	2.1962	−2.60	0.009	0.00333	4.49–0.246
	BMI	0.186	0.0859	2.16	**0.030**	1.20426	1.018–1.425

mGPS, modified Glasgow prognostic score; BMI, body mass index; SE, standard error; OR, odds ratio; CI, confidence interval.

The relationship between mGPS and preoperative complications was examined, and there was no statistically significant difference between the mGPS groups in the elderly patients with gastric cancer treated with perioperative FLOT ($p \geq 0.05$) (Table 5).

Table 5. The relationship between mGPS and preoperative complications in elderly patients with gastric cancer treated with perioperative FLOT.

	mGPS 0 (n = 41)			mGPS 1 (n = 15)			mGPS 2 (n = 15)			
	None	G1–2	G3–4	None	G1–2	G3–4	None	G1–2	G3–4	p-Value
Dose deferral	29 [70.7]	12 [29.3]	-	8 [53.3]	7 [46.7]	-	12 [80.0]	3 [20.0]	-	0.162
G-CSF prophylaxis	6 [14.6]	35 [85.4]	-	2 [13.3]	13 [86.7]	-	-	15 [100.0]	-	0.320
Grade 3/4 toxicity	30 [73.2]	11 [26.8]	-	11 [73.3]	4 [26.7]	-	13 [86.7]	2 [13.3]	-	0.294
Dose reduction	26 [63.4]	15 [36.6]	-	10 [66.7]	5 [33.7]	-	12 [80.0]	3 [20.0]	-	0.295
Fatigue	9 [22]	32 [78.0]	-	5 [33.3]	9 [60.0]	1 [6.7]	1 [6.7]	14 [93.3]	-	0.135
Hand-foot syndrome	29 [70.7]	12 [29.3]	-	9 [60.0]	6 [40.0]	-	13 [86.7]	2 [13.3]	-	0.306
Neutropenia	23 [56.1]	14 [34.1]	4 [9.8]	7 [46.7]	4 [26.7]	4 [26.7]	11 [73.3]	4 [26.7]	-	0.150
Anemia	19 [46.3]	21 [51.2]	1 [2.4]	8 [53.3]	7 [46. 7]	-	4 [26.7]	10 [66.6]	1 [6.7]	0.593
Thrombocytopenia	28 [68.3]	13 [31.7]	-	12 [80.0]	3 [20.0]	-	9 [60.0]	6 [40.0]	-	0.610
Febrile neutropenia	33 [80.5]	8 [19.5]	-	11 [73.3]	4 [26.7]	-	12 [80.0]	3 [20.0]	-	0.702
Mucositis	23 [56.1]	17 [41.5]	1 [2.4]	8 [53.3]	7 [46.7]	-	12 [80.0]	2 [13.3]	1 [6.7]	0.293
Diarrhea	13 [31.7]	27 [65.9]	1 [2.4]	4 [26.7]	11 [73.3]	-	3 [20.0]	11 [73.3]	1 [6.7]	0.846
Neuropathy	21 [51.2]	19 [46.3]	1 [2.4]	8 [53.3]	7 [46.7]	-	8 [53.3]	7 [46.7]	-	0.937
Nausea	7 [17.1]	33 [80.5]	1 [2.4]	3 [20.0]	12 [80.8]	-	1 [6.7]	13 [86.6]	1 [6.7]	0.773
Vomiting	15 [36.6]	25 [61.0]	1 [2.4]	6 [40.0]	9 [60.0]	-	7 [50.0]	7 [50.0]	1 [6.7]	0.845
Stomatitis	29 [70.7]	11 [26.8]	1 [2.4]	12 [80.0]	3 [20.0]	-	11 [73.3]	4 [26.7]	-	0.887
Allergic complications	36 [87.8]	5 [12.2]	-	12 [80.0]	3 [20.0]	-	14 [93.3]	1 [6.7]	-	0.223
Thrombosis	36 [87.7]	5 [12.2]	-	14 [93.3]	1 [6.7]	-	15 [100.0]	-	-	0.355
Renal toxicity	36 [87.8]	5 [12.2]	-	15 [100.0]	-	-	14 [93.3]	1 [6.7]	-	0.767
Hepatic toxicity	36 [87.8]	5 [12.2]	-	15 [100.0]	-	-	14 [93.3]	1 [6.7]	-	0.355

mGPS, modified Glasgow prognostic score; G, grade; G-CSF, granulocyte-colony stimulating factor.

4. Discussion

In the current study, we initially explored the association between pretreatment mGPS and prognosis and overall survival in geriatric patients with gastric cancer treated with perioperative FLOT. Our results revealed that elevated pretreatment mGPS is associated with poor overall survival of gastric cancer in elderly patients treated with perioperative FLOT compared with patients with a normal mGPS. To the best of our knowledge, this is the first study focusing on the role of mGPS in predicting prognosis and survival after perioperative FLOT treatment in elderly patients with gastric cancer. We used mGPS, a two-dimensional measure of malnutrition and systemic inflammation, including positive and negative acute phase reactants (CRP and albumin), as a prognostic indicator. mGPS, which has been shown to predict poor prognosis in solid cancers, may be suitable for use in elderly cancer patients as it is also an indicator of inflammation and malnutrition associated with immunosenescence [25,26]. Recently, preoperative mGPS has been reported to be a novel and reliable predictor for overall survival and disease-free survival in surgical non-small cell lung cancer [18]. A meta-analysis of 41 clinical trials involving 18348 patients with gastric cancer found that higher mGPS was associated with poorer overall survival [27]. Similarly, the predictive effect of mGPS on overall survival has been demonstrated in different cancer types, such as pancreatic cancer [28], esophageal cancer [29], hepatocellular carcinoma [30], renal cell carcinoma [31], and lung cancer [32]. Consistent with the literature, our current study evinced that geriatric patients treated with perioperative FLOT and with elevated pretreatment mGPS experienced poorer overall survival than those with normal mGPS. A study conducted on metastatic gastric cancer patients in Turkey revealed that mGPS was superior to PNI, cachexia index, prognostic index, neutrophil-lymphocyte ratio, and sarcopenia index in predicting mortality [33]. A recent study reported that a high mGPS was a significant prognostic factor for overall survival in elderly non-small cell lung cancer patients treated with anti-programmed death 1 (PD-1) blockade (HR: 0.31 [95% CI: 0.13–0.71], $p < 0.01$) [25]. Similarly, in elderly patients with gastric cancer treated with perioperative FLOT, we showed that those with mGPS 0 had significantly longer median overall survival compared with patients with mGPS 2 (HR: 6.59 [95% CI: 2.08–20.85], $p = 0.001$).

Next, we compared the general characteristics of participants according to their mGPS group, and there was no difference in terms of age, gender, and BMI. According to tumor characteristics, there was a significant difference only in the N stages but not in the T stages. The T-stage describes the size and extent of the tumor, while the N-stage describes lymph node involvement [34]. Lymph node staging (N0/N1/N2/N3a/N3b) has been shown to be one of the most important prognostic factors in resected gastric cancer [35]. When the mGPS 0 group was compared with both the mGPS 1 and mGPS 2 groups, a significant difference emerged in the N staging of the patients. As the mGPS scores increased, the number of lymph nodes containing cancer also increased (Table 1). We compared biochemical parameters according to mGPS groups, and there were natural differences in serum CRP and albumin levels, which are part of the mGPS calculation. As expected, higher serum CRP levels and lower albumin levels were observed in the highest mGPS group. Preoperative CEA and C19-9 levels, which indicate tumor burden, were higher in the highest mGPS group than in the mGPS 0 group. High levels of these two tumor markers, which have been shown to be associated with poor prognosis in gastric cancers [36–38], may indicate the presence of a more aggressive tumor.

Malnutrition is a predictor of unfavorable clinical results in older cancer patients. Several tools for screening malnutrition are available, though the optimal screening tool for this specific group remains unidentified. Based on the specific screening tool used (Nutrition Risk Screening 2002 (NRS 2002), Malnutrition Universal Screening Tool (MUST), and Mini Nutrition Assessment Short Form (MNA-SF)), the prevalence of malnutrition in older adults diagnosed with gastrointestinal cancer ranges from 20% to 52% [39]. Along with these nutritional screening tools, objective indices such as PNI, CONUT, NRI, and GNRI are used to evaluate the nutritional status of elderly cancer patients [40]. Although the prognostic importance of mGPS in cancer patients has been demonstrated in many studies, it is not among the diagnostic markers specific to malnutrition [40,41]. In our study, we showed that the PNI score was statistically significantly lower in the patients with the highest mGPS. Therefore, our findings suggested that mGPS and PNI are two tools with similar results for assessing the risk of malnutrition in elderly patients with gastric cancer receiving perioperative FLOT therapy. Another tool, the GNRI, was developed especially for the assessment of malnutrition risk in elderly patients [24]. In our study, we found that although GNRI scores tended to increase as mGPS increased, the difference was not significant.

In addition to PNI and GNRI, anthropometric markers such as weight loss and BMI are also used in the assessment of nutritional status in elderly cancer patients [40]. BMI is a simple tool that can be easily applied in clinical settings, but it does not detect the difference between fat and muscle mass and is an indicator of the total mass. Also, it should be noted that there is no consensus on any cut-off point of BMI to define obesity in the elderly [42]. In our study, the mean BMI of the mGPS 2 group was 26.4 ± 4.99 kg/m^2. We also found that each unit increase in BMI increased the risk of falling into the mGPS 2 group by 20%. (Table 4). This relationship can be explained by the serum CRP level, one of the components of mGPS. The pathophysiological mechanism linking high BMI to elevated CRP levels involves the release of pro-inflammatory cytokines (such as IL-6) by adipose tissue, which triggers the expression and release of CRP by hepatocytes in the liver [43]. Kawamoto et al. [44] demonstrated that a BMI $\geq$ 25 kg/m^2 was an independent risk factor for elevated CRP levels in adults between the ages of 65–74.

Previously, several studies have indicated that the preoperative nutritional status stands as an independent predictor of postoperative complications in geriatric cancer patients [45–48]. In a study conducted on geriatric patients with rectal cancer in Turkey, it has been reported that the mGPS and the CAR can predict serious postoperative complications and mortality [49]. Another study showed that mGPS was an independent predictor of major postoperative complications in elderly patients treated with radical cystectomy for urothelial bladder cancer [50]. A recent prospective study demonstrated that gastrointestinal cancer patients with higher mGPS were associated with postoperative complications, including requirements for blood transfusion, superficial surgical site infections, and sepsis [51]. Although the use of neoadjuvant chemotherapy in patients who underwent liver resection due to colorectal liver metastases was more prevalent in the mGPS $\geq$ 1 group than in those with mGPS 0, no relationship was found between preoperative mGPS and postoperative severe complications [52]. Similarly, our findings did not reveal any relationship between pretreatment mGPS and preoperative complications following neoadjuvant FLOT treatment (Table 5). No difference was observed between mGPS groups in terms of treatment-related serious side effects and chemotherapy response in geriatric patients receiving perioperative neoadjuvant FLOT treatment. In addition, similar radiological and pathological responses were observed in all three groups. In terms of toxicity, all patients' treatments were completed, and chemotherapy was not discontinued due to serious side effects. Therefore, we speculated that perioperative neoadjuvant FLOT therapy was an effective and safe treatment regimen in geriatric gastric cancer patients. On the other hand, the limited number of cases in our study could potentially explain this result. Apart from this, our single-center retrospective study has its own limitations, such as some selection and information bias. In this study, we defined the elderly patient as $\geq$65 years of age, and due to our limited number of cases, we could not evaluate our results according to age ranges. On the other hand, due to the increase in life expectancy and the increase in the population of individuals over the age of 80, prospective, multicenter studies with larger samples are needed.

5. Conclusions

The benefit of neoadjuvant treatments in the geriatric age group is more limited due to the physiological and metabolic consequences of aging. Due to the low expectation of efficacy and high toxicity of neoadjuvant therapy in elderly patients, patients who could potentially benefit may be overlooked and deprived of the expected benefit. Nevertheless, this study evinced that high pretreatment mGPS was associated with poor overall survival in geriatric patients with gastric cancer treated with perioperative FLOT. In conclusion, pretreatment mGPS may be a simple and useful tool to predict mortality in elderly patients with gastric cancer treated with neoadjuvant FLOT. Determining prognostic factors in elderly gastric cancer patients is important not only for managing the treatment of patients but also for providing long-term benefits to patients, and further studies are required.

Author Contributions: Conceptualization, E.M., E.B., S.S. and B.S.; methodology, E.M., E.B., S.S., B.M. and B.S.; software, E.M., E.B. and B.M.; validation, E.M., E.B., S.S. and B.M.; formal analysis, E.M., E.B. and B.M.; investigation, E.M. and E.B.; resources, E.M. and E.B.; data curation, E.M., E.B. and B.M.; writing—original draft preparation, E.M. and E.B.; writing—review and editing, E.M., E.B. and S.S.; visualization, E.M., E.B. and B.M.; supervision, B.S.; project administration, E.M. and E.B. All authors have read and agreed to the published version of the manuscript.

Funding: This research received no external funding.

Institutional Review Board Statement: The study was conducted in accordance with the Declaration of Helsinki and approved by the Non-Interventional Clinical Research Ethics Committee of Cukurova University (132:54; 07.04.2023).

Informed Consent Statement: Informed consent was obtained from all subjects involved in the study.

Data Availability Statement: Data are available on request from the authors.

Conflicts of Interest: The authors declare no conflict of interest.

References

1. Sung, H.; Ferlay, J.; Siegel, R.L.; Laversanne, M.; Soerjomataram, I.; Jemal, A.; Bray, F. Global Cancer Statistics 2020: GLOBOCAN Estimates of Incidence and Mortality Worldwide for 36 Cancers in 185 Countries. *CA Cancer J. Clin.* **2021**, *71*, 209–249. [CrossRef] [PubMed]
2. Balakrishnan, M.; George, R.; Sharma, A.; Graham, D.Y. Changing Trends in Stomach Cancer Throughout the World. *Curr. Gastroenterol. Rep.* **2017**, *19*, 36. [CrossRef] [PubMed]
3. Cunningham, D.; Allum, W.H.; Stenning, S.P.; Thompson, J.N.; Van de Velde, C.J.H.; Nicolson, M.; Scarffe, J.H.; Lofts, F.J.; Falk, S.J.; Iveson, T.J.; et al. Perioperative Chemotherapy versus Surgery Alone for Resectable Gastroesophageal Cancer. *N. Engl. J. Med.* **2006**, *355*, 11–20. [CrossRef] [PubMed]
4. Ychou, M.; Boige, V.; Pignon, J.P.; Conroy, T.; Bouché, O.; Lebreton, G.; Ducourtieux, M.; Bedenne, L.; Fabre, J.M.; Saint-Aubert, B.; et al. Perioperative chemotherapy compared with surgery alone for resectable gastroesophageal adenocarcinoma: An FNCLCC and FFCD multicenter phase III trial. *J. Clin. Oncol.* **2011**, *29*, 1715–1721. [CrossRef] [PubMed]
5. Al-Batran, S.E.; Homann, N.; Pauligk, C.; Goetze, T.O.; Meiler, J.; Kasper, S.; Kopp, H.G.; Mayer, F.; Haag, G.M.; Luley, K.; et al. Perioperative chemotherapy with fluorouracil plus leucovorin, oxaliplatin, and docetaxel versus fluorouracil or capecitabine plus cisplatin and epirubicin for locally advanced, resectable gastric or gastro-oesophageal junction adenocarcinoma (FLOT4): A randomised, phase 2/3 trial. *Lancet* **2019**, *393*, 1948–1957. [CrossRef] [PubMed]
6. Kattan, M.W.; Karpeh, M.S.; Mazumdar, M.; Brennan, M.F. Postoperative nomogram for disease-specific survival after an R0 resection for gastric carcinoma. *J. Clin. Oncol.* **2003**, *21*, 3647–3650. [CrossRef]
7. Han, D.S.; Suh, Y.S.; Kong, S.H.; Lee, H.J.; Choi, Y.; Aikou, S.; Sano, T.; Park, B.J.; Kim, W.H.; Yang, H.K. Nomogram predicting long-term survival after d2 gastrectomy for gastric cancer. *J. Clin. Oncol.* **2012**, *30*, 3834–3840. [CrossRef]
8. van den Ende, T.; Ter Veer, E.; Mali, R.M.A.; van Berge Henegouwen, M.I.; Hulshof, M.; van Oijen, M.G.H.; van Laarhoven, H.W.M. Prognostic and Predictive Factors for the Curative Treatment of Esophageal and Gastric Cancer in Randomized Controlled Trials: A Systematic Review and Meta-Analysis. *Cancers* **2019**, *11*, 530. [CrossRef]
9. Li, X.; An, B.; Zhao, Q.; Qi, J.; Wang, W.; Zhang, D.; Li, Z.; Qin, C. Combined fibrinogen and neutrophil-lymphocyte ratio as a predictive factor in resectable colorectal adenocarcinoma. *Cancer Manag. Res.* **2018**, *10*, 6285–6294. [CrossRef]
10. Han, C.L.; Meng, G.X.; Ding, Z.N.; Dong, Z.R.; Chen, Z.Q.; Hong, J.G.; Yan, L.J.; Liu, H.; Tian, B.W.; Yang, L.S.; et al. The Predictive Potential of the Baseline C-Reactive Protein Levels for the Efficiency of Immune Checkpoint Inhibitors in Cancer Patients: A Systematic Review and Meta-Analysis. *Front. Immunol.* **2022**, *13*, 827788. [CrossRef]
11. Carr, B.I.; Ince, V.; Bag, H.G.; Usta, S.; Ersan, V.; Isik, B.; Yilmaz, S. CRP is a superior and prognostically significant inflammation biomarker for hepatocellular cancer patients treated by liver transplantation. *Clin. Pract.* **2021**, *18*, 1626–1632.
12. Toyokawa, T.; Muguruma, K.; Yoshii, M.; Tamura, T.; Sakurai, K.; Kubo, N.; Tanaka, H.; Lee, S.; Yashiro, M.; Ohira, M. Clinical significance of prognostic inflammation-based and/or nutritional markers in patients with stage III gastric cancer. *BMC Cancer* **2020**, *20*, 517. [CrossRef] [PubMed]
13. Ozveren, A.; Erdogan, A.P.; Ekinci, F. The inflammatory prognostic index as a potential predictor of prognosis in metastatic gastric cancer. *Sci. Rep.* **2023**, *13*, 7755. [CrossRef]
14. Nogueiro, J.; Santos-Sousa, H.; Pereira, A.; Devezas, V.; Fernandes, C.; Sousa, F.; Fonseca, T.; Barbosa, E.; Barbosa, J.A. The impact of the prognostic nutritional index (PNI) in gastric cancer. *Langenbeck's Arch. Surg.* **2022**, *407*, 2703–2714. [CrossRef]
15. Aoyama, T.; Komori, K.; Nakazano, M.; Hara, K.; Tamagawa, H.; Kazama, K.; Hashimoto, I.; Yamada, T.; Maezawa, Y.; Segami, K.; et al. The Clinical Influence of the CONUT Score on Survival of Patients with Gastric Cancer Receiving Curative Treatment. *In Vivo* **2022**, *36*, 942–948. [CrossRef] [PubMed]
16. Saal, J.; Bald, T.; Eckstein, M.; Ralser, D.J.; Ritter, M.; Brossart, P.; Grünwald, V.; Hölzel, M.; Ellinger, J.; Klümper, N. Integrating On-Treatment Modified Glasgow Prognostic Score and Imaging to Predict Response and Outcomes in Metastatic Renal Cell Carcinoma. *JAMA Oncol.* **2023**, *9*, 1048–1055. [CrossRef]
17. Nagashima, Y.; Funahashi, K.; Kagami, S.; Ushigome, M.; Kaneko, T.; Miura, Y.; Yoshida, K.; Koda, T.; Kurihara, A. Which preoperative immunonutritional index best predicts postoperative mortality after palliative surgery for malignant bowel obstruction in patients with late-stage cancer? A single-center study in Japan comparing the modified Glasgow prognostic score (mGPS), the prognostic nutritional index (PNI), and the controlling nutritional status (CONUT). *Surg. Today* **2023**, *53*, 22–30. [CrossRef]

18. Yang, C.; Ren, G.; Yang, Q. Prognostic value of preoperative modified Glasgow prognostic score in surgical non-small cell lung cancer: A meta-analysis. *Front. Surg.* **2022**, *9*, 1094973. [CrossRef]
19. Nozoe, T.; Iguchi, T.; Egashira, A.; Adachi, E.; Matsukuma, A.; Ezaki, T. Significance of modified Glasgow prognostic score as a useful indicator for prognosis of patients with gastric carcinoma. *Am. J. Surg.* **2011**, *201*, 186–191. [CrossRef]
20. Brierley, J.D.; Gospodarowicz, M.K.; Wittekind, C. *TNM Classification of Malignant Tumours*; John Wiley & Sons: Hoboken, NJ, USA, 2017.
21. Dindo, D.; Demartines, N.; Clavien, P.A. Classification of surgical complications: A new proposal with evaluation in a cohort of 6336 patients and results of a survey. *Ann. Surg.* **2004**, *240*, 205–213. [CrossRef]
22. McMillan, D.C.; Crozier, J.E.; Canna, K.; Angerson, W.J.; McArdle, C.S. Evaluation of an inflammation-based prognostic score (GPS) in patients undergoing resection for colon and rectal cancer. *Int. J. Colorectal Dis.* **2007**, *22*, 881–886. [CrossRef] [PubMed]
23. Ding, P.a.; Yang, P.; Sun, C.; Tian, Y.; Guo, H.; Liu, Y.; Li, Y.; Zhao, Q. Predictive Effect of Systemic Immune-Inflammation Index Combined with Prognostic Nutrition Index Score on Efficacy and Prognosis of Neoadjuvant Intraperitoneal and Systemic Paclitaxel Combined with Apatinib Conversion Therapy in Gastric Cancer Patients with Positive Peritoneal Lavage Cytology: A Prospective Study. *Front. Oncol.* **2022**, *11*, 791912. [CrossRef]
24. Bouillanne, O.; Morineau, G.; Dupont, C.; Coulombel, I.; Vincent, J.P.; Nicolis, I.; Benazeth, S.; Cynober, L.; Aussel, C. Geriatric Nutritional Risk Index: A new index for evaluating at-risk elderly medical patients. *Am. J. Clin. Nutr.* **2005**, *82*, 777–783. [CrossRef] [PubMed]
25. Tanaka, T.; Yoshida, T.; Masuda, K.; Takeyasu, Y.; Shinno, Y.; Matsumoto, Y.; Okuma, Y.; Goto, Y.; Horinouchi, H.; Yamamoto, N.; et al. Prognostic role of modified Glasgow Prognostic score in elderly non-small cell lung cancer patients treated with anti-PD-1 antibodies. *Respir. Investig.* **2023**, *61*, 74–81. [CrossRef] [PubMed]
26. Diakos, C.I.; Charles, K.A.; McMillan, D.C.; Clarke, S.J. Cancer-related inflammation and treatment effectiveness. *Lancet Oncol.* **2014**, *15*, e493–e503. [CrossRef]
27. Kim, M.R.; Kim, A.S.; Choi, H.I.; Jung, J.H.; Park, J.Y.; Ko, H.J. Inflammatory markers for predicting overall survival in gastric cancer patients: A systematic review and meta-analysis. *PLoS ONE* **2020**, *15*, e0236445. [CrossRef]
28. Wu, D.; Wang, X.; Shi, G.; Sun, H.; Ge, G. Prognostic and clinical significance of modified glasgow prognostic score in pancreatic cancer: A meta-analysis of 4629 patients. *Aging* **2021**, *13*, 1410–1421. [CrossRef]
29. Wang, Y.; Chen, L.; Wu, Y.; Li, P.; Che, G. The prognostic value of modified Glasgow prognostic score in patients with esophageal squamous cell cancer: A Meta-analysis. *Nutr. Cancer* **2020**, *72*, 1146–1154. [CrossRef]
30. Chen, H.; Hu, N.; Chang, P.; Kang, T.; Han, S.; Lu, Y.; Li, M. Modified Glasgow prognostic score might be a prognostic factor for hepatocellular carcinoma: A meta-analysis. *Panminerva Med.* **2017**, *59*, 302–307. [CrossRef]
31. Hu, X.; Wang, Y.; Yang, W.X.; Dou, W.C.; Shao, Y.X.; Li, X. Modified Glasgow prognostic score as a prognostic factor for renal cell carcinomas: A systematic review and meta-analysis. *Cancer Manag. Res.* **2019**, *11*, 6163–6173. [CrossRef]
32. Zhang, Y.; Chen, S.; Chen, H.; Li, W. A comprehensive analysis of Glasgow Prognostic Score (GPS)/the modified Glasgow Prognostic Score (mGPS) on immune checkpoint inhibitor efficacy among patients with advanced cancer. *Cancer Med.* **2023**, *12*, 38–48. [CrossRef] [PubMed]
33. Demirelli, B.; Babacan, N.A.; Ercelep, Ö.; Öztürk, M.A.; Kaya, S.; Tanrıkulu, E.; Khalil, S.; Hasanov, R.; Alan, Ö.; Telli, T.A.; et al. Modified Glasgow Prognostic Score, Prognostic Nutritional Index and ECOG Performance Score Predicts Survival Better than Sarcopenia, Cachexia and Some Inflammatory Indices in Metastatic Gastric Cancer. *Nutr. Cancer* **2021**, *73*, 230–238. [CrossRef] [PubMed]
34. Díaz Del Arco, C.; Ortega Medina, L.; Estrada Muñoz, L.; García Gómez de Las Heras, S.; Fernández Aceñero, M.J. Pathologic Lymph Node Staging of Gastric Cancer. *Am. J. Clin. Pathol.* **2021**, *156*, 749–765. [CrossRef] [PubMed]
35. Zhou, Y.X.; Yang, L.P.; Wang, Z.X.; He, M.M.; Yun, J.P.; Zhang, D.S.; Wang, F.; Xu, R.H. Lymph node staging systems in patients with gastric cancer treated with D2 resection plus adjuvant chemotherapy. *J. Cancer* **2018**, *9*, 660–666. [CrossRef] [PubMed]
36. Cai, Q.; Zhou, W.; Li, J.; Ou, X.; Chen, C.; Cai, S.; He, W.; Xu, J.; He, Y. Association of Preoperative Serum Carcinoembryonic Antigen and Gastric Cancer Recurrence: A Large Cohort Study. *J. Cancer* **2021**, *12*, 397–403. [CrossRef]
37. Shibata, C.; Nakano, T.; Yasumoto, A.; Mitamura, A.; Sawada, K.; Ogawa, H.; Miura, T.; Ise, I.; Takami, K.; Yamamoto, K.; et al. Comparison of CEA and CA19-9 as a predictive factor for recurrence after curative gastrectomy in gastric cancer. *BMC Surg.* **2022**, *22*, 213. [CrossRef]
38. Wang, W.; Chen, X.L.; Zhao, S.Y.; Xu, Y.H.; Zhang, W.H.; Liu, K.; Chen, X.Z.; Yang, K.; Zhang, B.; Chen, Z.X.; et al. Prognostic significance of preoperative serum CA125, CA19-9 and CEA in gastric carcinoma. *Oncotarget* **2016**, *7*, 35423–35436. [CrossRef]
39. Ye, X.-J.; Ji, Y.-B.; Ma, B.-W.; Huang, D.-D.; Chen, W.-Z.; Pan, Z.-Y.; Shen, X.; Zhuang, C.-L.; Yu, Z. Comparison of three common nutritional screening tools with the new European Society for Clinical Nutrition and Metabolism (ESPEN) criteria for malnutrition among patients with geriatric gastrointestinal cancer: A prospective study in China. *BMJ Open* **2018**, *8*, e019750. [CrossRef]
40. Bullock, A.F.; Greenley, S.L.; McKenzie, G.A.G.; Paton, L.W.; Johnson, M.J. Relationship between markers of malnutrition and clinical outcomes in older adults with cancer: Systematic review, narrative synthesis and meta-analysis. *Eur. J. Clin. Nutr.* **2020**, *74*, 1519–1535. [CrossRef]
41. White, J.V.; Guenter, P.; Jensen, G.; Malone, A.; Schofield, M. Consensus Statement of the Academy of Nutrition and Dietetics/American Society for Parenteral and Enteral Nutrition: Characteristics Recommended for the Identification and Documentation of Adult Malnutrition (Undernutrition). *J. Acad. Nutr. Diet.* **2012**, *112*, 730–738. [CrossRef]

42. Ji, T.; Li, Y.; Ma, L. Sarcopenic Obesity: An Emerging Public Health Problem. *Aging Dis.* **2022**, *13*, 379–388. [CrossRef] [PubMed]
43. Ellulu, M.S.; Patimah, I.; Khaza'ai, H.; Rahmat, A.; Abed, Y. Obesity and inflammation: The linking mechanism and the complications. *Arch. Med. Sci.* **2017**, *13*, 851–863. [CrossRef] [PubMed]
44. Kawamoto, R.; Kusunoki, T.; Abe, M.; Kohara, K.; Miki, T. An association between body mass index and high-sensitivity C-reactive protein concentrations is influenced by age in community-dwelling persons. *Ann. Clin. Biochem.* **2013**, *50*, 457–464. [CrossRef] [PubMed]
45. Kushiyama, S.; Sakurai, K.; Kubo, N.; Tamamori, Y.; Nishii, T.; Tachimori, A.; Inoue, T.; Maeda, K. The Preoperative Geriatric Nutritional Risk Index Predicts Postoperative Complications in Elderly Patients with Gastric Cancer Undergoing Gastrectomy. *In Vivo* **2018**, *32*, 1667–1672. [CrossRef] [PubMed]
46. Li, L.; Wang, H.; Yang, J.; Jiang, L.; Yang, J.; Wu, H.; Wen, T.; Yan, L. Geriatric nutritional risk index predicts prognosis after hepatectomy in elderly patients with hepatitis B virus-related hepatocellular carcinoma. *Sci. Rep.* **2018**, *8*, 12561. [CrossRef] [PubMed]
47. Furuke, H.; Matsubara, D.; Kubota, T.; Kiuchi, J.; Kubo, H.; Ohashi, T.; Shimizu, H.; Arita, T.; Yamamoto, Y.; Konishi, H.; et al. Geriatric Nutritional Risk Index Predicts Poor Prognosis of Patients After Curative Surgery for Gastric Cancer. *Cancer Diagn. Progn.* **2021**, *1*, 43–52. [CrossRef]
48. Yin, J.; Qu, J.; Liang, X.; Wang, M. Prognostic significance of controlling nutritional status score for patients with gastric cancer: A systematic review and meta-analysis. *Exp. Ther. Med.* **2023**, *25*, 202. [CrossRef]
49. Manoglu, B.; Sokmen, S.; Bisgin, T.; Semiz, H.S.; Görken, I.B.; Ellidokuz, H. Inflammation-based prognostic scores in geriatric patients with rectal cancer. *Tech. Coloproctol.* **2023**, *27*, 397–405. [CrossRef]
50. Mari, A.; Muto, G.; Di Maida, F.; Tellini, R.; Bossa, R.; Bisegna, C.; Campi, R.; Cocci, A.; Viola, L.; Grosso, A.; et al. Oncological impact of inflammatory biomarkers in elderly patients treated with radical cystectomy for urothelial bladder cancer. *Arab. J. Urol.* **2020**, *19*, 2–8. [CrossRef]
51. Mahalingam, S.; Amaranathan, A.; Sathasivam, S.; Udayakumar, K.P. Correlation of Preoperative Anemia Subtypes with Tumor Characteristics, Systemic Inflammation and Immediate Postoperative Outcomes in Gastrointestinal Cancer Patients—A Prospective Observational Study. *J. Gastrointest. Cancer* **2023**. [CrossRef]
52. Sahakyan, M.A.; Brudvik, K.W.; Angelsen, J.-H.; Dille-Amdam, R.G.; Sandvik, O.M.; Edwin, B.; Nymo, L.S.; Lassen, K. Preoperative Inflammatory Markers in Liver Resection for Colorectal Liver Metastases: A National Registry-Based Study. *World J. Surg.* **2023**, *47*, 2213–2220. [CrossRef] [PubMed]

MDPI

St. Alban-Anlage 66

4052 Basel

Switzerland

www.mdpi.com

Nutrients Editorial Office

E-mail: nutrients@mdpi.com

www.mdpi.com/journal/nutrients

www.ingramcontent.com/pod-product-compliance
Lightning Source LLC
LaVergne TN
LVHW070900170726
843515LV00005B/687